Optimal Wellbeing of Ageing Wild Animals in Human Care

Sabrina Brando • Sarah Chapman
Editors

Optimal Wellbeing of Ageing Wild Animals in Human Care

Springer

Editors
Sabrina Brando
Director of AnimalConcepts
Teulada, Spain

Sarah Chapman
Chapman Zoo Consultancy
Birmingham, UK

ISBN 978-3-031-30658-7 ISBN 978-3-031-30659-4 (eBook)
https://doi.org/10.1007/978-3-031-30659-4

Cover photo of Henry the Tuatara taken by Lindsay Hazley - Senior Living Species Officer. Henry lives in Invercargill, New Zealand and has his own Kaitaki (guardians), the Ngati Koata who are the guardians of the Tuatara.

This Springer imprint is published by the registered company Springer Nature Switzerland AG
The registered company address is: Gewerbestrasse 11, 6330 Cham, Switzerland

Paper in this product is recyclable.

This book is dedicated to all ageing animals everywhere and the people who care for them. Winnie the Pooh was wise (as always); let's all take his advice to heart and listen to animals as well as talk to them.

Foreword

'Death and taxes' are the two certain things in life according to Benjamin Franklin's letter of 1789. I think Franklin missed out the most important part of death, which is the life that preceded it. To paraphrase the protagonist Nathan Algren from the 2003 film *The Last Samurai*, when questioned about how the Last Samurai died by Emperor Meiji, he responded: 'I will tell you how he lived.' It is the quality of life of a human or an animal that is important, and of course, we wish for a humane and quick death.

Humans have been fixated on longevity for thousands of years, and the longevity argument has been used to criticise the keeping of animals in captivity, especially zoo animals. It is certainly true that when the first modern zoos appeared in the 1800s, the animals had short lives. Very little is known about the variety and rate of ageing of different species. Only a few species such as naked mole rats and some species of bat seem to have very slow ageing processes. Some very long-lived species such as Greenland sharks (400 years) may age very slowly but they age, and little is known about life expectancy in wild animals in general.

Since the Industrial Revolution in the UK, the life expectancy of the average citizen has more than doubled due to improvements in health care, sanitation, and nutrition, amongst others. I am sure our human ancestors two hundred years ago would have been happy to know that their descendants would live on average much longer than them but would not have predicted all the wellbeing concerns and problems that can come with a long life. Currently many animals from a variety of species in zoos are now outliving their wild counterparts, with a few notable exceptions such as elephants. Whilst many would agree that improvements in human and animal lifespans are in general a good thing, this ignores the important point of **how** we age. There are still knowledge gaps and a lack of understanding of the animals needs and preferences in terms of what constitutes physical and psychological wellbeing for an individual.

The common diseases associated with human ageing are arthritis, dementia, diabetes, hypertension, and cognitive decline, amongst others. Other common consequences of ageing are a deterioration in hearing and sight. In addition to physical changes humans may also experience loss of confidence and reduced social support and interaction, which can also be true for other animals. Many of the conditions are related to the 'wear and tear' of life, but, importantly, how lives are lived will affect whether we age in a healthy manner and have a good quality of life in advanced years. Thus, it is now common to hear wellbeing experts talk about healthy ageing.

This edited book features chapters by an array of global specialists and seeks to address ageing in zoo animals, and specifically healthy ageing. The idea of the book is first to be proactive in caring for zoo animals and promote healthy ageing through a variety of approaches such as diet, exercise, cognitive stimulation, and a good human–animal relationship. The book also acknowledges that aspects such as pain cannot always be avoided, and that treatments may be necessary, including medication and/or physiotherapy. It highlights the importance of a challenging environment which, e.g., encourages the use of their whole body, to maintain strength and coordination. Additionally, animals need a sense of agency and the ability to look after themselves, e.g., through a complex enriched environment with a wide variety of opportunities. It is important that we take a proactive approach to address problems due to ageing. Animals should be motivated through social and environmental factors, as well as participating in positive reinforcement training. Ethical care should be at the heart of and the lens through which ageing animal programmes are developed and how decisions are made. Part of the caring process is realising when a life is no longer a 'life worth living', and when euthanasia is the best option. Another highlight of the book is the opportunity and responsibility to communicate stories about the lives of ageing animals, including disease and death. The book comes full circle by providing space for grieving and celebrating life. Animals, unlike humans, may avoid taxes but death befalls all living beings; more importantly, it is how they live and how they age that will affect their quality of life (i.e., wellbeing). This book is an important starting point for everyone interested and or involved in the care and wellbeing of ageing zoo animals.

University of Salford
Salford, UK

Robert Young

Preface

In 1894, Governor Grey-Wilson of the British Overseas Territory, the Island of St Helena, an isolated volcano in the mid-South Atlantic, did us all a great favour. He wrote a letter to steamship captain Pascal Oliver telling him that Jonathan had landed in 1882 and was reported to be fully grown. Jonathan is a Seychelles Giant Tortoise (***Aldabrachelys gigantea hololissa***). That alone makes him a very rare survivor of a once thought-to-be extinct species. This species is now known to be a subspecies of the closely related Aldabra from the atoll of the same name. Based on full maturity being approximately 50 years of age, the landing date means he hatched in around 1832, making him currently at least 190 years old. He is now in the Guinness World Records as the oldest living land animal in the world and the oldest ever recorded chelonian. Jonathan is blind, with cataracts, and has lost his sense of smell. Twelve years ago, the island's first permanent veterinarian saw that his beak was soft and crumbly, so that not only did he struggle to graze, but he was biting blindly into soil and leaf mould. The veterinarian began a regime of hand feeding once weekly, and was astonished to watch Jonathan rejuvenate, even to the point of regrowing his beak. It transpired that he was not just suffering from a shortage of calories but vital vitamins, minerals, and trace elements.

After mouth swabs were taken to acquire his DNA, researchers, in the USA, unravelled his genome and his methylome, looking into why giant tortoises are so resistant to cancer, for the benefit of humankind. Jonathan is the tortoise that just keeps on giving. To the islanders, known as Saints, he is an icon, a symbol of endurance and persistence in the face of change. Pandemics, world wars, kings and queens, and even empires have come and gone, and every human being on this planet was born within his lifetime. And for the tourists, he is a 'must-see', grazing in the paddock below the stately Plantation House, built by the East India Company, set in beautifully groomed historic gardens. Long may he continue to delight us with his remarkable presence.

Welcome to this book dedicated to ageing animals and those who care about them. With so many wonderful animals of all kinds living longer under human care in zoos, sanctuaries, aquariums, and alike, there is a need to expand our knowledge and deepen our understanding of ageing animals. Ageing is the process of becoming older. As an animal ages, they can be referred to as geriatric, older, elderly, or aged. Our approach should reflect meeting the needs and preferences of individuals in a respectful and compassionate manner. The purpose of the book is to bring together information, ideas, and viewpoints as foundations to promote positive wellbeing. What we do must be in the best interest of the animal and be at the heart of their care. The terms animal welfare and animal wellbeing have both been used over the years to describe the state of the animal. The terminology within this book will be interchangeable within the chapters depending on the authors' preference. In the authors' experience, the changing needs and preferences of many ageing animals in human care are often overlooked or even disregarded and being labelled as 'just old'. The key message is that the wellbeing of ageing animals should not be dismissed or overlooked but prioritised.

In 2016, AnimalConcepts organised in collaboration with the San Francisco Zoo in the USA the first seminar on the care and wellbeing of ageing wild animals which was followed by another collaborative seminar with Chapman Zoo Consultancy and the Allwetterzoo Münster in Germany in 2017. Both seminars had a diverse group of speakers contributing their knowledge, experience, case studies, and evidence-based research to further the field of caring for ageing animals. The events were attended by animal care staff including caregivers, veterinary professionals, students, researchers, and those with an interest in wild animal care and wellbeing. We are delighted that Springer Nature recognised the need for a book on this topic and for the invitation to cocreate. We are grateful that almost all seminar speakers were available to write chapters, and organisations and caregivers for the stories on individual animals. Although each of the chapters in this book can be read independently, we encourage them to be read together to provide a holistic overview of the care of ageing wild animals under human care.

Each chapter starts with a story of an individual animal to highlight for whom we ultimately do all this effort, for ageing animals everywhere. Chapter "Holistic Approaches to Optimal Wellbeing of Ageing Wild Animals" by Sabrina Brando and Chapman provides an overview of various approaches to the monitoring and assessment of animal wellbeing including quantitative and qualitative methods. Understanding emotional expressions and mental states in animals is an important and growing field in animals in human care, including personality and individual differences. They include the focus on positive animal–animal and human–animal interactions playing a central role in positive wellbeing. It highlights the need for a holistic approach considering a wide spectrum of physical and psychological aspects, providing choice and control through the lens of 24/7 across lifespan.

Compassion for animal elders in the last phase of their life is central in Chapter "Ageing Gracefully: Compassion for Nonhuman Animal Elders' by Joachim

Nieuwland and Frank Meijboom who explore the ethical considerations for caring for ageing animals with genuine curiosity. Those who take care of ageing animals are confronted with various moral considerations, especially now that nonhuman animals grow old within zoos. Ageing animals undergo changes which keeps them from flourishing which can lead to suffering and a change in standing. Moral deliberation helps to engage with the moral issues of taking care and compassion appears especially promising to care for animals, so they continue to exert agency over the later stages of their lives.

Good, consistent, and meaningful record keeping is fundamental to ageing animal care and can inform the types of choice and control possible for the individual. In Chapter "The Importance of Meaningful Record Keeping in Caring for Ageing Wild Animals', Max Norman, Sarah Chapman, and Sabrina Brando outline the various methods available for use by care animal staff. The ability to monitor, collate, and review an animal's information throughout his or her lifetime, based on physical and psychological indicators, is vital to the ability to provide good quality care. This chapter outlines the type of records that should be kept and how these can be documented, including quantitative and qualitative measures of behaviour. It focuses on the importance of reliable and accessible data for the ongoing monitoring and assessment of ageing animal wellbeing, including predictive and proactive management practices.

Ageing animal wellbeing assessments should inform environmental and habitat designs. Chapter "Facility Design for Diasabled and Ageing Wild Animals" by Jon Coe blends science with empathy through descriptions and examples of multidisciplinary designs of animal facilities around the world. The importance of designed features benefiting both animals and animal carers is emphasised. Subjects include integration of management and facility planning, a model design and animal co-design process, features which may delay the onset of health problems, and general recommendations including flexible spaces, suitable surfaces, horizontal and vertical circulation, animal rotation, and aquatic/aerobic exercise. Animal/computer interaction (ACI) opportunities are touched on, including animal control of ambient conditions and round-the-clock personalised access to food, enriching opportunities, and alternative spaces.

The use of environmental enrichment is commonplace in zoos, aquariums, sanctuaries, and other facilities, and Chapter "Environmental Enrichment for Ageing Zoo Animals" by Julian Chapman illustrates how we can use enrichment to enhance the lives of animals. Various types of enrichment are described as well as how enrichment programmes can be created by, e.g., using a four-week timetable and to ensure that enrichment is not overused but is always provided. This chapter describes practical ways to adapt enrichment for the older animal, as capabilities change, to reduce the likelihood of frustration, increase engagement, ensure safety, and assist in maintaining physical and psychological wellbeing.

Positive reinforcement promotes psychological wellbeing. Understanding animal learning and training can greatly support animal care and good animal wellbeing through all life stages, including when they are ageing as they often require special care. In Chapter "The Role of Learning and Training in Caring for Ageing Animals", Sabrina Brando and Debbie Marrin describe the opportunities and challenges involved in proactive ageing animal training and how training can be used to diagnose, monitor, and treat individuals with age-related health concerns with positive reinforcement and good human–animal interactions. They highlight the importance of training staff for end-of-life decisions requiring organisational, leadership, and team support. Designing for age-specific and adapting environments will facilitate care and treatments, making it easier and more comfortable for animals to participate in training sessions.

All foods, also those used in training sessions, must be accounted for and considered. Some age-related changes may have an impact on the animal's nutritional status, through changes in nutrient requirements, ability to find, consume, and digest food, and ability to absorb and assimilate nutrients from the diet. Chapter "Supporting Geriatric Zoo Animal Welfare through Nutrition" by Francis Cabana and Amy Plowman concentrates on nutrition and they describe the positive changes seen in ageing animals' gait such as improvements in lameness and rising from rest, following changes in diets within two zoos. They also discuss how different aspects of the diet can influence ageing physiological changes such as changes in bone mass, degenerative joint disease, declining brain function, and gastrointestinal changes. Suggestions for geriatric diet modifications are included.

Weight and body condition scoring are some of the most important monitoring tools. Within Chapter "Veterinary Care of Ageing Zoo Animals: Description of a Proactive Approach" on veterinary care, Sarah Chapman and Phillipa Dobbs describe a proactive approach to the assessment of ageing animals. Individuals were identified as aged if they were 75% through their expected natural longevity. They highlight that common conditions found in older animals are often hidden and may only come to light once at an advanced stage. These conditions can be painful and proactive assessment can enable early diagnosis and allow treatments and management practices to be instigated. 'Hands-off' methods such as positive reinforcement training and behavioural observations, followed by 'hands-on' methods such as a clinical assessment, should be combined for a holistic understanding of the wellbeing of the individual.

Conditions experienced by ageing animals are often painful which is a serious welfare concern. In Chapter "Pain: Physiology, Recognition, and Management in Zoo Animals ", Heather Bacon discusses pain management setting the scene by explaining the different neurological pathways involved. Barriers to effective management of pain across the many and varied taxa that zoos house include our current understanding of pain physiology and the recognition of pain states in different species. Additionally, the therapeutic approach to pain management may be a challenge depending on species. This chapter outlines the physiology of pain, types of pain, pain recognition, and the principles of preventing and managing pain experiences which will improve the animal's quality of life.

Environmental modification and a wide variety of therapies and treatments can assist in reducing pain. Physiotherapy and occupational therapy are an intrinsic part of the treatment and management of numerous types of disease. In Chapter "Physiotherapy and Management of the Musculoskeletal Health of Ageing Wild Animals in Human Care", Matthew Shackleton and Louise Lefrere describe alternative therapies providing in-depth information regarding the theory and practicalities of utilising physiotherapy techniques in ageing animals. They make suggestions on adaptations to facilitate a thorough assessment and how best practice zoo protocols can be utilised to successfully apply low-stress rehabilitation. Protocols follow the human model of exercise prescription, lifestyle and environmental adaptation, and behavioural health. The focus should be creating an environment for lasting change rather than just treating symptoms. Case studies, with some examples of physiotherapy being used with wild animal patients, are included.

Sometimes even after assessments, physiotherapy, and treatments we have to accept that we have exhausted all available options to provide a good quality of life. A wide range of age-related pathologies have been identified in zoo animals which can present subtly and are typically debilitating, are painful, and negatively impact on the overall welfare of the animal. In Chapter "Euthanasia of Geriatric Zoo Animals: Decision Making and Procedure", Sarah Chapman, James Chatterton, and Julian Chapman highlight the importance of end-of-life decision-making which can be an emotive subject. Being able to plan procedures and ensure the relevant stakeholders are involved is fundamental in the euthanasia process. Modern zoos make end-of-life decisions based on detailed welfare assessments with an overall focus on preventing unnecessary suffering of the animals involved. The decision-making process, practical arrangements, management of the individual and other animals, and public communication, including cases and scenarios, are described.

The evidence of subtle to devasting effects of different pathologies can be traced back and observed through the bodies of ageing animals. In Chapter "The Longevity Legacy: The Challenges of Old Animals in Zoos", Andrew Kitchener summarises the principal skeletal and dental pathologies of aged zoo mammals. Data are presented on pathologies from zoo specimens in National Museums Scotland that were collected over the last 30 years in a range of large mammals, including bears, big cats, great apes, babirusas, and pygmy hippopotamuses. The causes of these pathologies and the possible role of enrichment and habitat modification in reducing their prevalence are discussed. He raises awareness of these conditions and how painful they can be. Proactive monitoring for the possible presence of these pathologies requires regular and effective assessments of the individual to minimise negative welfare states.

The impact of pathologies leading to euthanasia of an elderly animal can in turn have an effect on those left behind. The behavioural responses of animals to the death of companions, also called mourning-like behaviours, have been described in various species. In Chapter "Mourning-Like Behaviour in a Malayan Sun Bear", Friederike Schmitz and Simone Schehka describe behaviours observed in Josy, a Malayan sun bear (*Helarctos malayanus*) following the death of her companion. Many caregiving staff have reported witnessing changes in animal's behaviour when other animals have died, either suddenly or following a longer illness or deterioration. The appreciation that animals may have feelings of grief and show mourning behaviour is an important factor in their care. Whilst the literature in this area is scarce and is an underrepresented topic of discussion, it is an important element when managing older animals in zoos.

There often is a deep emotional investment in and connection to an animal's life, including with elderly animals. This experience of caring for that animal comes with a range of positive feelings such as joy, love, and trust—and, inevitably, a range of negative feelings such as sorrow, pain, and grief when the animal dies. In Chapter "Caring for Elderly Wild Animals: The Human Experience", Sabrina Brando, Mickey Gjerris, Nicola Field, and Lynette Hart highlight the need for compassionate awareness of different human–animal bonds. Fostering a culture of care and respect allows for humans to be at their best for the animals they care for. Providing safe spaces for all to share in their mutual experiences, including opportunities for mourning and celebrating animals as well as supporting mental health, is discussed.

Stories about celebration of life and grief can reveal our connection with the animals in our care. Older animals are often the best-known individuals in a facility and often have a special place in the hearts of visitors. Chapter "The View from Beyond the Fence: Ageing Zoo Animals and Communicating with the Outside World" by Philip Knowling discusses public communications and emphasises the importance of being open and honest with the public. Communicating about the wellbeing of ageing animals provides insights into their lives and excellent educational opportunities. It can inspire the public and shape opinions and is a chance to strengthen and highlight a facility's commitment to animal care and wellbeing. Working with journalists, as well as the use of social media, helps reach a wider audience by sharing stories of the animals and the people who care about them.

This book illuminates ageing animals and puts forth the importance of a holistic approach interconnecting all aspects of their care. The ultimate aim is supporting a good life and a good death. The late global spiritual leader Thich Nhat Hanh describes connection as interbeing. '*To inter-be and the action of interbeing reflects reality, we inter-are with one another and with all life.*' Our relationship and care for animals, past, present, and future, is at the heart and the purpose of this book. Together with all contributors we invite you to take action for the benefit of ageing animals.

Teulada, Spain
Birmingham, UK
8 January 2023

Sabrina Brando
Sarah Chapman

Acknowledgements

The book has been brought to life through the contributions by knowledgeable and enthusiastic authors who translated their expertise and practices in order to further the understanding and improve the implementation of a diverse animal care and wellbeing programme for ageing animals.

We would like to thank the following organisations, staff, and external individuals for their stories and photos livening the pages of this book. They are listed in alphabetically order.

Allwetterzoo, Münster, Germany
Animals Asia
Léa Audier, France
Roger Belis Manyós, Fundació Mona, Spain
Max Block, Los Angeles, USA
Jody Carrigan, Zoo Atlanta, USA
Phillipa Dobbs, Twycross Zoo, UK
Molly Feldman, Animals Asia, China
Kerry Hill, Wildfowl and Wetlands Trust, UK
Dr Joe Hollins, Plantation House, Saint Helena
Malgosia Kaczmarska, Zoo d'Amneville, France
Toronto Zoo
Esmay van Strien

We would like to thank friend and colleague Professor Robert Young for supporting this book and writing the foreword.

We would like to acknowledge the support of the staff at Springer, especially Annette Klaus who supported the publication proposal and liaised with the publication team on our behalf.

Last but not least, we would like to thank the production team for creating the book from all the diverse content and media.

Sarah would like to thank her family and friends for their support during this project, especially her husband Julian, and their dogs Meg and Bea for their moral support and warmth.

Sabrina would like to thank her friends and colleagues, especially all who shared photos and stories which brighten this book, and elderly chimpanzee La Vieille at Tchimpounga Sanctuary in the Republic of the Congo who is one of the inspirations for this book.

Caring for elderly animals in human care is a responsibility bringing a range of joys and sorrows; as editors we are grateful to all those who have contributed to the co-creation of this resourceful and engaging book.

Contents

About the Authors

Heather Bacon OBE, BSc (Hons), BVSc, CertZooMed, SFHEA, MRCVS, is a veterinarian, the head of the School of Veterinary medicine at the University of Central Lancashire and a DEFRA-appointed UK zoo inspector. She has worked in the UK and internationally on zoological, veterinary education, conservation, and animal welfare projects for over 15 years.

Sabrina Brando is the director of AnimalConcepts, uniting people and organisation on 6 continents through interconnected online platforms combining animal, people, and planetary wellbeing. She is a psychologist with an MSc in Animal Studies and currently pursues a PhD at the University of Stirling in Scotland focusing on human wellbeing at an individual, leadership, and organisational level in zoos and aquariums. Since 2009, Sabrina is an instructor on the Primate Care Training Program for the Pan African Sanctuary Alliance; in 2019, she became Research Associate with the Smithsonian Institution and is a graduate of the Inner MBA class of 2022. She is a Certified Compassion Fatigue Professional and Optimize Eudaimonia Coach.

Sabrina teaches at various universities and colleges as well as working with many zoos, aquariums, sanctuaries, and other facilities worldwide, on topics such as animal wellbeing, behaviour, environmental enrichment, animal training, habitat design, and the human–animal relationship. AnimalConcepts has organised over 500 events worldwide since 2004, and Sabrina is a frequent invited and keynote speaker. She serves on zoo expert groups and collaborates on a wide variety of research projects. Sabrina serves as a reviewer on academic journals, contributes chapters to animal wellbeing books, and has extensively published popular and academic articles. She also works with governments writing policies and guidelines.

AnimalConcepts 'One Care' approach is based on optimal wellbeing for all, based on the premise that compassionate and holistic self-care (individuals) and we-care (teams, leadership, and organisational) approaches combined will support the commitment and desire of serving animals, people, and the environment.

Francis Cabana is one of the few wildlife nutritionists today. With his PhD in slow loris nutrition and ecology, his specialty has been with primates although his 14+ years of zoo experience has led him to be a jack of all trades. He has worked at zoos in Canada, the UK, Indonesia, and Singapore and given nutrition workshops and seminars all over the world, building capacity in the nutrition welfare sector.

Jon Coe is a landscape architect, environmental planner, and zoo designer with fifty-six years of professional experience including one hundred sixty planning and design projects for zoos, botanic gardens, animal sanctuaries, and nature parks in fifteen countries on six continents. He has published widely including over seventy scholarly papers, book chapters, and conference papers and lectured in twenty countries. https://joncoe.net/

Julian Chapman began his career in zoos in 1978 at Whipsnade Zoo. Over the next 35 years, he progressed to senior keeper roles at Paignton Zoo and Twycross Zoo where he worked with and learnt from many academics and veterinarians. Environmental enrichment was always his great passion, and he co-founded Shape-Regional, part of the Shape of Enrichment organisation. He is currently the co-director of Chapman Zoo Consultancy.

Sarah Chapman qualified as a vet in June 2000 and is an experienced zoo veterinary surgeon having worked with mammals, birds, reptiles, amphibians and invertebrates in a number of large zoos in the United Kingdom. She worked at Woburn Safari Park then at Paignton Zoo as Associate Veterinarian. She was Head of Veterinary Services for Twycross Zoo working with all four types of great apes and large mammals including big cats and hoof stock especially elephants. She was an Honorary Assistant Professor of the University of Nottingham Veterinary and developed the zoo's residency programme. In 2015, she became a RCVS recognised specialist in zoo and wildlife medicine having completed her certificate and diploma in this field. She has also completed a master's degree in Wild Animal Health working on Gorilla pathology as her dissertation. She has spent time teaching in China and Romania, speaking at international conferences and in Cameroon doing veterinary work with rescued primates. She is currently co-director of Chapman Zoo Consultancy providing specialist services to zoological collections and mentoring and training to veterinary and zoo professionals. She is an inspector for local authorities with respect to animal welfare and zoo licensing and has undertaken international work including zoo welfare assessments in Thailand, Sri Lanka and Vietnam. Her main interests include ageing animal care and providing optimal welfare for any animal under human care.

James Chatterton MSc, DipECZM(ZHM), MRCVS, Dr. James Chatterton is the Veterinary Services Manager at Auckland Zoo (New Zealand), having worked as a veterinarian since 2000 and in zoological medicine for over 13 years. He is a recognised specialist in zoological medicine (Zoo Health Management) by both the EBVS and RCVS and is veterinary advisor to ZAA's Ungulate TAG and the Kākāpō and Takahē Recovery Group.

Phillipa Dobbs , BSc (hons), BVetMed (hons), Cert AVP (ZM), MRCVS, started her career in zoo medicine in 2003 at Paignton Zoo. She gained a degree in Animal Science at Plymouth University and qualified in July 2011 from the Royal Veterinary

College. She has worked as a vet at Twycross Zoo since 2012. She attained her RCVS CertAVP (zoological medicine) and completed the ECZM residency (ZHM).

Mickey Gjerris (M.Th., PhD) is an associate professor in bioethics at the University of Copenhagen. For +20 years he has wondered about different aspects of the ethical relationship between humans and nature, focusing on areas such as animals, agriculture, biotechnology, climate change, and biodiversity. He is a frequent speaker at public events, media expert, and hugger of trees and consenting animals.

Lynette Hart , MA, PhD, is Professor of Anthrozoology and Animal Behavior and Vice Chair in the Department of Population Health and Reproduction at UC Davis School of Veterinary Medicine. Assistance dogs and dealing with animal death are amongst her long-term interests.

Andrew Kitchener is Principal Curator of Vertebrates at National Museums Scotland. His research interests include geographical variation and taxonomy of mammals, hybridisation, functional morphology and biomechanics, Scottish zooarchaeology, and the effects of captivity on endangered species. He is a member of the Scottish Animal Welfare Commission and leads the IUCN Cat Specialist Group's Cat Classification Task Force.

Phil Knowling is a writer and artist born in the South West of England. He was in charge of media and communications at a major zoo for 18 years and worked in conservation for over 20 years. He is the author of several books on local subjects, wrote for a county magazine, and is a published cartoonist.

Louise Lefrere BSc (Hons), PGDipVetPhys, PGCE, PGCert, MIRVAP(VP), qualified as Veterinary Physiotherapist at the University of Nottingham School of Veterinary Medicine and Science. She previously studied undergraduate animal behaviour and uses positive reinforcement training with a variety of animal species. Louise is a qualified lecturer, and alongside treating domestic and wild animals, she teaches workshops for veterinary and zoo professionals as part of Shackleton Veterinary Physiotherapy.

Debra Marrin started her animal care career in 1975. She has worked with a large variety of marine, exotic, wild, and domestic animals. She is currently at the San Francisco Zoo where she is the Director of Training and Behavioral Husbandry. Teaching animals to participate voluntarily in their own health care has been one of her main focuses for decades.

Franck L. B. Meijboom is Associate Professor of Ethics of Human-Animal interactions. He studied theology and ethics at the Universities of Utrecht (NL) and Aberdeen (UK). As Associate Professor he is affiliated to the Faculty of Veterinary Medicine and the Ethics Institute (Faculty of Humanities) of Utrecht University. Additionally, he is Head of the Centre for Sustainable Animal Stewardship.

Joachim Nieuwland is Assistant Professor at the Faculty of Veterinary Medicine, where he teaches veterinary ethics. He studied veterinary medicine at the University of Utrecht (NL) and philosophy at Tilburg University (NL) and received his PhD degree from Leiden University (NL).

Nicola (Nic) Field (BSc, MSc, FHEA) is the Co-Founder of Global Animal Welfare, an international NGO building capacity in wildlife sanctuaries. She lectures in animal behaviour and welfare. She has more than 25 years' experience working with captive wildlife internationally including Asia, Europe, and Africa including more than a decade with former bile-farmed bears rescued by Animals Asia.

Max Norman is a graduate of the Royal Veterinary College's MSc Wild Animal Biology programme with their postgraduate research project reviewing nonhuman primate personality and an avid advocate for evidence-based animal care practice. They have a particular interest in nutrition as well as comparative psychology and how psychology can be used to benefit animal welfare.

Amy Plowman was based at Wild Planet Trust (Paignton Zoo, Newquay Zoo, Living Coasts in the UK) for over 20 years conducting zoo nutrition research and formulating, monitoring, and evaluating zoo diets, especially for primates. She is currently Head of Conservation and Science for the Bumblebee Conservation Trust but still an active member of the European Zoo Nutrition Group.

Simone Schehka studied Biology at the FU Berlin and graduated in 2009 with a first-class degree in Ethology at the TiHo Hannover. She started her zoo career in 2009 as zoological assistant at Erlebniszoo Hannover. Since 2011, she worked as curator of mammals at the Allwetterzoo Münster. In July 2020, she assumed the role of CEO at Allwetterzoo Münster.

Friederike Schmitz graduated with a first-class degree in biology/ethology. She has a strong interest in animal training and has worked with a variety of species including marine mammals, bears, primates, and birds. Since 2013, she has been working as a keeper at the Allwetterzoo Münster, with a focus on animal welfare assessment.

Matthew Shackleton BSc (Hons), MScVetPhys, PGCert, PTLLS, MIRVAP(VP), has a master's degree in veterinary physiotherapy from the University of Nottingham School of Veterinary Medicine and Science, where he now works as a teaching associate. Matthew has a background in zoo-keeping and completed research into Komodo dragon locomotion. He treats wild and domesticated animals and conducts CPD for veterinary and zoo professionals as part of Shackleton Veterinary Physiotherapy.

Holistic Approaches to Optimal Wellbeing of Ageing Wild Animals

Sabrina Brando and Sarah Chapman

Tatzi. Tatzi is a male Asian Black Bear (*Ursus thibetanus*) who arrived at the Chengdu Bear Rescue Centre (CBRC) in October 2009 and was estimated at that time to be in his mid to late teens. He is now well into his geriatric years, and while he continues to thrive, his life has changed in many ways. He has always been a sweet and gentle soul with two main concerns: food and friends. He still loves to go outside and will eagerly gobble up any treat we have to offer. However, his social life is not what it once was as many of his former playmates have passed away. This combined with deteriorating mobility means Tatzi now lives in a smaller enclosure space with just two other bears. While he does not play as much as he used to, he gets a lot of extra attention through his involvement

S. Brando (✉)
AnimalConcepts, Teulada, Spain
e-mail: sbrando@animalconcepts.eu

S. Chapman
Chapman Zoo Consultancy, Birmingham, UK

S. Brando, S. Chapman (eds.), *Optimal Wellbeing of Ageing Wild Animals in Human Care*, https://doi.org/10.1007/978-3-031-30659-4_1

in a cooperative care program. As an older bear, he needs help wearing down his claws as well as close monitoring of a thyroid condition. His eagerness to participate (for perhaps honey or condensed milk) means we can now do conscious nail trims and blood draws, both of which keep him physically and mentally healthy in the later stages of his life

Abstract

Promoting good wellbeing in ageing animals benefits from a holistic approach considering a wide spectrum of physical and psychological aspects, through the lens of 24/7 across lifespan. A particular focus on the hours care staff is not usually at a facility is key, and professional zoos endeavour to provide animals with choice and control to meet their own needs and preferences when they want to. Ageing animals require proactive care such as screening, frequent behavioural observations, and assessing e.g., locomotion, social interaction, and feeding to detect early signs of age-related changes which could be of concern. This chapter discusses a variety of approaches for monitoring and assessment of animal wellbeing, which can support the changing needs and preferences of ageing animals in human care which are often overlooked or even disregarded and being labelled as 'just old'. The key message is that the wellbeing of ageing animals should not be dismissed or overlooked but prioritised so that their additional needs and preferences are addressed and managed respectfully and compassionately.

Keywords

Holistic · Wellbeing · Assessment · Ageing · Zoo · Compassion · Ethics · Monitoring · Individual · Social · 24/7 across lifespan

1 Introduction

We have an ethical responsibility to provide animals in zoos with environments that allows them to experience good welfare 24/7 across lifespan (Brando and Buchanan-Smith 2018). Professional zoos and aquariums, sanctuaries and other facilities caring for wild animals (henceforth zoos) have seen a significant evolution, today promoting 'cradle-to-grave', birth to death (Seidensticker and Doherty 1996), and '24/7 across lifespan' (Brando and Buchanan-Smith 2018). Animal welfare science and practice focuses on the wellbeing of individuals. Animal welfare is defined as 'an animal's collective physical, mental, and emotional states over a period of time, and is measured on a continuum from good to poor' (Association of Zoos and Aquariums [AZA 2023]). When considering the primary components of the welfare, the Association of Zoos and Aquariums' (AZA) Animal Welfare Committee further explains this as 'an animal typically experiences good welfare when healthy, comfortable, well-nourished, safe, able to develop and express species-typical relationships, behaviours, and cognitive abilities, and not suffering from unpleasant states such as pain, fear, or distress. Because physical, mental, and emotional states may be dependent on one another and can vary from day to day, it is important to consider these states in combination with one another over time to provide an assessment of an animal's overall welfare status'. There are many definitions of animal welfare: Animal wellbeing comprises the psychological and physical experiences, including all needs and most preferences, of an individual animal as perceived by herself/himself and the ability to exert agency over one's life to a meaningful extent. A wide variety of experiences, ranging from positive through negative, affect animal wellbeing. A variety of inputs may all act at the same time and may be synergistic rather than simply additive in their consequences. Therefore, the evaluation of wellbeing should consider all measurable factors and how they interact. The focus should be on monitoring and assessing animals based on their experiences and on promoting agency, choice, control and predominantly positive wellbeing (Brando 2022). The terms 'animal welfare' and 'animal wellbeing' have both been used over the years (Moberg and Mench 2000), to describe the state of the animal, and are used in this chapter and throughout the book interchangeably. Contemporary zoos today have professionalised their approaches to the whole of life care, including expanding programs to support animals through the last phase of life, which is also called senescence, the process of becoming old and showing the effects of being old. It also comprises specific attention to death.

The term 'professional' refers to excellence in animal care and wellbeing programs, incorporating the latest best practice, rooted in science and evidence-based approaches. Being professional means having an up-to-date theoretical and practical knowledge of animal welfare–related topics (Brando and Coe 2022). Professional also means having an auto-critical approach as well as a dynamic approach as science and best practices are evolving and changing depending on the available information, which might require to breaking institutional habits (Olle 2018). The ability to challenge the status quo through the review of emerging scientific evidence and ethical considerations and frameworks (De Mori et al. 2019) remains necessary and forms an important aspect in the process of changing long-held beliefs or practices. Being professional means to continue asking whether this is the best that we can be and whether this is what is in the best interest of the animal 24/7 and an 'animal-first' approach and to achieve the goals of education, research, conservation and recreation (Brando and Coe 2022). Professional animal welfare programs today strive to operate on an evidence-based approach as set out by Maple (2008) for the 'empirical zoo' and include evidence-based animal welfare programs (Melfi 2009). These programs reflect best practice processes of care and conducting research, including the importance of good human–animal relationships and interactions (Hosey and Melfi 2012). Choice and control should also revolve around human–animal interactions and relationships, including staff, the general public and others who interact or are in the direct or indirect space of the animals.

2 Choice and Control

Modern and professional zoos and aquariums strive to provide their animals with agency: the ability to exert choice and control over doing what they want and when they want to do it and to meet their own needs and preferences when they choose. The benefits of choice and control for animal wellbeing have been well described since the 1990s (Chamove and Anderson 1989; Snowdon and Savage 1989) and more recently in reviews by Brando (2009) and Allard and Bashaw (2019). A lack of control over one's environment has been empirically described as leading to behavioural and physiological wellbeing concerns (Mineka and Hendersen 1985; Perdue et al. 2014). There is also evidence demonstrating that simply providing animals with the ability to make their own choices, regardless of whether they are acted upon or not, is rewarding and beneficial to animals (Owen et al. 2005; Leotti et al. 2010; Kurtycz et al. 2014). Giving more agency over their own lives is thus a common goal of many welfare approaches, allowing animals to self-maintain with decreased dependence on their human caregivers (Brando and Buchanan-Smith 2018; Coe 2018; Allard and Bashaw 2019).

Modern animal wellbeing concepts encompass all aspects of welfare, including physical, behavioural and psychological domains, and are increasingly emphasising the importance of positive emotional states (Mellor and Beausoleil 2015) and the centrality of the animal's feelings to their wellbeing (Veasey 2017). Such schools of thought have developed from those of early modern zoos, which predominantly focused on the survival of the animal, towards coping, and ultimately into the modern values of providing animals opportunities to thrive and promoting optimal animal welfare. Indeed, while we have the Five Domains Model (Mellor et al. 2020), we also have other modern frameworks such as the '5 Cs + N' welfare concept (Coe 2018) that focuses on an animal's ability to achieve competence, to have choice, to take control, to engage in complexity and challenges and to experience novelty. The provision of complex and enriching environments that offer animals plentiful opportunities to exert choice and control is a core component of professional and modern animal care. Progress through research and the implementation of best practices in areas such as veterinary care, nutrition, habitat design, animal training and environmental enrichment are keys to promoting, monitoring and assessing animal welfare, including at times outside regular working hours. Particular and specific attention should be drawn to back-of-house (BOH) or off-exhibit areas, where animals may variably be locked inside overnight or at other times. On average, an animal caregiver's working day consists of 8 hours during which zoos are open. Consequently, animals may spend as much as two-thirds of their day inside BOH areas without 24/7 open access – equating to a total of 243 days out of a year (Brando and Coe 2022).

The concept of providing animals choice and control should be used proactively, with an element of common sense and compassion. The choices given should be meaningful from the perspective of the animal, and care staff must accept the choices made by the animals. For example, if we provide an animal with a choice between two non-preferred options – such as two social partners it does not particularly get along with – have we truly provided a choice that serves to improve their wellbeing? If all the choices we have provided in a circumstance lead to negative outcomes, does that truly feel like control from the animal's perspective (Brando and Coe 2022)?

3 Animal Welfare Approaches

Animal welfare science has an extensive history, with some schools of animal welfare thought originating decades if not centuries in the past. Although the term 'animal welfare' has not always been used to describe the concept, humans have long been aware of the effects that environment, nutrition, housing, handling, social structure and interactions with humans can have on animals (Hemsworth and Coleman 2010). In the past, animal welfare approaches predominantly focused on an animal's ability to lead a life that can be perceived as natural (Duncan and Petherick 1991), how the animal feels (Dawkins 1998; Duncan and Petherick 1991) and approaches such as the biological functioning approach (Broom 1986). Such approaches are now seen as being dynamically integrated components of whole animal welfare. In the modern days, animal welfare science is approached from a holistic perspective that considers and encompasses all the physiological, behavioural and psychological components on an additive basis, wherein each component compounds to create an overarching view of the animal's overall wellbeing.

There is an increasing emphasis on the importance of positive states and the feelings of the animals from their perspective (Wemelsfelder et al. 2001; Wemelsfelder 2007; Mellor and Beausoleil 2015; Mellor 2016; Veasey 2017; Brando and Buchanan-Smith 2018). Positive welfare is not something that we can give to animals, but instead something that each individual animal experiences. A holistic approach to promoting optimal welfare for all animals at an individual level is essential to enhancing the experience of wellbeing from the perspective of the animals. Animal welfare assessment for zoo-housed animals has commonly used frameworks and approaches used for farm animals, such as the four key welfare principles and 12 welfare criteria proposed by Welfare Quality (2009) and adapted to 14 points with an increased focus on behaviour and positive emotions for captive wild animals by Brando and Buchanan-Smith (2018). There is also the Five Domains Model (Mellor and Beausoleil 2015), which forms the core of many animal welfare strategies around the world including that of the World Association of Zoos and Aquariums (WAZA) (Mellor et al. 2015). The Five Domains Model stresses that animal welfare assessment should follow an integrated and a holistic approach, including the four physical domains of nutrition, environment, physical health and behaviour and the mental domain. Modern reviews of the Five Domains Model have put an increased emphasis on the importance of human–animal interactions and their ability to have both positive and negative effects on the animals (Mellor et al. 2020).

4 Assessing Animal Welfare

Animal welfare science concerns individuals, in terms of the whole animal: We look at the inputs (care) that we provide them, at the independent variables that may impact their experience and also at the outputs, or the dependent variables that give us information regarding how the animals are reacting to their lived experience across time and context. There exist multiple bespoke software systems for monitoring and evaluating animal behaviour such as ZIMS Care and Welfare, ZOOMONITOR, WelfareTrak and TRACKS, which can be used by animal caregivers to understand animal welfare over time.

Scientific monitoring of animal behaviour has been conducted for many years (Kleiman 1992), although it is a relatively new concept to consistently use animal-based measures to assess animal welfare from a whole animal perspective. Animal-based assessment is mostly used for specific aspects of animal care, such as the effectiveness of environmental enrichment or techniques for reducing stereotypic and other undesirable behaviours. Practical and modern animal welfare assessment should integrate outcome-based measures of animal behaviour to provide more insights into individual welfare states (Meagher 2009). For example, technologies such as dataloggers can be used to record information such as locomotion, sleeping, habitat use and social preferences, which can be used to gauge animal welfare and assess changes that may indicate concerns or opportunities for optimal wellbeing. Other observable behaviours that may give us insights into internal welfare states include play, rest, vigilance and vocalisations; we may also assess indirect evidence such as wounding, the quality of nests constructed, the use of enrichment and other devices and stool quality. Using both inputs and outputs to measure animal welfare is essential to achieving a complete and an integrated welfare assessment approach, as discussed by Whitham and Wielebnowski (2009) and Kagan and Veasey (2010).

Maple (2008) set out an evidence-based approach for the 'empirical zoo' that highlights evidence-based animal welfare programs as central to reflecting best practice of all processes, including care, research, exhibit design (Fuller et al. 2016), nutrition (Hulbert et al. 2017), environmental enrichment (Bashaw et al. 2016), positive reinforcement training (Spiezio et al. 2017; Brando and Norman 2023) and human–animal relationships and interactions (Hosey and Melfi 2012). Such programs have made advances as a result of decades of research and the development of new best practices, although there are still many areas of concern in need of further exploration such as night-time welfare and confronting back-of-house traditions (Brando and Coe 2022). Indeed, an increased focus on off-hours care is an essential next step in optimising animal wellbeing in zoos and aquaria; animals require and deserve as much choice, complexity and agency during these hours as they do during the day when staff are present on site, and their welfare needs may be different at night from during the day (Brando et al. 2023).

The observations, knowledge and skills of animal care staff who work with the animals daily are integral to holistic zoo animal welfare approaches. The caregivers who work directly with the animals know them better than anyone else, balancing the art of compassion with the science of animal care. Daily keeper reports, for example, might qualitatively describe an animal's behaviour that day, which together with the incorporation of objective quantitative assessments can be used to capture the individual's perspective, subjective experiences and their emotions and feelings (Whitham and Wielebnowski 2013). In respect of the importance of the caregiver's experience and knowledge, Kagan et al. (2015) presented a zoo animal welfare assessment framework based on the combined values of science, compassion and sustainability. Kagan's framework is centred on a universal outline consisting of (1) Institutional Philosophy & Policy, (2) Programmatic Structure & Resources, (3) Execution and (4) Evaluation to ensure great welfare for each individual animal. Kagan also highlighted the importance of training staff to conduct thorough and holistic animal welfare assessments; "Staff must understand the difference between inputs (what is provided for animals) and outputs (what animals experience) and why assessing each is important." The integration of qualitative and quantitative assessment is particularly necessary when considering welfare during off-hours when caregivers are not present, allowing for improved wellbeing from the perspective of the animals without necessitating increased staff presence overnight.

Professional animal welfare programs today strive to operate on an evidence-based approach as set out by Maple (2008) for the 'empirical zoo' and to include evidence-based animal welfare programs (Melfi 2009). These programs reflect best practice processes of care and conducting research to,for example, exhibit design (Fuller et al. 2016), nutrition (Hulbert et al. 2017), the use of environmental enrichment (Bashaw et al. 2016) and positive reinforcement training (Spiezio et al. 2017; Brando and Norman 2023) to manage animals in their care and the importance of good human–animal relationships and interactions (Hosey and Melfi 2012). Providing complex and enriching environment that offers choice and control is a core building block of professional animal care. These programs have made advances through research and implementation of best practices in veterinary care, nutrition, habitat design, animal training and environmental enrichment. However, welfare at night, when staff are not present or only a few staff are present (off-hours), is not usually discussed. This is worrying because animals also require and deserve the choice, complexity and control of their environment during these hours, and their needs may also differ during these hours. Humans, like most other primates, are diurnal. As such, humans involved in animal welfare tend to focus on changes made during the daylight hours. Video cameras and recording systems make behavioural observation of primates during off-hours feasible, by analysing recordings or via live feeds such that staff can monitor primates in real time from off-site (Spiezio et al. 2017; Brando and Norman 2023). Assessing animals individually promotes focusing on what is in the best interest of that individual.

5 Ageing Animals First

The animal's first approach, as proposed and described by Lincoln Park Zoo, is not only helpful in determining the needs of animals in different programs and activities, but also when caring for ageing animals. An 'animal-first' approach highlights the attention to promoting optimal animal welfare as a condition for all the goals zoos have (Brando and Herrelko 2021) and animal wellbeing at the core of them. The cradle-to-grave approach requires the prioritisation of caring for ageing animals and questions from the animal-first approach including the following: (1) Does the animal have choice? (2) Does the animal have control? (3) Can the animal remain in the comfort of his/her habitat? Instead of asking about a program, we can ask whether the care and decisions are taken based on the benefit(s) to the animal.

5.1 Attention to Life Stages: Senescence and Death

The natural history of an animal, its biology, ecology and diet, sensory systems, natural habitat, social structure, major life history events, activity patterns and human–animal interactions are among the many topics considered when developing species-specific animal welfare programs. Looking at the life cycle of a species, we find different life stages commonly divided into birth, baby, juvenile, adolescence, reproductive age, senescence and death. When we consider different life stages, we can identify key features and considerations likely to be of importance to the welfare of the species and zoom in on the needs and preferences of an individual. Table 1 below provides examples highlighting these features across a wide range of species. To manage a species appropriately in captivity, it is important to find out about each of these key considerations and develop a management plan accordingly (Brando and Buchanan-Smith 2018).

5.2 A Spotlight on the Social Life of Ageing Animals

There are many aspects to consider in an animal's life as discussed throughout this chapter. In this section, we will shine a spotlight on the social life of animals and how it pertains to ageing animals. It is important to note that species and individual animals are social in different ways, impacting types of behaviours, distance between animals, needs for areas to socialise and sleep, among others. Understanding the social needs of a species and preferences of an individual allows us to support animals in a phase where many changes can occur and prevent undesired and needless separation and isolation. Social interactions help group-living animals manage socio-environmental challenges and are central to survival and reproductive success. Social interactions also contribute to feelings of contentment, security and joy. Social behaviours and relationships can change across the lifespan, a phenomenon referred to as 'social ageing', and age-dependent changes in social behaviours can modulate how fitness changes with age (Siracusa et al. 2022). This in turn can have effects on behaviour and social interactions within the group and with human care staff. Siracusa et al. (2022) suggested that explanations for social ageing can be categorised into three groups: changes in sociality that (a) occur as a result of senescence; (b) result from adaptations to ameliorate the negative effects of senescence; and/or (c) result from positive effects of age and demographic changes. Considering these aspects that can affect changes in sociality allows for a more holistic

Table 1 Examples of key features at different stages of an animal's life, specifically senescence and death. Adapted from Brando and Buchanan-Smith (2018)

Senescence
Social behaviour: Similar affiliative social behaviour in old and younger adult chimpanzees implies continued need for social housing with advancing age (Baker 2000).
Physical decline, reduced mobility and pain: Dental problems and osteo-spondyloarthroses are common in animals in zoo and appear to increase with age in bears, big cats, great apes, babirusas and pygmy hippos (Kitchener and Macdonald 2002). See also Chapters "Veterinary Care of Ageing Zoo Animals: Description of a Proactive Approach", "Pain: Physiology, Recognition, and Management in Zoo Animals" and "The Longevity Legacy: The Challenges of Old Animals in Zoos".
Behaviour and physical aspects: Changes in behaviour and physical characteristics in octopus are indicators of senescence (Anderson et al. 2002).
Nutrition: Ageing rate is higher under free-ranging conditions than in captivity. Differences in ageing rates between captive and free-ranging ruminants increased as species were more specialised on grass diets (Lemaître et al. 2013). See also Chapter "Supporting Geriatric Zoo Animal Welfare through Nutrition".
Cognitive decline in memory: Ageing negatively affects performance in cognitive tests in rhesus monkeys (Herndon et al. 1997).
Sex-specific actuarial and reproductive senescence: Males tended to have a higher juvenile mortality and a faster senescence than females, while females had more pronounced reproductive senescence, that is, decline of reproductive parameters with age (Tidière et al. 2021).
Death
Social aspects of sickness and dying: The responses of a group of chimpanzees to a dying female included pre-death care, close inspection and testing for signs of life at the moment of death, all-night attendance by the deceased's adult daughter, cleaning the corpse and later avoidance of the place where death occurred. In some cases, it may be more humane if the elderly apes are allowed to die naturally in a supported social setting than separating them for veterinary care or euthanasia (Anderson et al. 2010). See also Chapter "Mourning-Like Behaviour in a Malayan Sun Bear".
Euthanasia: Post-mortem findings on correct end-of-life decisions can be validated by a scoring system to evaluate physical condition and quality of life in geriatric zoo mammals (Föllmi et al. 2007). See also Chapters "Veterinary Care of Ageing Zoo Animals: Description of a Proactive Approach" and "Euthanasia of Geriatric Zoo Animals: Decision Making and Procedure".

understanding in ageing animal care and wellbeing program development. Using the Five Domains Model, mentioned earlier in this chapter, can assist in anticipating the types of experiences animals may have now and in the future and requires reflection on proactive measures to support positive social ageing. Environments, outdoor and indoor, can be designed and/or refurbished for group-living including specific needs and preferences, for example, quiet areas to get away from others, comfortable resting and interacting areas and adapted treatment areas allowing animals to live together for as long as possible. All these efforts prevent individuals to be moved away from a complex habitat and indoor areas with other, to being separated off and/or isolated from others. This is of particular importance in an organisational culture where euthanasia is not acceptable nor practiced or is illegal by law.

6 Summary

This chapter described a variety of approaches for monitoring and assessing animal wellbeing with a special attention given to ageing animals. Other chapters within this book provide more details regarding the various aspects of individual animal care, which all feed into a holistic approach to promoting optimal wellbeing. In the authors' experience, the changing needs and preferences of many ageing animals in human care are often overlooked or even disregarded and being labelled as 'just old'. The key message is that the wellbeing of ageing animals should not be dismissed or overlooked but prioritised so that their additional needs and preferences are addressed and managed respectfully and compassionately.

References

Allard SM, Bashaw MJ (2019) Empowering zoo animals. In: Kaufman AB, Bashaw MJ, Maple TL (eds) Scientific foundations of zoos and aquariums: their role in conservation and research. Cambridge University Press, Cambridge, pp 241–273

Anderson RC, Wood JB, Byrne RA (2002) Octopus senescence: the beginning of the end. J Appl Anim Welf Sci 5(4):275–283

Anderson JR, Gillies A, Lock LC (2010) Pan thanatology. Curr Biol 20(8):R349–R351

AZA Animal Welfare Committee. https://www.aza.org/animal_welfare_committee. Accessed 6 January 2023

Baker KC (2000) Advanced age influences chimpanzee behavior in small social groups. Zoo Biology: Published in affiliation with the American Zoo and Aquarium Association 19(2):111–119

Bashaw MJ, Gibson MD, Schowe DM, Kucher AS (2016) Does enrichment improve reptile welfare? Leopard geckos (Eublepharis macularius) respond to five types of environmental enrichment. Appl Anim Behav Sci 184:150–160

Brando SICA (2009) Exploring choice and control opportunities applied in enrichment and training. Presentation, Paignton

Brando S (2022) Marine mammals. In: Routledge handbook of animal welfare. Routledge, London, pp 293–308

Brando S, Norman M (2023) Handling and training of wild animals: evidence and ethics-based approaches and best practices in the modern zoo. Animals 13(14):2247

Brando S, Buchanan-Smith HM (2018) The 24/7 approach to promoting optimal welfare for captive wild animals. Behav Process 156:83–95

Brando S, Coe J (2022) Confronting back-of-house traditions: primates as a case study. J Zool Bot Gard 3(3):366–397

Brando S, Herrelko ES (2021) Wild animals in the city: considering and connecting with animals in zoos and aquariums. In: Animals in our midst: the challenges of co-existing with animals in the anthropocene. Springer, Cham, pp 341–360

Brando S, Vitale A, Bacon M (2023) Promoting good nonhuman primate welfare outside regular working hours. Animals 13(8):1423

Broom DM (1986) Indicators of poor welfare. Br Vet J 142(6):524–526

Chamove AS, Anderson JR (1989) Examining environmental enrichment. In: Segal EF (ed) Housing, care and psychological Well-being of captive and laboratory primate. Noyes Publications, New Jersey, pp 183–202

Coe JC (2018) Embedding environmental enrichment into zoo animal facility design. Retrieved from http://www.zoodesignconference.com/wp-content/uploads/2017/10/03.-Coe-Embedding-enrichment.pdf

Dawkins MS (1998) Evolution and animal welfare. Q Rev Biol 73(3):305–328
de Mori B, Ferrante L, Florio D, Macchi E, Pollastri I, Normando S (2019) A protocol for the ethical assessment of wild animal–visitor interactions (AVIP) evaluating animal welfare, education, and conservation outcomes. Animals 9(8):487
Duncan IJH, Petherick JC (1991) The implications of cognitive processes on animal welfare. J Anim Sci 69(12):5017–5022
Föllmi J, Steiger A, Walzer C, Robert N, Geissbühler U, Doherr MG, Wenker C (2007) A scoring system to evaluate physical condition and quality of life in geriatric zoo mammals. Anim Welf 16:309–318
Fuller G, Raghanti MA, Dennis PM, Kuhar CW, Willis MA, Schook MW, Lukas KE (2016) A comparison of nocturnal primate behavior in exhibits illuminated with red and blue light. Appl Anim Behav Sci 184:126–134
Hemsworth PH, Coleman GJ (2010) Human-livestock interactions: the stockperson and the productivity of intensively farmed animals. CABI, Wallingford
Herndon JG, Moss MB, Rosene DL, Killiany RJ (1997) Patterns of cognitive decline in aged rhesus monkeys. Behav Brain Res 87(1):25–34
Hosey G, Melfi V (2012) Human–animal bonds between zoo professionals and the animals in their care. Zoo Biol 31(1):13–26
Hulbert AJ, Hunt KA, Rose PE (2017) A multi-zoo investigation of nutrient provision for captive red-crested turacos. Zoo Biol 36(2):152–160
Kagan R, Veasey J (2010) Challenges of zoo animal welfare. Wild mammals in captivity: principles and techniques for zoo management. University of Chicago Press, Chicago, pp 11–21
Kagan R, Carter S, Allard S (2015) A universal animal welfare framework for zoos. J Appl Anim Welf Sci 18(sup1):S1–S10
Kitchener, A., & Macdonald, A. A. (2002). The longevity legacy: the problem of old animals in zoos
Kleiman DG (1992) Behavior research in zoos: past, present, and future. Zoo Biol 11(5):301–312
Kurtycz LM, Wagner KE, Ross SR (2014) The choice to access outdoor areas affects the behavior of great apes. J Appl Anim Welf Sci 17(3):185–197
Lemaître JF, Gaillard JM, Lackey LB, Clauss M, Müller DW (2013) Comparing free-ranging and captive populations reveals intra-specific variation in aging rates in large herbivores. Exp Gerontol 48(2):162–167
Leotti LA, Iyengar SS, Ochsner KN (2010) Born to choose: the origins and value of the need for control. Trends Cogn Sci 14(10):457–463
Maple TL (2008) Empirical zoo: opportunities and challenges to a scientific zoo biology. Zoo Biology: Published in affiliation with the American Zoo and Aquarium Association 27(6):431–435
Meagher RK (2009) Observer ratings: validity and value as a tool for animal welfare research. Appl Anim Behav Sci 119(1–2):1–14
Melfi VA (2009) There are big gaps in our knowledge, and thus approach, to zoo animal welfare: a case for evidence-based zoo animal management. Zoo Biology: Published in affiliation with the American Zoo and Aquarium Association 28(6):574–588
Mellor DJ (2016) Updating animal welfare thinking: moving beyond the "five freedoms" towards "a life worth living". Animals 6:21
Mellor DJ, Beausoleil NJ (2015) Extending the 'five domains' model for animal welfare assessment to incorporate positive welfare states. Anim Welf 24(3):241
Mellor DJ, Hunt S, Gusset M (2015) Caring for wildlife: the world zoo and aquarium animal welfare strategy. WAZA Executive Office, Gland
Mellor DJ, Beausoleil NJ, Littlewood KE, McLean AN, McGreevy PD, Jones B, Wilkins C (2020) The 2020 five domains model: including human–animal interactions in assessments of animal welfare. Animals 10(10):1870
Mineka S, Hendersen RW (1985) Controllability and predictability in acquired motivation. Annu Rev Psychol 36(1):495–529
Moberg GP, Mench JA (eds) (2000) The biology of animal stress: basic principles and implications for animal welfare. CABI

Olle CD (2018) Breaking institutional habits: a critical paradigm for social change agents in psychology. Couns Psychol 46(2):190–212

Owen MA, Swaisgood RR, Czekala NM, Lindburg DG (2005) Enclosure choice and Well-being in giant pandas: is it all about control? Zoo Biology: Published in affiliation with the American Zoo and Aquarium Association 24(5):475–481

Perdue BM, Evans TA, Washburn DA, Rumbaugh DM, Beran MJ (2014) Do monkeys choose to choose? Learn Behav 42(2):164–175

Seidensticker J, Doherty JG (1996) Integrating animal behavior and exhibit design. In: Kleinman D et al (eds) Wild mammals in captivity. University of Chicago Press, Chicago, pp 180–190

Siracusa ER, Higham JP, Snyder-Mackler N, Brent LJ (2022) Social ageing: exploring the drivers of late-life changes in social behaviour in mammals. Biol Lett 18(3):20210643

Snowdon CT, Savage A (1989) Psychological Well-being of captive primates: general considerations and examples from callitrichids. In: Housing, care and psychological Well-being of captive and laboratory primates. Noyes Publications, Park Ridge, NJ, pp 75–88

Spiezio C, Vaglio S, Scala C, Regaiolli B (2017) Does positive reinforcement training affect the behaviour and welfare of zoo animals? The case of the ring-tailed lemur (Lemur catta). Appl Anim Behav Sci 196:91–99

Tidière M, Müller P, Sliwa A, Siberchicot A, Douay G (2021) Sex-specific actuarial and reproductive senescence in zoo-housed tiger (Panthera tigris): the importance of sub-species for conservation. Zoo Biol 40(4):320–329

Veasey JS (2017) In pursuit of peak animal welfare; the need to prioritize the meaningful over the measurable. Zoo Biol 36(6):413–425

Welfare Quality (2009). http://www.welfarequalitynetwork.net/en-us/reports/assessment-protocols. Accessed 6 January 2023

Wemelsfelder F (2007) How animals communicate quality of life: the qualitative assessment of behaviour. Anim Welf 16:25–31

Wemelsfelder F, Hunter TE, Mendl MT, Lawrence AB (2001) Assessing the 'whole animal': a free choice profiling approach. Anim Behav 62(2):209–220

Whitham JC, Wielebnowski N (2009) Animal-based welfare monitoring: using keeper ratings as an assessment tool. Zoo Biology: Published in affiliation with the American Zoo and Aquarium Association 28(6):545–560

Whitham JC, Wielebnowski N (2013) New directions for zoo animal welfare science. Appl Anim Behav Sci 147(3-4):247–260. Chicago

Aging Gracefully: Compassion for Nonhuman Animal Elders

Joachim Nieuwland and Franck Meijboom

Margo. Margo is a 52-year-old chimpanzee (*Pan troglodytes*) living at Artis Zoo in the Netherlands. She came from the wild and has had a long life including as a high-ranking female and the mother of four babies. She is loved and cared for by the troop, giving her the nest she chooses and treating her with respect. Only the teenagers like to tease her by poking her with sticks; she does her best to make them stop. She used to run after the one poking her to grab the stick and break it in half, but now that she is older, she asks others for help. Margo likes to interact with the caretakers when food and enrichment are delivered, and some of her favourite foods are endive and bananas. For enrichment, she likes the tubes with some peanut butter, honey, or a green smoothie that she needs to manipulate with a stick to reach and eat. She loves building large nests with blankets and sheets.

J. Nieuwland (✉) · F. Meijboom
Utrecht University, Utrecht, The Netherland
e-mail: j.nieuwland@uu.nl; F.L.B.Meijboom@uu.nl

S. Brando, S. Chapman (eds.), *Optimal Wellbeing of Ageing Wild Animals in Human Care*, https://doi.org/10.1007/978-3-031-30659-4_2

Margo has a kind personality and comforted her daughter Saphira when Sabra died. Her best friends are Leentje and Quincy, and she is still seen interacting and grooming with other group members

Abstract

Many nonhuman animals (hereafter animals) grow old within zoos. Aging animals undergo innumerable bodily and mental changes, some of which lead to suffering and keep them from flourishing while others move them towards other roles, opportunities, and standing. Those who take care of aging animals are confronted with various moral considerations of what it takes to care for these animals. Increased human intervention in the later stages of the lives of animals can make it more difficult to find one's bearings. Conspecifics in the wild gradually become less of a guide for shaping the lives of geriatric animals in confined spaces. Perhaps to take care is to be responsive to the individual animal, both to understand how they have been shaped by living their lives within a zoo and to explore and make available ways for them to continue to exert agency over the later stages of their lives. Moral deliberation helps to engage with the moral issues of taking care. Compassion comprises a vital part of moral deliberation and appears especially promising to care for animals in the later stages of their lives.

Keywords

Nonhuman animals · Zoo · Aging · Animal elders · Animal welfare · Animal agency · Wisdom · Compassion

1 Introduction: Caring about Animal Elders

The start and end of animal life often attract a lot of attention. A newborn polar bear or elephant in a zoo is quickly picked up by media and the public. Likewise, the end of animal life often raises discussion, especially when death comes unexpected or is unwanted, for instance, in the case of the killing of the gorilla Harambe, a 17-year-old western lowland gorilla (*Gorilla gorilla gorilla*). To protect the 3-year-old boy that had entered the enclosure, zoo personnel had to quickly decide upon the best course of action, resulting in the fatal shooting of Harambe. This common emphasis on the start and end of animal life in the confinement of zoos makes the stages in between, especially the aging animal, a blind spot. Do we pay enough attention to the frailty of animals of old age in zoo enclosures? In empirical terms, aged animals confront us with questions of changes in, among others, health, individual and social behaviour, and feeding behaviour. In this chapter, however, we focus on the moral reasons to focus on the aging animals. The process of aging raises moral considerations. While aged animals generally need more care and attention, what this involves is not always self-evident. What does it take to take care?

In this chapter, we first discuss the concept of aging. Next, we focus on the impact of aging in terms of animal welfare and animal agency. As a background for these considerations, we bring out the cultural setting in which these animal elders find themselves, as well as the changes to how animals are viewed within human cultures. From that point onwards, we develop a view of morality and moral deliberation in the context of taking care of aging animals. We highlight compassion as highly relevant for taking care of aging animals, as it fosters a genuine curiosity for how animal elders experience their lives alongside the motivation to take care in whatever way appears the most appropriate.

2 Aging

Aging refers to the relation between an individual and time, usually concerning specifically the later stages in life. An individual's age relates her or him statistically to the life span of a species, population, or another subset of individuals, for instance, a community of animals confined in a zoo setting. In that sense, it is a relative term, subject to alteration. The baseline for being old as a human being, for instance, has dramatically shifted over the last two centuries. A similar shift has introduced differences between many but not all animal species when comparing confinement and the wild (Tidière et al. 2016).

Aging implies change. Most basically, aging tracks the progress of time, which might be interpreted in terms of a linear progression as ever-continuing change. The present appears to slide into the past, always edging into the future. With the flow of time, physical bodies change just as much as mental experience. Such changes are not necessarily for better or worse. However, when an individual reaches the later phases of her or his life span, senescence—the processes of physiological deterioration—may lead to disability and disease, potentially hindering individual flourishing (Crews 2007).

When animals grow old, like humans, they inevitably undergo various changes in their bodies, cognitive functioning, and behaviour (Krebs et al. 2018). These changes affect individuals in various ways across time, broadening welfare considerations across the life span of individuals (Brando and Buchanan-Smith 2018). A slow onset of arthritis, for instance, could make it increasingly difficult to roam the territory or maintain social status. Aging not only directly negatively affects welfare, but it also brings up the question to us what it implies for animals to age gracefully. In other words, it invites one to think about what wellbeing and flourishing mean for animals in the final stages of their lives.

2.1 Welfare of Aging Animals

Caring for aging animals calls for a responsiveness to the individual and the specific challenges that are raised simply by growing old. What this entails in practice, however, requires context-sensitive deliberation among those responsible for taking care. In some cases, especially where suffering is profound and not expected to diminish in due time, euthanasia should be considered. Here, we understand euthanasia as the ending of a life being genuinely in the interest of the individual and with respect for animal welfare in the process (Fawcett 2013). In other cases, carefully starting up treatment can improve the welfare of aging animals, allowing them to continue to age. All these cases, whether considering euthanasia or determining treatment options, are permeated by moral concern for the individual elderly animal.

Care for the deteriorating wellbeing of an elderly animal sets these animals apart from the life they would have lived outside of human captivity. To the extent that the wild provides inspiration or a guide for shaping the lives of captive aging animals, this natural standard may become more and more difficult to approximate, if so desired. Increasingly individually tailored medical care and adaptation of the environment provide further opportunities for animals to become more geriatric. Paradoxically, while zoos strive to approximate the wild, one is inescapably confronted with the care for their lives in a way unparalleled in the wild. The older the animals get, and/or the more they are cared for, the less natural or wild their lives appear to become. This calls for careful reflection on what makes a (sufficiently) good animal life, into these later stages of life.

Animal welfare functions as an important (though not the sole) guide in such reflection. Especially as animal welfare (just like a conception of what makes a good life) involves an evaluative judgment, there are different views on what it entails specifically. That is, one's values determine in part what comprises animal welfare. David Fraser distinguishes between three prevalent value-laden assumptions often discernable in various animal welfare conceptions. These include (1) overall health and functioning, (2) experiential (affective) states or feelings, and (3) living according to one's nature in terms of species-specific functioning within an appropriate environment (Fraser 2009).

People will differ in their emphasis on these perspectives based on the values that they endorse. For some, aging diminishes animal welfare as soon as it limits natural living. Others look at the way in which mental and bodily changes affect the animal and limit his or her ability to adapt to the prevailing circumstances. Moreover, they might consider modifying the environment further away from the natural ideal, to adapt the environment to the altered abilities of the individual animals (Krebs et al. 2018).

2.2 Wisdom in Agency

There is more to the question of aging gracefully. Apart from the moral values at play in conceptions of animal welfare as indicated by Fraser, should aging animals have opportunities to make decisions about their own lives? Should one foster their agency, to allow and facilitate individual intentional action, that is, acting with a goal in mind (Delon 2018)? What range of opportunities is available to animal elders to shape their own lives?

Agency opens the door to freedom. It might do so to the extent of freeing animals from the confinement and captivity altogether. Less sweeping, at least it offers the freedom to explore a range of opportunities. Doing so requires (1) a supportive environment and (2) individual ability. At the intersection of the individual and her or his environment, one once again must place her or his values on the table. If one values agency, what range of opportunities, in terms of both scope and content, should be available to animals?

The agency of aging animals itself reveals rather autobiographically the life story of an individual. Tracing back the historical and socioecological roots of an animal offers some insights into the agency he or she displays. Agency reflects how the individual has interacted with his or her surroundings and learned—and in some cases being trained, for instance to make possible veterinary treatment (Krebs et al. 2018)—to extend the range of intentional behaviour and shape one's own life (Donaldson and Kymlicka 2015; Donaldson and Kymlicka 2016).

Animal agency invites one to take an overview of the whole life of an animal: to both identify and promote opportunities for animals to make meaningful choices about their own lives, throughout their lives. It invites to consider what opportunities these animals have and, importantly, should have to shape their own lives. Rather than rendering an individual a mere member of a species, attention for agency aligns with respecting individual personality (Cole and Fraser 2018). It makes one wonder to what extent animals can and want to exert choice and control over their lives (Allard and Bashaw 2019). Especially when it comes to geriatric animals, for whom few natural benchmarks exist of what a good life entails, considering the animals' own ability to shape their own daily lives through their decisions may provide a key component of caring for aging animals (Krebs et al. 2018).

Individuals are intimately attuned to their environment to the extent that it becomes difficult to definitively demarcate one from the other. Interdependence between individual and environment shows up as the accumulated knowledge and skill set of an individual at a particular point in time, as these can (hypothetically) be traced back to the unique range of learning opportunities experienced over time. It calls for a biographical understanding of individual animals. Such an understanding provides a viewpoint to assess the skills and knowledge that animals have acquired throughout their lives. Aging animals may very well, for instance, lack various skills that are invaluable to living outside the confines of zoo enclosures, as they have not encountered the learning opportunities to acquire these skills. Their knowledge and skills will reflect social learning in a confined setting (Swaisgood 2010). This includes their capacity to deal with novel choices, opportunities, enrichments, abilities to control, etc. Depending on their familiarity with opportunities provided, promoting agency could also backfire, as geriatric animals living in zoos could lack the ability to meaningfully engage with such kind of unfamiliarity (Donaldson and Kymlicka 2016; Krebs et al. 2018).

2.3 Cultures of Animal Elders

The interdependence between individual animals and their social and ecological context brings us to another way of looking at aging animals, as members of a particular culture. As indicated, social learning situated in a particular socioecological setting results in a particular set of skills and abilities. Different populations of a particular species may exhibit differences in their behaviours due to these manifold differences found in terms of both intraspecies interaction and ecological backdrop (Swaisgood 2010). As social learning takes place in social groups that—despite their similarities—vary from each other in myriad ways, cultural differences are to be expected.

Some might push back at this point to argue that culture is sole human provenance. Much depends on how one defines culture. However, if one takes culture as “the inheritance of an array of behavioural traditions through social learning from others” (Whiten 2021), it does capture the way in which many nonhuman animals learn from their conspecifics.

This prompts one to consider the cultural setting of aging animals. Different populations in the wild display relevant cultural differences (Brakes et al. 2021). All the ways in which geriatric animals have interacted with conspecifics, not to forget animals of other species and humans, cannot but result in a diversity of cultures across zoos. Discovering the intricacies of animal culture in more natural settings invites us to consider the ways in which zoo settings restrict and/or promote the development of distinct zoo cultures and to ask, for instance: “What is the role of animal elders in their respective culture?”

When one asks questions about the culture of aging animals that have been living in zoos for a long time, potentially troubling aspects of zoo life could arise. If one wants to explore the moral dimensions of taking care of aging animals, these could enrich such reflection and engagement. For instance, it might point out that breeding strategies in place, including the ending of individual lives for reasons other than being genuinely in the interest of the animal, or translocation could disrupt pathways for social learning and show how human-animal interaction is shaping behaviors in particular ways (Hosey 2013). In addition, husbandry configurations themselves might fail to replicate features of a natural socioecological context, which affects social learning (Swaisgood 2010). The range and richness of culture may be impoverished for these reasons, for some a shadow of the cultures that have developed outside of human captivity. Such questions together invite engagement with the broader context of how these aging animals have lived their lives before becoming the elders that they are now.

3 Animals in Human Culture

Zoos reflect the human culture in which they exist. If one looks at the history of zoos, it becomes apparent that zoos themselves are interwoven with cultural views, both of animals and in more general. Zoos today are descendants of the early menageries where animals were put on display primarily to entertain the privileged class of society, where exoticism and rarity functioned as status symbols. Today, many zoos have been and are transitioning away from the remnants of earlier exoticism to become centres that endorse conservation, education, and research and focus on animal care and welfare. The gradual transitioning to conservation and education correlates with an increase in public concern about the natural world (Boachá Sampaio et al. 2020).

An awareness of nature increasingly informs zoo practice. The natural lives of animals in their natural habitats provide a standard of species-specific potential, urging zoos to develop facsimiles of natural habitats. Alongside the call for nature, one hears animal voices. In many countries, animals gain recognition as sentient beings. Animal welfare and intrinsic value have become guiding principles, aligning with individuated care. Growing acknowledgment of the moral status of animals translates into more stringent demands on how humans ought to treat them, raising all kinds of considerations, some of which raised earlier in this chapter: Should ailing animals (of age) receive intensive veterinary treatment? Should animals in zoos age beyond their capacity to live a more-or-less natural life? Does death harm an animal when he or she has no anticipation nor suffering from it being brought about? Should animals be subjected to the human gaze? Do animals benefit from interaction with humans, and if so, in what way? In what way can these considerations help to bolster and foster care for elderly animals?

3.1 What Is Morality?

Elderly animals, both frail and wise, prompt those who take care of them to consider what it takes to care for them. These considerations are thoroughly moral, as they require one to establish the needs and interests of these aging animals to flesh out what it means to take care. As those tasked with the responsibility of taking care of aging animals in zoos are inevitably faced with a plethora of moral decisions, can philosophy inform and support moral deliberation?

Philosophers have made various suggestions to address the question of what one should do. For instance, ethical theorists map out possibilities for moral action. These theories invite one to look at the world from a well-crafted, conceptual framework, to intuit and judge its fecundity in the face of a moral problem. When put next to each other, ethical theories may end up at different sides, fueling the question: Who is right? Here, adopting a single theoretical viewpoint may overlook the insights of other theories or those not articulated in competing theories. Perhaps ethical theories are best viewed as vehicles for our moral imagination to expand our views and moral perception to better address the complexities of the moral problems one faces. Their value lie in making sense with what one believes to be warranted upon reflection, enriching rather than ossifying moral views of the world. If these theories fail to do so, they might be counterproductive to moral engagement (McKenna and Light 2004).

Caring for aging animals in zoos will generally map onto the culture in the background. The cultural backdrop may harbour deep-seated ethical or even metaphysical beliefs informing such care, for instance, views about euthanasia, the role of medicine and science, the relation between individual autonomy and collective responsibility, and the position of humans in relation to animals and nature writ large.

Deliberation about caring for aging animals may require engaging with these underlying beliefs about societal norms and values and wondering how they relate to the caring for elderly animals.

Engaging with moral beliefs involves engaging with emotions. Perhaps, as philosopher Jesse Prinz suggests, morality arises out of ongoing emotional engagement of individuals with the prevailing social fabric of their lives. The interaction between the individual and her or his social and cultural environment attunes emotional responses in a particular way. Throughout one's life, emotional states, such as anger, guilt, shame, disgust, and admiration, are prompted in certain contexts, inhibited in others, resulting in a particular set of sentiments, one's moral perception of the world (Prinz 2006).

3.2 Caring for Others

If morality is emotional at its core, moral deliberation not only requires an open mind but also the ability to feel and to explore experiences. What lies beneath the principles that guide action and shape moral perception? How do you feel about each possible course of action to take care of an aging animal, and how have former experiences informed your view? Of course, opening to one's feelings and experience is not always easy, nor should it be demanded. An atmosphere of mutual respect and acceptance among colleagues could foster meaningful exploration of how one feels in all honesty about the moral issue at hand. Exploring your own experiences and feelings while attending to how others feel about the situation, if done in a respectful manner, can help to gain a deeper understanding of what matters for yourself and for others (Misak 2000). Through empathy, "feeling what one takes another person to be feeling" (Prinz 2011), one can look at the moral problem at hand from a different perspective, perhaps to the enrichment of one's own view of the matter. Moreover, empathy could also make you resonate with the experiential world of the animals in your care (Gruen 2013).

Despite the possible benefits of empathy, one needs to be wary of the ways in which it could interfere with moral agency. For instance, empathy could make you focus on the experiences of those you bond with, blinding you to others—both humans and animals—with whom you lack this way of bonding (Kasperbauer 2015). In other words, empathy might open you up to the experience of those elderly animals that happen to kindle your affection, while leaving you much less interested in the wellbeing of those that lack such affection. Affection and interest appear to join forces. Empathy could also misfire, attributing mental states to others that misalign with their actual feelings. This is something particular at risk when approximating the experience of members of another species, where the projection of human characteristics—anthropomorphism—eschews the reality of animals themselves (Gruen 2013). While empathy provides a potential way for humans to engage with the animals in their lives, we need to be careful not to overdo it. However, there is also a risk of swinging the pendulum too far the other side. As Frans de Waal puts it, we need to avoid not only anthropomorphism, but also anthropodenial, which he defines as "a blindness to the humanlike characteristics of other animals, or the animal-like characteristics of ourselves" (de Waal 1999). Given the extent to which humans share their evolutionary history with other animals, we need to be careful not to overlook the biological overlap between humans and other animals. To escape the perils of anthropomorphism and anthropodenial, de Waal suggests heuristic anthropomorphism, where identification with another animal can enrich scientific practice, by fostering empathetic thinking within the realm of developing testable hypotheses. Empathy could also prove too powerful, too emotionally exhausting, if one vicariously experiences the suffering of others. Relying on empathy alone makes one vulnerable to empathic distress, "a strong aversive and self-oriented response to the suffering of others, accompanied by the desire to withdraw from a situation in order to protect oneself from excessive negative feelings" (Singer and Klimecki 2014).

Getting in touch with the feelings underneath moral judgement not only deepens understanding but also helps to avert moral distress. Moral distress occurs when individuals find it difficult to act upon their core beliefs because of circumstance, for instance, peer pressure (Jameton 2013). In response to such distress, psychological numbing could function as a coping mechanism to offset the failure to live up to one's own moral standards, resulting possibly in burnout in the long run and blunted moral agency (Oh and Gastmans 2015; Rushton 2017). Moral deliberation as sketched here effectively counters numbing for its engagement with feelings as a basis for aligning one's actions with one's beliefs. Genuine moral deliberation requires regular honest and respectful exchange of perspectives. Such open deliberative effort promotes the exploration of one's own values and those of others, providing the basis for aligning relevant values with the practice of taking care. As such, moral deliberation helps to cultivate moral resilience, the ability to engage with moral complexity and difficulty in a way that seeks for possibilities rather than succumbing to defeat. In the process of learning to deal with moral difficulty, one also bolsters one's own moral integrity (Rushton 2017).

3.3 Compassion for Animal Elders

To fruitfully engage in moral deliberation, it appears, one should not feel too much (risking empathic distress) or too little (risking psychological numbing). To navigate between these emotional extremes, some find guidance in compassion (Halifax 2011; Singer and Klimecki 2014). Compassion can be defined broadly as the "capacity to attend to the experience of others, to feel concern for others, to sense what will serve others, potentially to be able to be of service" (Halifax 2013). More narrowly, it combines the affective recognition of the suffering of others with a motional state to help (Halifax 2013). Luckily, as compassion involves primarily a recognition rather than a resemblance of the emotional state of others, it is not associated with empathic distress (Singer and Klimecki 2014). As such, compassion is well equipped to propel moral action, as it circumvents vicarious suffering while nonetheless attending to it (Bloom 2017).

Compassion invokes a heightened sensitivity for the suffering of other individuals (and oneself) coupled with the aspiration to alleviate, diminish, or eliminate it. As such, it provides a motivational attitude that has been viewed as highly relevant to the practice of caring for others (Halifax 2011). In the process of aging, geriatric animals will be affected by changes to their bodies and minds. Part of taking care is to recognise (nuanced) indications of suffering and adversity, to handle being confronted with suffering, and to work towards alleviating suffering and fostering flourishing. For precisely these reasons, compassion appears profoundly relevant to the care for aging animals in zoos. It encourages curiosity and interest for the lives of elderly animals, to acquire a fine-grained view of their lived experience and to support them in ways that help them overcome or deal with the adversities, both great and small, that they face in the last phases of their lives. While compassion throws the interest of other sentient beings into sharp relief, aligning with the consonant call for the recognition of the intrinsic value of animals, what it entails at the practical level will be a question of honest, open deliberative engagement in practice. Here, both individual efforts (for instance, engaging in mindfulness or other forms of contemplative practices) and organisational initiative help to foster compassion for each other (Kirby et al. 2017).

4 Conclusions: Listening to Animal Elders

Aging animals have a story to tell. To take care of them requires that one listens, to the biographies of their lives and to the ways in which they tell us how they want to live their lives now that they have become elders. As they diverge further away each day from the lives they would have lived in a natural habitat, it becomes even more important to listen. The geriatric phase of the lives of animals will often resemble untrodden territory, especially for it being increasingly shaped by human intervention. Novelty and uncertainty together prompt an array of moral questions for those that take care of them. Compassion proves especially promising as a way of navigating these concerns, as it fosters a genuine curiosity for how elderly animals experience their lives alongside the motivation to take care in whatever way appears the most appropriate. In other words, it allows one to be responsive to their authentic interests, to enable these elders to age gracefully.

References

Allard SM, Bashaw MJ (2019) Empowering zoo animals. In: Kaufman AB, Bashaw MJ, Maple TL (eds) Scientific foundations of zoos and aquariums: their role in conservation and research. Cambridge University Press, Cambridge, pp 241–273

Bloom P (2017) Empathy and its discontents. Trends Cogn Sci 21(1):24–31. https://doi.org/10.1016/j.tics.2016.11.004

Boachá Sampaio M, Schiel N, Souto A (2020) From exploitation to conservation: a historical analysis of zoos and their functions in human societies. Ethnobiology and Conservation 9:1–32. https://doi.org/10.15451/ec2020-01-9.02-1-32

Brakes P, Carroll EL, Dall SRX, Keith SA, McGregor PK, Mesnick SL, Garland EC (2021) A deepening understanding of animal culture suggests lessons for conservation. Proc R Soc B Biol Sci 288(1949):20202718. https://doi.org/10.1098/rspb.2020.2718

Brando S, Buchanan-Smith HM (2018) The 24/7 approach to promoting optimal welfare for captive wild animals. Behav Process 156:83–95

Cole J, Fraser D (2018) Zoo animal welfare: the human dimension. J Appl Anim Welf Sci 21(sup1):49–58. https://doi.org/10.1080/10888705.2018.1513839

Crews DE (2007) Senescence, aging, and disease. J Physiol Anthropol 26(3):365–372. https://doi.org/10.2114/jpa2.26.365

de Waal FBM (1999) Anthropomorphism and anthropodenial: consistency in our thinking about humans and other animals. Philos Top 27(1):255–280

Delon N (2018) Animal agency, captivity, and meaning. Harv Rev Philos 25:127–146. https://doi.org/10.5840/harvardreview201892519

Donaldson S, Kymlicka W (2015) Farmed animal sanctuaries: the heart of the movement? Politics Anim 1(1):50–74

Donaldson S, Kymlicka W (2016) Rethinking membership and participation in an inclusive democracy: cognitive disability, children, animals. In: Arneil B, Hirschmann N (eds) Disability and political theory. Cambridge University Press, Cambridge, pp 168–197. https://doi.org/10.1017/9781316694053.009

Fawcett A (2013) Euthanasia and morally justifiable killing in a veterinary clinical context. In Animal Death; University of Sydney: Sydney, Australia, pp. 205–220

Fraser D (2009) Assessing animal welfare: different philosophies, different scientific approaches. Zoo Biol 28(6):507–518. https://doi.org/10.1002/zoo.20253

Gruen L (2013) Entangled empathy. In: The politics of species. Cambridge University Press, Cambridge, pp. 223–231. Retrieved from https://doi.org/10.1017/CBO9781139506755.023

Halifax J (2011) The precious necessity of compassion. J Pain Symptom Manag 41(1):146–153. https://doi.org/10.1016/j.jpainsymman.2010.08.010

Halifax J (2013) G.R.a.C.E. for nurses: cultivating compassion in nurse/patient interactions. J Nurs Educ Pract 4(1). https://doi.org/10.5430/jnep.v4n1p121

Hosey G (2013) Hediger revisited: how do zoo animals see us? J Appl Anim Welf Sci 16(4):338–359. https://doi.org/10.1080/10888705.2013.827916

Jameton A (2013) A reflection on moral distress in nursing together with a current application of the concept. J Bioeth Inq 10(3):297–308. https://doi.org/10.1007/s11673-013-9466-3

Kasperbauer T (2015) Rejecting empathy for animal ethics. Ethical Theory Moral Pract 18(4):817–833. https://doi.org/10.1007/s10677-014-9557-1

Kirby JN, Steindl SR, Doty JR (2017) Compassion as the highest ethic. In: Practitioner's guide to ethics and mindfulness-based interventions. Springer International Publishing, Cham, pp 253–277. https://doi.org/10.1007/978-3-319-64924-5_10

Krebs BL, Marrin D, Phelps A, Krol L, Watters JV (2018) Managing aged animals in zoos to promote positive welfare: a review and future directions. Animals (Basel) 8(7):116. https://doi.org/10.3390/ani8070116

McKenna E, Light A (2004) Animal pragmatism: Rethinking human-nonhuman animal relations. Indiana University Press, Bloomington

Misak C (2000) Truth, politics, morality. Routledge, London. https://doi.org/10.4324/9780203162286

Oh Y, Gastmans C (2015) Moral distress experienced by nurses. Nurs Ethics 22(1):15–31. https://doi.org/10.1177/0969733013502803

Prinz J (2006) The emotional basis of moral judgments. Philos Explor 9(1):29–43. https://doi.org/10.1080/13869790500492466

Prinz JJ (2011) Is empathy necessary for morality? In: *Empathy: philosophical and psychological perspectives*. Oxford University Press, Oxford. https://doi.org/10.1093/acprof:oso/9780199539956.003.0014

Rushton CH (2017) Cultivating moral resilience. Am J Nurs 117(2 Suppl 1):S11–S15. https://doi.org/10.1097/01.NAJ.0000512205.93596.00

Singer T, Klimecki OM (2014) Empathy and compassion. Curr Biol 24(18):R875–R878. https://doi.org/10.1016/j.cub.2014.06.054

Swaisgood RR (2010) The conservation-welfare nexus in reintroduction programmes: a role for sensory ecology. Anim Welf 19(2):125–137

Tidière M, Gaillard J, Berger V, Müller DWH, Bingaman Lackey L, Gimenez O et al (2016) Comparative analyses of longevity and senescence reveal variable survival benefits of living in zoos across mammals. Sci Rep 6(1):36361. https://doi.org/10.1038/srep36361

Whiten A (2021) The burgeoning reach of animal culture. Science (American Association for the Advancement of Science) 372(6537):eabe6514. https://doi.org/10.1126/science.abe6514

The Importance of Meaningful Record-Keeping in Caring for Ageing Wild Animals

Max Norman, Sarah Chapman, and Sabrina Brando

M. Norman (✉)
AnimalConcepts, Essex, UK
e-mail: mnorman@animalconcepts.eu

S. Chapman
Chapman Zoo Consultancy, Birmingham, UK

S. Brando
AnimalConcepts, Teulada, Spain
e-mail: sbrando@animalconcepts.eu

S. Brando, S. Chapman (eds.), *Optimal Wellbeing of Ageing Wild Animals in Human Care*, https://doi.org/10.1007/978-3-031-30659-4_3

Machi. Machi, a female Western Lowland Gorilla (*Gorilla gorilla gorilla*) came to Zoo Atlanta with her mother, Choomba, in 1988 from Emory University's Yerkes National Primate Research Center. She lived at Zoo Atlanta until 2013, when she moved to Zoo Knoxville on a recommendation from the Association of Zoos and Aquariums' (AZA) Gorilla Species Survival Plan® (SSP). Western lowland gorillas are considered geriatric after the age of 40. In recent times, Machi has been experiencing challenging bouts of arthritis. This is not unusual for gorillas who experience many of the same issues ageing humans do. Zoo Atlanta is a leader in working with geriatric apes. The Zoo's Animal Care and Veterinary Teams have made impressive advances in geriatric care over the years. The Gorilla Care Team focuses on creating habitats and social environments that allow the gorillas to thrive, and gorilla facilities have been redesigned with age in mind. Given Zoo Atlanta's established leadership and long-term experience in this area, Zoo Atlanta and Zoo Knoxville collaborated on Machi's move to Atlanta in 2021. Machi translates to March in Swahili, and she was given that name because she was born on March 1, 1976. She is now 46 years old

Abstract

Good, consistent and meaningful record-keeping is fundamental to all aspects of animal care. The ability to monitor, collate and review an animal's information throughout his or her lifetime is vital to the ability to provide good quality care. Knowledge of previous care and management techniques, both good and suboptimal, allows animal caregivers to adjust and improve the care provided to support the wellbeing of the animal as optimal as possible. This chapter outlines the type of records that should be kept and how these can be documented, including quantitative and qualitative measures of behaviour. It focuses on the importance of reliable and accessible data for the on-going monitoring and assessment of ageing animals' wellbeing, including predictive and proactive management practices.

Keywords

Records · Consistency · Meaningful · Data · Wellbeing · Zoo animals

1 Introduction

The effective management of animals starts with a fundamental concept—we need to know the individual animal. We need to know not only who they are, but also their background; from day-to-day observations to their life history, who their social group and parents and grandparents are. We need to know what they were like yesterday to know if they are behaving differently today; we need to know what they ate for dinner last night to know if that could have impacted their health the next morning. This much is true whether we are talking about a companion dog that lives with us, or if we are a veterinarian treating that dog at any stage in his or her life. This knowledge is vital to being able to make decisions effectively and successfully about short- and long-term management and understanding what the risks may be in terms of wellbeing, both physically and psychologically. The primary goal is to monitor wellbeing of an individual animal and promote predominantly positive states throughout life. It can also be used for effective conservation breeding programs that account for the life history and genetics of all animals involved at a population level among record-keeping for other goals.

2 Types of Records

In an ideal world, we would have comprehensive information about every aspect of an animal's life; if we could have all information available to us, all the time, it would certainly make many management decisions, for both the individual and the population, much easier. We could analyse in detail any aspect of the animals' lives that we wanted to if we had all possible information recorded on a database, but of course, collecting, collating and synthesising that data takes a great deal of time. Database management is a time-consuming job, and larger facilities, such as zoos and aquariums, may need to hire staff whose sole job is to enter and manage data on their animals (Hosey et al. 2013). Therefore, it is imperative to consider based on the time, staff and, in some cases, money available, to collect the data we need and what data we are logistically able to collect while still meeting the relevant criteria. In some jurisdictions, there are legal requirements for minimum information that is held about individual animals. In the UK, the Department for Environment, Food and Rural Affairs (DEFRA) is responsible for setting out the responsibilities of zoos according to legislation and set out in the Secretary of State's Standards of Modern Zoo Practice 2012 (SSSMZP) that individual records should contain information as indicated in Table 1. While the requirements may vary from one jurisdiction to another or in different contexts, the general types of information that is needed is by and large similar to those set out here. This information can be broadly categorised as life history, such as births, deaths and transfers, and daily changes or care, such as diet, enrichment, and other daily inputs. It includes outputs, also known as animal-based indicators, such as measurable behaviour and physiological parameters which incorporate medical information and physical records.

Other types of information should also be collected to have a holistic perspective and overview of the animal's life and wellbeing. These can include social life, such as play, friends and family, psychological wellbeing, such as joy, curiosity and lack of fear and anxiety. Using the lens of the Five Domains Model can assist in creating complete record-keeping protocols and processes. This information is useful not only if a governing authority needs to review the information, but also for a multitude of other reasons. If the animal moves elsewhere, all information on the past health and other events is available to their new caregivers. If a veterinarian and care staff are concerned about the animal's health, small changes from their behaviour in the months prior may be helpful in making a diagnosis. Data can also be used for retrospective studies on animal behaviour and management; however, past records are not always comprehensive enough to make meaningful conclusions in all cases (Hosey et al. 2012). Nonetheless, historical records have been used to make a great effect in studies of lifespan in several species (Weiss et al. 2013; Altschul et al. 2018).

2.1 Life History

When we speak of life history, we are largely referring to major life events that occur throughout an animal's life cycle: when and where they are born, who their parents and other relatives are, the time and manner of death and events that can be

Table 1 Types of information required to be kept on individual zoo animal records, adapted from the Secretary of State's Standards of Modern Zoo Practice (Department for Environment, Food and Rural Affairs, 2012), with definitions of the type of information included

Information required	Definition
Individual identity	May include a 'given' name, but in many cases may also include an individual identity number to prevent confusion if the name is the same as another of the species or the name changes if moved to another collection.
Scientific name	Accepted binomial name of the species to which the animal belongs.
Origin	Information on where the animal was born (captive or wild), who the parents are (if known), previous locations if the animal has been transferred, and so forth.
Dates of transfer	In the case that an animal is transferred between facilities during his or her life, accurate records of where he or she was moved to and when, date of exit, etc. to allow for tracking.
Date of birth	Date, or estimated date, of birth or hatching.
Sex	May not be possible for all species; if unknown, this information should also be kept.
Distinctive markings	Useful for individual identification purposes; may include individual features such as fur patterns, scars, uniquely shaped or marked body parts that differ from others of the species and so forth; can also include physical identification markers such as tags, rings, transponders and other devices.
Clinical data	All data on veterinary visits, including notes taken on said visits, as well as information on any treatments given including dosages of medications and dates administered.
Behavioural and life history data	Records should be kept on any notable changes in behaviour or other events that may be significant. For example, caregivers can provide notes of daily observations.
Date of death	When applicable.
Post-mortem results	All findings of post-mortems to a deceased animal.
Escapes (where applicable)	Records should be kept of any escapes, where to, when, with accurate information on why the escape occurred, any damage or injury caused to property or people and measures taken to prevent repeat escapes. Escapes should also include aspects on how the animal is faring and if she or he survived the event.
Food and diet	All information on nutrition provided to the animals including types of food and supplements. For carnivores, additional records may need to be kept on the weight and type of meat fed.

anticipated based on biology and ecology (Brando and Buchanan-Smith 2018). We are also referring to any transfers between organisations during their lifetime, such as when zoos transfer animals for breeding, wellbeing or other management purposes. This information is useful for tracking purposes, particularly where disease transfer may be a concern (Hosey et al. 2012), and for monitoring breeding programmes and improved social life and wellbeing in more appropriate environments. Detailed information to be submitted before an animal transfer can be compiled through the use of a pre-transfer checklist; see Fig. 1, for an adapted version for ageing animals.

If we know one or more of an animal's parents or grandparents suffer or suffered from a condition known to be hereditary, this information can be used not only to inform breeding decisions involving that animal but also to make decisions regarding preventative care later in his or her life. Demographic information on the life history of a species as it lives under human care is further useful to those caring for the same species in the future. Lifespan and other life history traits can vary between animals living in the wild and those living in captivity; comprehensive records on those who came before can, therefore, be used to inform our decisions when caring for others of the same species. For example, we can use mortality records from the past to better understand what diseases commonly afflict certain species. As an example, Lamglait et al. (2015) conducted a retrospective study of mortality at Réserve Africaine de Sigean on captive African wild dogs (*Lycaon pictus*). If we know that a certain species, in this case African wild dogs, is prone to tumours later

Ageing Animal Pre-Transfer Checklist	
	Animal ID and house name
	Sex confirmation: M/F
	Weight and body condition score
	Photos and videos of the animal
	Medical history, diagnostic images e.g., radiographs and current medications
	Pre-transfer testing results
	Diet sheet including special requirements and any supplements
	Water provision (bowl, auto drinker, etc.)
	Behavioural history and related information
	Breeding history and related information
	Social history and information on friends and family
	Animal training information and videos including any special requirements
	Environmental enrichment information including any items to avoid
	Descriptions and photos of enrichment items that worked well
	Photos of the habitat – outdoors and indoors
	Information on human-animal interactions, preferred care staff
	Journey information (crate, box, substrate, temperature, length)
	Legal paperwork (CITES, export/import, licenses)
	Any other relevant information e.g., notes on how medication is given, special notes on personality

Fig. 1 Pre-transfer checklist for information required prior to transfer of an ageing animal between facilities

in life, we can implement early screening processes to catch and treat illness early and potentially improve quality of life in their senior years.

2.2 Daily Changes

Daily events that occur in an animal's life may not seem significant at the time they occur but are nonetheless important to record. For example, an animal may seem lethargic one day and appear normal for all the days that follow; even though nothing came of the initial observation, that information may have been crucial when a veterinarian needed to be called or if the same lethargy was demonstrated later. As an example, a dead fox was found in a tiger's enclosure one morning, and while nothing came of this event, another animal entering an enclosure could cause illness or injury to the inhabitants and, further, alerted staff to weaknesses in the fence that needed addressing. Collecting this type of information on a daily basis—what did the animal eat today, what were the faeces like, how did he or she interact with the care staff (or not), how interested was the animal in their enrichment and so on—is crucial for understanding if any changes that come about in behaviour or health are a cause for concern or the consequences of a known event from the days prior. It is also important to make note of important events that occurred outside of what is observed from the animal. For example, perhaps the routine had differed compared to what is considered a normal day, or there was construction work happening near the enclosure. These changes in what is "normal" outside of the habitat that can nonetheless impact an animal's daily experience may be useful if, for example, the construction work disturbed the animal enough to influence their behaviour or health the next day. We have to keep in mind that other animals perceive the world in many different ways we humans cannot, and meaningful records should also reflect seasonal or other fluctuating needs, such as types of required lighting for birds or reptiles or differing heating requirements of ill or elderly animals. We can use the information gleaned from these events to prepare for the future. If we know that an animal is afraid of fireworks due to common sense or past records, we can make reasonable adjustments—playing peaceful music, bringing them inside earlier, providing more bedding to build a safer nest and so on, all to help them feel less afraid when we know there will be fireworks.

2.3 Physical and Psychological Information

Recording information on the physical and psychological health of animals is essential for a multitude of reasons. Of course, it is vital to understand on the day-to-day basis how an animal is doing, whether he or she is healthy, whether any medical treatment he or she is undergoing is working and so forth. It is also useful in establishing parameters that indicate good or poor physical and mental health in a species or in an individual. For example, some zoos regularly collect samples of faeces for testing to collect baseline information such as stool quality, hormone levels and parasite burden, which can all be used to monitor not only the health status of an animal but also other areas such as reproductive status. Knowing what baseline is 'normal' for an individual or a species is an essential part of recognising when something is *not* normal and a cause for concern. Further, historical data collected form an important resource bank that can be used to understand the health of other individuals in the future, including the samples collected from conspecifics living in the wild. Historical information on animal disease and treatments is especially useful when we are talking about exotic species for whom there is not a lot of medical literature or case studies surrounding their veterinary care. Often when a wild animal living in a zoo falls ill, there may only be a small selection of species-specific literature available to assist in making a diagnosis and determining the best course of treatment; it, therefore, often falls to collating data from similar species to make these decisions. Whenever an animal falls ill, undergoes surgery or receives a medication, accurate records on diagnosis, prognosis and treatment are invaluable not only in tracking that individual animal, but also in building an evidence base that can be used to improve the way we care for animals experiencing the same affliction in the future. This knowledge base can also be applied to the treatment of rescued or otherwise unwell free-living animals in a conservation context.

Biometrics record-keeping refers to the collection of body measurements, such as body weight, body length, wingspan, tail length, ear length and so forth. Regular weighing of animals is an essential aspect of any animal wellbeing monitoring program, as it can reveal a lot of information about an animal's health status; it can inform us whether their diet is appropriate, whether the animal is getting enough exercise, how the animal's appetite is doing and so forth. Collecting weights is especially useful for young and developing animals and for mothers who are expecting young, as it can be a reliable indicator of whether the animal is developing at the expected pace for its species. For some species, such as bears and reindeer, the quantity and quality of food available in the wild vary seasonally, and some animals may drop their food intake to almost zero in some seasons naturally. Of course, in zoos where climate is far more controlled and there is not a risk of poor or different food availability in certain species, these changes in wild conspecifics are not always mimicked in husbandry practices. Seasonal weight changes should nonetheless be monitored as some animals may have natural fluctuations in their physiological processes as a result of changes in the season, and this would be reflected in their weight. Body condition scoring (BCS) is also an important tool to use for certain species and can provide information regarding loss of muscle, decrease or increase in general condition or abnormal increases in the parts of the body, for example, fluid in the abdomen. Regular photographs and video footage are recommended to ensure consistency in data collection and to have the BCS linked to the weight of the individual.

All these are examples of records that predominantly focus on the physical and measurable aspects of animal health. The field of psychological wellbeing in animals, as well as cognitive wellbeing, is growing but is not documented or described in great detail or reflected in contemporary record-keeping today, and hopefully, this will continue to change. These types of documentations reflect an effort to collecting and documenting not only the measurable but also the meaningful. This can include a wide variety of details of individual behaviours and ways of being and recognising the centrality of the feelings of animals which are currently challenging to measure directly. While all these are challenges in record-keeping, they are not a reason not to try. Physiological and health-orientated measures have long emerged as popular metrics for assessing welfare because they are quantifiable; however, their popularity in animal welfare assessment has led to them having a disproportionate influence on animal management to the detriment of animal welfare in numerous instances (Veasey 2017). Psychological experiences and positive wellbeing states, as well as concerns and poor wellbeing states, should also be priorities in record-keeping. This can be particularly important in ageing animals where changes in, for example, environment and physical wellbeing can have positive or negative effects on mental states.

3 Record-Keeping Systems

The National Research Council (NRC) set out ten criteria that gathering information on animal health should meet to be effective: (1) accessibility, (2) accuracy, (3) appropriateness, (4) comprehensibility, (5) comprehensiveness, (6) consistency, (7) relevance, (8) reliability, (9) timelessness and (10) usefulness (NRC 2004). These criteria apply whether we are talking about simple pen-and-paper record systems or more complex database software. Although there is, of course, a question of how we keep and maintain such records and which is the best method to use. In small, private organisations, or when we are speaking of companion animals at home, it may be more straightforward to keep records with pen and paper. However, there are critical issues with paper-based systems. It is difficult to search and track paper systems, creating issues with traceability and accessibility. Paper is easily lost or misplaced, meaning that finding the relevant record(s) when needed can become a challenge. Of course, paper records can be digitised; we can scan paper trails into computers to make them searchable, allowing for both types of records to be kept and stored in data repositories, but interrogating the data will not be possible and is not recommended. This is particularly challenging when there are many animals, or groups of animals, such as those housed in zoos and aquariums. Such was noted in Hosey et al.'s (2013) comprehensive book on zoo animal management; the authors rightfully highlight that some zoo organisations can be responsible for the records of several thousands of individuals, sometimes across multiple different sites, representing some hundreds of different species. Not only this, but as part of wider population management for many species, animals can be transferred between zoos at any stage of their lives for breeding and other reasons, and their new homes will need accurate records to best understand how to care for the animals when they arrive.

Many zoos today use centralised electronic information systems such as Zoological Information Management System (ZIMS) for population management and Medical ZIMS (previously known as MedARKS) for veterinary records, developed specifically for their use (NRC 2004). ZIMS also has a Care and Welfare module that can assist facilities in tracking and scoring animal-based indicators such as interaction with environment, social life, sleeping and alertness. These databases contain multi-institutional records on animals housed around the world, with the animals' unique records following them whenever they move. Other systems have historically been used depending on the institution; for example, the Australasian Regional Association of Zoological Parks and Aquaria (ARAZPA) developed the Regional Animal Species Collection Plan (REGASP) software for institutional and regional collection planning in Australia and New Zealand's zoos and aquariums. Some collections have their own in-house database systems, although they may still export these data to a central database such as ZIMS (Dubois et al. 2003). In the modern day, most records kept by zoos are well-kept and up-to-date including in the developing countries (Ward et al. 2020).

3.1 Collecting Data

From observations, many zoos still rely on pen-and-paper for recording information on daily occurrences such as notes on behaviour or daily husbandry changes. Not all animal caretakers have regular access to a computer or laptop during their workday, and a diary or notebook kept in the staff meeting or break room continues to be a quick and straightforward method not only to make notes but also for other members of staff to quickly reference the events of previous days without needing to spend time searching up information on a computer. For these records to be useful across time and context, it is important that the records eventually find their way into a centralised system such as ZIMS. For example, they must be scanned or typed up at the end of a working day or week by a member of staff to be entered by the zoo's database manager for them to be useful and for looking at trends and making evidence-based decision-making.

Some types of data are more challenging to collect. While faeces can be collected easily, the laboratory tests needed are not always available to zoos on-site and may need to be sent for external testing if parameters such as hormone levels are desired. For other types of information, such as collecting blood samples or even something as simple as getting the weight of animals, animals will often need to be trained to participate—showing the relevant body part or stepping on to scales—to avoid the need for an animal to be unnecessarily stressed or anaesthetised for the procedure. Training animals to participate in their own care is an essential component of an effective husbandry program to enable choice and control in their own care. There can be challenges to this, of course (see also Chapter "The Role of Learning and Training in Caring for Ageing Animals"). For example, it can be difficult to get the individual weights of a mother and baby primate when the baby is still dependent on its mother; we can make estimates on their individual weights based on historical information on the mother, but separating the two to get a weight unless it were completely necessary would be stressful for both the animals. Collecting data should also include aspects on, for example, an animal's social life, enrichment engagements and the human–animal interaction.

4 Monitoring Wellbeing of Ageing Animals

Keeping and monitoring records are even more important when we are working with animals in their senior years. Especially for long-lived animals, we may not always have been the ones caring for them; we may not know what 'healthy' looked like in their younger years, or it was so long ago that it may be difficult for us to remember. The physiological and psychological changes that potentially indicate deteriorating wellbeing may happen so gradually that we would not necessarily notice them if we did not have records. When gradual changes such as slow loss of body weight, declining muscular function and confidence or an increase in anxiety are the only signs we may have that an animal is beginning to decline in wellbeing later in life, it is imperative that we have systems in place for monitoring such information. By keeping records throughout the lives of the animals we care for, as well as comparing them with the records kept from other individuals of the same species, it becomes possible to establish baselines and what the individual idiosyncrasies are. A drop in weight for an ageing animal may be the only indicator that there is a health concern or the start of an illness. This is due to the fact that wild animals can hide signs of illness until late on in the disease process (see also Chapter "Veterinary Care of Ageing Zoo Animals: Description of a Proactive Approach"). Loss of appetite may be the first sign that an animal is ill, but the causal factor may have been present for much longer. The change or lack of use of the enclosure space is a common indicator that there could be a painful condition or reduced ability present; therefore, knowledge of the normal usage is vital to have as a baseline. Weather conditions are a useful piece of information to record as some health conditions are exasperated by colder weather, for example, joint disease leading to more obvious stiffness. Understanding what is going on today requires an understanding of what was going on yesterday, a week ago and a month ago. Detailed and meaningful record-keeping on the physical and psychological aspects of an animal life will assist in supporting a holistic animal care and wellbeing program for the elderly animals.

References

Altschul DM, Hopkins WD, Herrelko ES, Inoue-Murayama M, Matsuzawa T, King JE, Ross SR, Weiss A (2018) Personality links with lifespan in chimpanzees. elife 7:e33781

Brando S, Buchanan-Smith HM (2018) The 24/7 approach to promoting optimal welfare for captive wild animals. Behav Process 156:83–95

Dubois S, Johnson K, Smith B (2003) Building better zoological information systems for zoos and aquariums. Available online at http://www.zims.org/

Hosey G, Hill SP, Lherbier ML (2012) Can zoo records help answer behavioral research questions? The case of the left-handed lemurs (*Lemur catta*). Zoo Biol 31:189–196. https://doi.org/10.1002/zoo.20402

Hosey G, Melfi V, Pankhurst S (2013) Zoo animals: behaviour, management, and welfare, 2nd edn. Oxford University Press, Oxford

Lamglait B, Trunet E, Leclerc A (2015) Retrospective study of mortality of captive African wild dogs (Lycaon pictus) in a French zoo (1974-2013). J Zoo Aquar Res 3(2):47–51. https://doi.org/10.19227/jzar.v3i2.124

National Research Council (NRC) (2004) Committee on a review of the Smithsonian Institution's National Zoological Park. Animal care and Management at the National zoo: interim report. Washington, DC: National Academies Press (US); 3, Record Keeping. Available from: https://www.ncbi.nlm.nih.gov/books/NBK207745/

Veasey JS (2017) In pursuit of peak animal welfare; the need to prioritize the meaningful over the measurable. Zoo Biol 36(6):413–425

Ward SJ, Williams E, Groves G, Marsh S, Morgan D (2020) Using zoo welfare assessments to identify common issues in developing country zoos. Animals 10(11):2101. https://doi.org/10.3390/ani10112101

Weiss A, Gartner MC, Gold KC, Stoinski TS (2013) Extraversion predicts longer survival in gorillas: an 18-year longitudinal study. Proc Royal Soc Biol Sci 280(1752):20122231

Facility Design for Ageing Wild Animals

Jon C. Coe

J. C. Coe (✉)
Jon Coe Design, Healesville, Victoria, Australia
e-mail: jon@joncoedesign.com

S. Brando, S. Chapman (eds.), *Optimal Wellbeing of Ageing Wild Animals in Human Care*, https://doi.org/10.1007/978-3-031-30659-4_4

Biji. Biji is the oldest Bornean Orangutan (*Pongo pygmaeus*) at Zoo Atlanta, born on 18 October 1970. Orangutans are solitary, spending the majority of their adult lives on their own rather than in groups, and Biji is nothing but independent. Taking care of a geriatric individual like Biji requires some extra work. Like some geriatric humans, she receives special medications and dietary items for her health. She is also specially trained for regular voluntary urine, blood pressure and blood collection so that we can monitor her wellbeing and make changes to her daily routine if needed. Despite her advanced age, Biji is one of the more playful orangutans at Zoo Atlanta. She seems to enjoy interacting with a huge variety of enrichment items. Some of her favourites include picture books, weaving toys, puzzles, water misters and bubbles

Abstract

The blending of science with empathy through multidisciplinary design of animal facilities and features can contribute to the wellbeing of disabled and elderly animals in zoos, sanctuaries and research facilities and may prevent or slow some common degenerative conditions. The importance of designed features benefiting both animals and animal carers is emphasized. Subjects include the integration of management and facility planning, a model design and animal codesign process, features that may delay the onset of health problems and general recommendations including flexible spaces, suitable surfaces, horizontal and vertical circulation, animal rotation and aquatic/aerobic exercise. Animal–computer interaction (ACI) opportunities are touched on, including animal control of ambient conditions and round-the-clock personalized access to food, enriching opportunities and alternative spaces, which are all automatically monitored for research and medical observation.

Keywords

Animal welfare · Ageing animals · Disabled animals · Zoo design · Sanctuary design · Codesign with animals

1 Introduction

This chapter overviews the multilayered subject of facility planning and is intended to support the aims, activities and programs suggested in other chapters. While evidentiary citations are provided, some content is based upon my own experience derived from theoretical and practical experiences in a field where art and science comingle, and empirical evaluations of built works lag far behind the emergence of new ideas and developments.

How can we use the design of physical features to improve the wellbeing of ageing animals in zoos, laboratories, sanctuaries and rescue centres? How do we integrate overlapping subjects in other chapters such as animal welfare, environmental enrichment, medical care and animal training into the design of animal-friendly built-in features? Could some features be activated by the animals themselves? And can we plan features that not only benefit ageing and infirm animals, but also delay early onset of common age-related infirmities? These are among the opportunities discussed in this chapter.

1.1 Planning and Design

Planning for lifelong wellbeing as described by Maple (2019) and Brando and Buchanan-Smith (2018) may apply to large and small animal populations. Well-managed zoos tend to have smaller populations of a given species with good population planning, resulting in wide age distribution and few ageing individuals. However, some medical research laboratories, and sanctuaries for past research and entertainment animals, have large primate populations. "Unlike the cooperative breeding program populations in zoos, the sanctuary population is creeping toward a future in which there will only be elderly animals" (Steve Ross per comm. 26 March 2021). Rescue centres for elephants, 'problem' tigers and leopards or native species rescued from fire, flood or urbanization may be facing similar emergencies. It is hoped that the materials to follow are useful to a broad range of animal care facilities with an international audience.

> *...most animals born and raised in a zoo will live out their lives in one. Planning for different life stages is therefore critical for the welfare of all animals living in zoos.* (Watters et al. 2009, p. 14.)

1.2 Integration of Facility and Management Design

Other chapters in this book respond to a wide range of animal management issues and opportunities, many of which concern the 'design' of animal and staff management programs and initiatives. It is essential that planning and design of physical features, as varied as buildings, water circulation, planting plans or Wi-Fi systems, be fully integrated with animal management systems, such as animal training, nutrition, health and enrichment programs. Therefore, the terms 'planning' and 'design' presume the built features are jointly planned with related operational planning described in the other chapters.

1.3 Universal Design: Planning for All Needs

Many of the design recommendations suggested for older animals also would benefit younger animals with disabilities and may even prevent or slow onset of harmful age-related conditions such as hoof, foot, joint, pelt or feather problems described in the other chapters (Chapter "Veterinary Care of Ageing Zoo Animals: Description of a Proactive Approach" on veterinary care, Chapter "Euthanasia of Geriatric Zoo Animals: Decision Making and Procedure" on euthanasia, and Chapter "The Longevity Legacy: The Challenges of Old Animals in Zoos" on longevity). Good design supports the needs and welfare of all stakeholders, young and old and non-human and human. '*Universal Design is the design of an environment so that it can be accessed, understood and used… by all people regardless of their age, size, ability or disability*' (Centre for Excellence in Universal Design [CEUD] 2022). This design approach for humans seems equally applicable to design for animals and animal carers. '*If an environment is accessible, usable, convenient and a pleasure to use, everyone benefits. Simply put, universal design is good design*' (National Disability Association [NDA] 2022).

1.4 Design Process

> *Design Thinking is an iterative process in which we seek to understand the user, challenge assumptions, and redefine problems in an attempt to identify alternative strategies and solutions…* (Dam and Siang 2020, third paragraph).

The following Plattner Stanford Process Model (Chebyniak 2019) for use in design for humans could well be applied to captive animal residents. While the steps given are sequential, the authors mention that the order may vary if each step is thoughtfully considered and integrated. I have added the final three steps and observations based upon my own experience.

1. *Empathize—with your users.* In the past, designers were taught to inventory and analyse information about human and animal clients, their needs, habitats and habits, with the implication of a rather clinical, empirical and impersonal approach. In my own work, I have found greatest success with an approach balancing science and empathy. In addition, it is beneficial to balance the emotional connections and insights of caregivers with the empirical observations of veterinarians, behaviour specialists and academics.
2. *Define—your users' needs, their problem and your insights.* It is said that proper problem definition is 90% of the solution. Also consider that a problem may have multiple interrelated causes and require several interrelated solutions, some combining built structures with management practices. An especially useful practice during this step is the documentation of baseline, preoccupation or pre-existing use by animal and human stakeholders. Some zoos have implemented annual welfare audits of each animal (Sherwin et al. 2018), which provide valuable input during this step. Licensed and accredited zoos and other institutions housing animals have periodic inspections, the reporting of which may help to identify and prioritize design needs and opportunities.

3. *Ideate—by challenging assumptions and creating ideas for innovative solutions.* Creativity is required at each stage, but here it is the focus. The temptation is simply to copy the solution of someone else. While previous solutions, including today's best practice, should be evaluated, 'outside-the-box' thinking is required to innovate substantial improvements rather than simply maintaining the status quo.
4. *Prototype—to start creating solutions.* While large-scale architectural structures are impractical to prototype, simple three-dimensional scale models can be useful in helping human users visualize and simulate practical day-to-day and emergency operations. Detailed fabrications, such as gate operators or animal-operated interfaces, should be prototyped at full size.
5. *Test.* Plans, models and prototype fabrications should be tested early and often. Do not wait until the final design to test basic conceptual directions. Test with a wide variety of stakeholders. Is it needed? Is it easy and intuitive for staff to use? Is it easy and intuitive for animals to use? French et al. (2020) described offering elephants a variety of prototypes for animal-activated enrichment devices and then further developing the elephant's preferred devices through multiple cycles of testing and refinement.
6. *Repeat.* While not in the Stanford model, I added this to emphasize the iterative and somewhat open-ended nature of good design. For larger architectural structures, these five steps would typically be repeated during the three stages: the vision stage, the preliminary design stage and the final design stage. For smaller built-in features design, prototyping and testing would continue until the desired result is realized.
7. *Post-occupancy evaluation.* How do you measure success? What lessons were learned? How did behaviour and welfare of animals and human stakeholders change because of the design and implementation process? Evaluation methods should be compatible with the open-ended and iterative design process and comparable with baseline or preoccupation research findings.
8. *Publish.* Sharing results, good and bad, formally or informally, is essential to on-going improvements in the welfare of all stakeholders.

Disney's Animal Kingdom also developed an effective six-step design process they refer to as 'SPIDER': (1) setting goals, (2) planning, (3) implementing, (4) documenting, (5) evaluating and (6) readjusting (Colahan and Breder 2003).

1.5 Design Team Building

In addition to the synergistic benefits of using an inclusive, multidisciplinary design group, team building and follow-through advantages may be gained when outside designers collaborate with long-term users. In my experience, a well-managed design process helps develop a constituency of supporters for effective and continued use of the features designed by the group. Participation may enhance a sense of shared authorship and ownership when compared to designs generated by others.

1.6 Codesign with Animals

This concept represents an emerging design frontier largely based upon the general notion of *becoming with* (Westerlaken and Gualeni 2016) or thinking of *animals as co-workers* (Kelling et al. 2012) and providing animal and human clients repeated opportunities to test prototypes using their species-typical *Umwelt*, dominant senses and individual preferences.

1.7 Related Literature on Design for the Elderly and Disabled Humans

Abundant literature exists on the design of facilities for infirm and ageing humans (e.g., George 2017). The International Council on Active Aging (ICAA) (2022) identifies seven dimensions of wellness in humans: (1) physical, (2) social, (3) spiritual, (4) intellectual, (5) emotional, (6) vocational and (7) environmental wellness. Most of these also appear relevant to animal welfare. How would we design animal environments to optimize animal access and agency in these seven dimensions?

1.8 Literature on Design for the Elderly and Disabled Animals

While some studies have used great apes and other primates for investigations of human ageing (Erwin et al. 2004), my literature search and contacts in wild animal management have uncovered little generalized guidance in our field. Hosey et al. (2009) provided good general design direction, but both domesticated and non-domesticated animals with disabilities, young or old, tend to be managed on a case-by-case basis. For example, Singapore Zoo has just lowered binturong (*Arctictis binturong*) resting platforms for two ageing animals (G. Laule pers. comm. 14 June 2021). Zoo Atlanta has recently modified facilities for their oldest male and female gorillas by lowering hammocks, shelves and benches; adding steps, ramps and rung handholds; and adding fire hose so gorillas can slide down rather than climbing down (J. Carrigan, pers. comm. 26 June 2021).

Other reports discuss how these animals have adapted to the existing conditions rather that how conditions have been modified to meet their special needs. Examples included studies of an amputee chimpanzee (Ang et al. 2017) and amputee and visually impaired rescued bears (Dallaire et al. 2012). Studies are underway at Chimp Haven in Keithville, Louisiana, to develop and validate a mobility index to identify emerging mobility problems developing in older rescued chimpanzees (S. Ross per.

Comm., 26 March 2021). Dr. Ross added that researchers have a fundamental knowledge of chimpanzees... *'which is important if we want to understand how potential limitations of aging are affecting them'*. For example, Ross et al. (2009) added to our understanding of chimpanzee spatial preferences between highly naturalistic and highly artificial enclosures. Browning and Maple (2019) also suggested a system for evaluating usable space in zoo exhibits. These may prove to be valuable tools for conducting baseline studies and later evaluating the immediate and long-term success of design efforts.

1.9 Designed Enrichment Features for Younger Disabled Animals that May Apply to Care for Older Animals

Studies of housing and enrichment accommodations for blind and amputee Asian black bears (*Ursus thibetanus*) at the Animals Asia Foundation China Bear Sanctuary in Chengdu, China, found that these individuals explored their outdoor environments and searched for scattered food less than able-bodied bears and were less active overall. However, several disabled bears showed novel adaptations to receiving and using enrichment features (Dallaire et al. 2012). This suggests that smaller areas with more specialized enrichment opportunities may be better suited to the needs of these animals and to ageing animals with similar disabilities. This is discussed further in Chapter "Environmental Enrichment for Ageing Zoo Animals" on environmental enrichment.

1.10 Design Features That May Benefit Older Animals and Delay Onset of Health Problems

1.10.1 Self-Care

Many mammal species use mud and dust baths to maintain healthy skin and fur. Many birds use dust and water baths (Mayntz 2020; Earth.com 2022). Many reptiles seek specific wavelengths and intensities of light and ranges of temperature (Baines et al. 2016). All species require sufficient quantity and quality of space, furnishings and motivation to develop and maintain adequate bone density and muscle tone through physical exercise (Panagiotou et al. 2021). Provision of abundant opportunities for appropriate self-care may prevent or delay future health problems.

1.10.2 Circadian Rhythms

'Many characteristics of animals' circadian rhythms change as they age, and changes to various sleep characteristics have been noted in elderly humans and animals' (Krebs et al. 2018). Based upon studies with humans and rodents, Panagiotou et al. (2021 abstract) stated: *'We emphasise that even moderate age-matched exercise is able to ameliorate several aging characteristics as far as sleep and circadian rhythms are concerned, independent of species studied'*. This finding suggests facilities and management programs encouraging lifelong exercise may

ameliorate circadian rhythm and sleep disruption as animals age, although a wider range of species, including nocturnal species, should be tested.

1.10.3 Suitable Surfaces

Surface materials for animal areas have long been selected for durability and ease of cleaning, primarily human objectives. While these considerations remain important, the trend has been to provide more animal-friendly surfaces such as deep mulch or biofloors (Chamove et al. 1982; Fiby 2020). Experienced elephant managers believe that confinement on hard, wet concrete flooring may be responsible for many long-term foot and joint problems (G. Creighton pers. comm., 5 May 2021). Soft, warm and resilient underfoot and deep surfacing of rounded sand particles or mulch are thought to reduce foot and joint problems and may prevent the onset of long-term problems as elephants age. Examples include Dublin Zoo, Denver Zoo, Copenhagen Zoo and Heidelberg Zoo; rhinoceroses and tapirs are at Magdeburg Zoo (Fiby 2020).

Dublin Zoo recommends a 2-m-deep sand bed with mounded sand pillows piled daily. It is believed that this leads to better quality of sleep for elephants of all ages (G. Creighton pers. com., 5 May 2021). Is it possible that better sleep throughout life would result in better quality of life and better long-term health? Biofloors for big cats and primates are usually 50 cm to 100 cm deep with under floor drainage (Fiby 2020) and so are most practical for installation in new facilities. Both long-lasting biofloors and frequently replenished sand and mulch mounds require material storage areas and convenient access for small tractors or other equipment. These must be included in facility designs. Of course, all accessible surfaces should provide safe traction for hand, foot, hoof, claw, scales and so on, without undue abrasion.

2 General Recommendations

2.1 Flexible Spaces

Most smaller animals in zoos and related facilities spend their entire lives in a single area, while some are moved between day-use areas, usually exhibits, and night containment areas called back-of-house, bedrooms, night dens and holding areas (Brando and Coe 2022). Some specialized facilities for great apes, such as those at Louisville Zoo in Kentucky and Lincoln Park Zoo in Chicago, Illinois, provide a series of indoor and outdoor activity areas and areas for temporary isolation and quarantine, often allowing the apes ad lib 24/7 access to most indoor areas.

Many species such as chimpanzees, bachelor gorillas and other group-living primates, carnivores and ungulates display highly changeable dominance systems. Thus, it is important to have flexible arrangements of accommodations, including areas where older animals and deposed dominant individuals may seek safety, with or without visual contact with their group (Coe et al. 2009; Watters et al. 2009; Brando and Buchanan-Smith 2018; Brando and Coe 2022). See Fig. 1 for an example.

Some zoos develop flexible off-display housing for powerful animals, suitable for large carnivores, for example, to provide humane life care for retired older

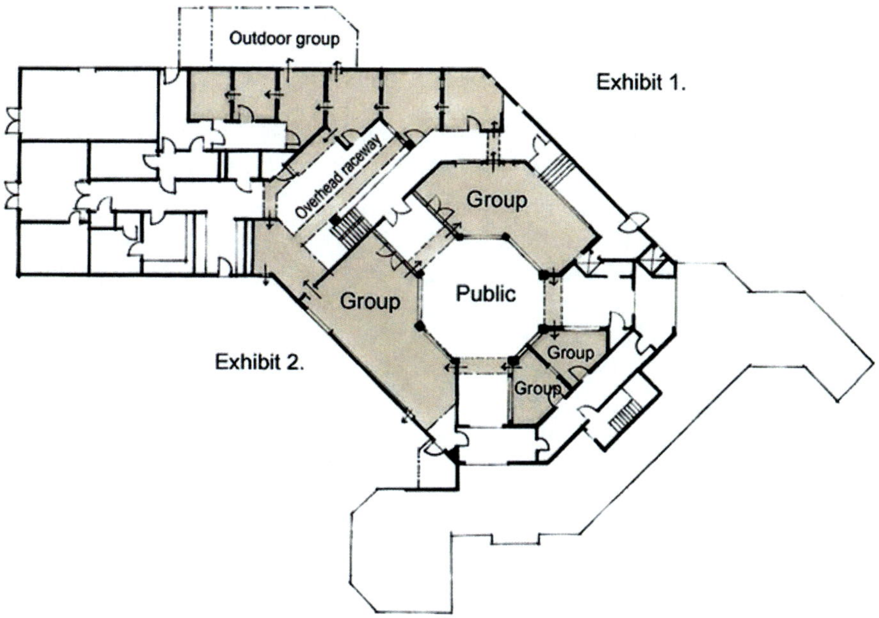

Fig. 1 Flexible floor plan for the Louisville Zoo gorilla facility showing the diversity of areas for apes. Image: J. Coe

animals and temporary housing for animals being held for shipping to other locations or waiting to be rehoused in facilities under construction.

2.2 Accessibility

Ageing and disabled animals, with their widely varied sensory and locomotion adaptations, may offer different accessibility challenges than humans. However, certain fundamentals may equally apply. For example, disabled and infirm individuals, youthful or elderly, may lack endurance and range of motion. Thus, circulation systems may be subdivided into small units, both horizontally and vertically, with convenient rest areas and perhaps bypasses for more physically able individuals. While there is little information on the circulation standards of blind or deaf animals, Dallaire et al. (2012) suggested that such animals found their way using other senses such as touch and smell. Spatial familiarity also seems likely to play a role.

2.3 Lighting

Lighting systems should be suited to diurnal, nocturnal or other species-appropriate levels, with separate light levels for cleaning and servicing by diurnal humans. Perhaps visually impaired animals of visually oriented species could benefit from motion-activated lighting to illuminate areas they occupy as they circulate or enter new areas.

2.4 Service Access

Animal circulation areas should be immediately accessible for carers to provide training, encouragement, medical treatment or other supports when needed, as well as for ease of service. However, for timid individuals or species, animal circulation areas should not be too-near areas of frequent human activity and include options allowing animals to feel sheltered.

2.5 Vertical Circulation

The benefits of overhead circulation, such as uninterrupted access to multiple areas, depend upon easy accessibility for both animals and carers. Vertical access for disabled animals may be improved by providing multiple small steps with frequent intermediate rest areas as shown in Fig. 2.

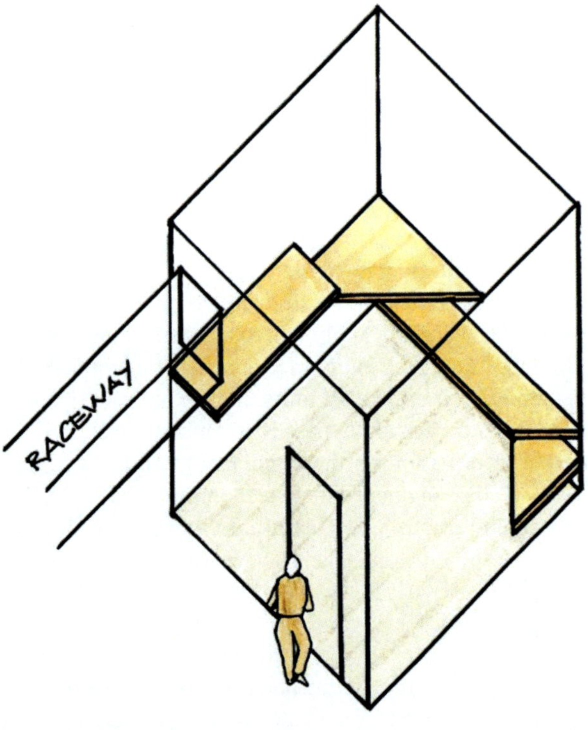

Fig. 2 Positioning of elevated rest areas can ease vertical circulation. Image: J. Coe

Provision of steps or ramps is a usual method considered for vertical circulation. In my experience, steps are suitable for many types of animals. However, heavy-bodied species like tapir, hippopotamus, rhinoceros and elephant, for example, may be reluctant to descend steps unless they are quite low and large. Ramps are useful for species like these and for carers moving wheeled equipment. However, it is becoming more common for carers to stay on lower accessible pathways and use wheeled lifts when safe access to higher levels is required.

I have heard the complaint that the extensive use of ramps for disabled and older animals requires excessive space and cost when fit individuals can easily ascend vertical passageways or furnishings. However, sometimes ramps or steps can be efficiently wrapped around central vertical shafts and integrated with safe stairs for carers and cleaners. Vertical circulation in the three-level chimpanzees' night building at the Los Angeles Zoo in California provides built-in handholds on the back wall for able-bodied apes. Resting platforms are available to older apes and facilitate cleaning. Vertical ape circulation is coordinated with stairways for care staff as seen in Fig. 3. Figure 4 shows the vertical circulation concept with rest platforms circling a central climbing pole, and Fig. 5 shows a three-level ramp and raceway concept proposed for rescued leopards in India.

2.6 Overhead Horizontal Circulation

Providing vertical circulation is accessible and comfortable, and horizontal overhead circulation could provide older and disabled animals access to a wide variety of physical and behavioural experiences, passing over lower mobility barriers. Some animals may feel more confident when above caregivers and especially when above crowds of visitors or perceived predator species. If so, this supports the use of elevated circulation areas and overhead raceways.

2.7 Animal Rotation and Circulation

Like ourselves, animals live in a multidimensional world, including accessible surface area, spatial volume and time, all of which are mediated by species and individual needs, characteristics and preferences. While most wild animals in human care spend their lives in one or two confined areas, free-ranging animals regularly visit multiple resource areas using well-defined horizontal and vertical pathways as described by Professor H. Hediger (1950 p. 14) and used as a defining feature of animal rotation (or alternation) facility designs and management systems in zoos and sanctuaries (Coe 2004, 2014).

Multiple areas connected by raceways, both on-ground and elevated, have now been used in several wild animal care facilities including the Philadelphia, Pennsylvania and Jacksonville Florida Zoos; sanctuaries such as Chimp Haven; and research facilities such as the Primate Research Center of Kyoto University. The

Fig. 3 Vertical circulation shaft for chimpanzees at Los Angeles Zoo also provides access for cleaners. Photo J. Coe

Center for Great Apes in Wauchula, Florida, was the first wildlife rescue sanctuary to develop extensive 'aerial trailways' to encourage movement and mental stimulation for animals at all ages, including ageing animals. They have 2.4 km of aerial trailways connecting 12 night houses and 19 outdoor enclosures, as well as their clinic and quarantine area. According to the founder Ms. Patti Ragan, '*The connecting trailways…give the groups of apes many choices for changing habitats with frequent rotations…*'. Fixed ladders are provided to assist older animals on accessing trailways. She believed at least six primate sanctuaries were using overhead raceways of various designs (P. Ragan, pers. comm., 16 May 2021). Images can be seen in Fig. 6.

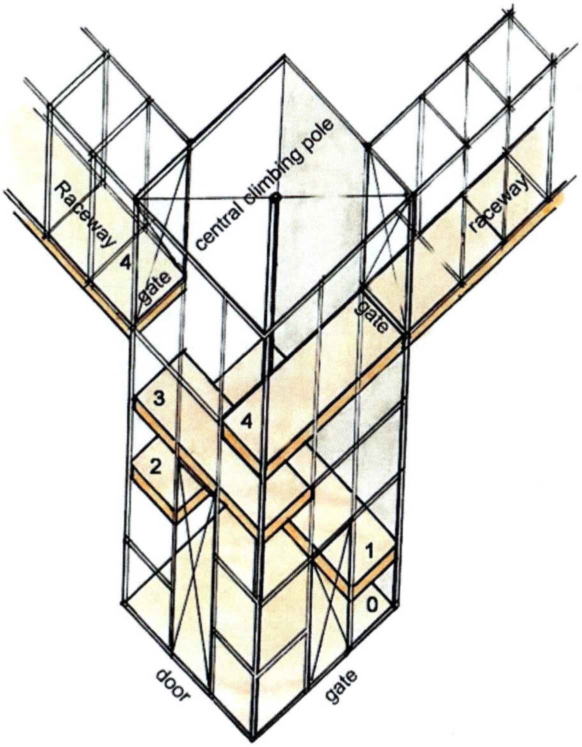

Fig. 4 Vertical circulation concept with rest platforms circling a central climbing pole. Image: J. Coe

2.8 Aquatic–Aerobic Exercise

Moderate physical exercise is known to benefit both physical and cognitive abilities in both ageing animals and humans (Kramer et al. 2006) and should be part of planning an animal care and welfare program in general to promote movement and physical exercise for animals, for example, good bone, muscle and core health (S. Brando personal communication 21/10/2022). Low-impact aquatic exercise is shown to benefit older humans (Bergamin et al. 2012) including conditioning exercises while swimming or partially submerged. These activities may also benefit many species of older animals in human care. Hydrotherapy using cool or warm water is used with lame horses (King 2016). Underwater equine treadmills are available and may be adaptable for use with similar species in zoos and sanctuaries. Melbourne Zoo, Victoria, Australia, has developed an innovative 'Reptile Gymnasium' in which various snakes, lizards and turtles are brought to swim against a gently water current. This can be seen in Fig. 7. Keeper Alex Mitchell said that the new facility was keeping the animals fit and active and, in turn, helping prevent illness and disease (Ausleisure, January 22, 2019).

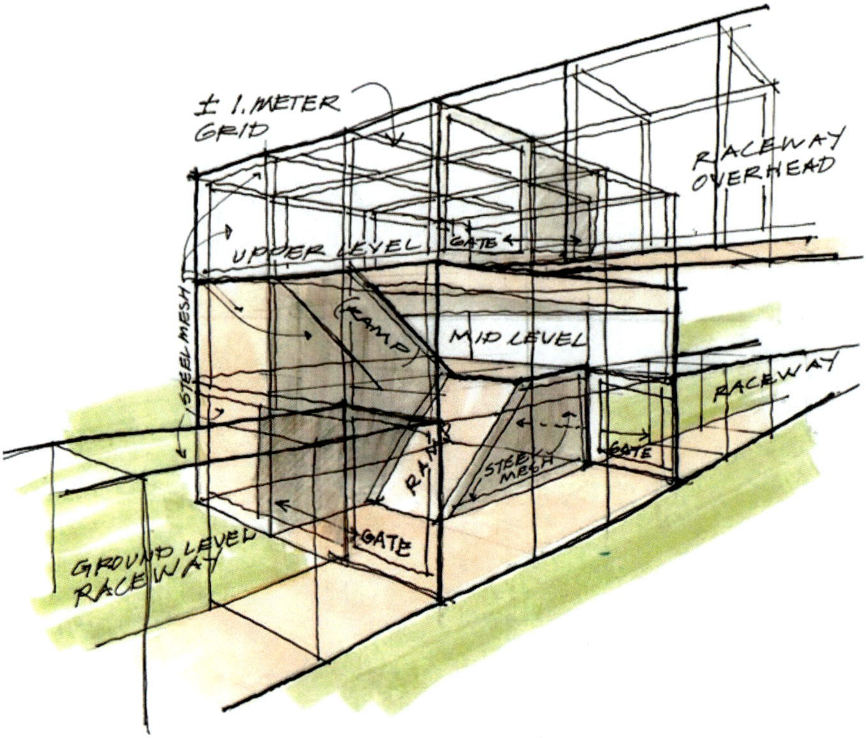

Fig. 5 Three-level ramp and raceway concept proposed for rescued leopards in India. Either ramps or steps could be used. Image: J. Coe

The elephant facility at Taronga Zoo, in New South Whales, was provided with a 60-m-long, 3-m-wide and 3-m-deep pool, with shallow accessible ends, designed to encourage aquatic–aerobic fitness and prevent lameness for the zoo's elephants. This can be seen in Fig. 8.

2.9 Animal–Computer Interaction (ACI) Benefits for Ageing Animals

Creative use of ACI technology, based upon solid animal welfare science, has the potential for greatly increasing managed animal choice, control and self-sufficiency and thus relative animal freedom in zoos, aquariums and sanctuaries. Greater automation can reduce human labour costs, but most are not allowed to impair the formation of positive carer–animal relationships (Coe and Hoy 2020).

Fig. 6 Aerial trailways allow great apes of all ages to 'take a walk through the park' at the Center for Great Apes. Photos: P. Ragan

Fig. 7 Reptiles exercise by swimming against a current of water at Melbourne Zoo's 'Reptile Gymnasium'. Photo: Zoos Victoria

2.10 Providing Older Animals with Increased Choice and Control Using Advancing Technology

The notion that animals housed in zoos and sanctuaries benefit from increased levels of choice and control over their own activities is now widely accepted (Snowdon and Savage 1989), as is the understanding that advanced age is often associated with

Fig. 8 An elephant pool designed for aquatic–aerobic fitness and image showing proposed fitness training. Photo and image: J. Coe

diminished agency. For example, older individuals may be less able to thermoregulate. Therefore, designed features allowing animals to individually control ambient conditions (temperature, ventilation and acoustics) should be especially useful for older animals. Older animals may be provided with increased bedding and other substrates or a variety of perches offering different ambient conditions. However, these opportunities may be usurped by younger, more dominant individuals, so care will have to be taken to provide sufficient options and quantities. Alternatively, easily accessible rest areas could be equipped with localized heaters, cooling mist or fans, which are activated by motion sensors and/or radio frequency identification (RFID) microchips by specific animals and not others. RFID and motion sensor–activated 'smart gates' can provide older individuals access to refuge or feeding areas for specialized diets and medication, safe from aggressive individuals or vermin (Coe and Hoy 2020). Animals themselves may be able to accommodate changing activity periods associated with ageing by using technology providing 24/7 access to their desired locations and features. Small RFID-controlled pet feeders and doors are commercially available (https://www.surepetcare.com).

2.11 ACI-Assisted Monitoring

Regular periodic health monitoring of younger animals is a normal part of animal management, in which animal–computer interactive devices may assist. While recovering, disabled and ageing animals may require frequent monitoring, especially those on veterinary treatment, for example, heart disease in great apes (S. Chapman personal comm. 31 August 2022). Carers at Zoo Atlanta, Georgia, have trained an orangutan to provide an electrocardiogram (EKG) using the commercially available KardiaMobile finger EKG, which can be seen in Fig. 9.

ACI features installed in the elephant facility at Dublin Zoo provide monitoring during times staff are not usually on duty. These features record videos of behaviours, times and locations when animals access feeders, sleeping mounds and even travel distance throughout the day and night. Caregivers can gain live access to these monitors and operate surveillance cameras or feeders 24/7 remotely from anywhere with a broadband service, including from aircraft during international travel (G. Creighton pers. com., 5 May 2021). ACI features can be designed to automatically record body weight and temperature and eventually even internal functions recorded by microchip implants, which are all recorded remotely.

Whitham and Miller (2016) provided an overview of monitoring systems such as accelerometers, global positioning system (GPS), bioacoustics, thermography and radio frequency identification systems.

2.12 Virtual Enrichment and Cognitive Research

Dr. Francine Dolins and her colleagues are developing virtual reality landscapes for laboratory apes and monkeys to explore as combined virtual enrichment and cognitive research initiatives. They also have developed a 'robo-bonobo-bot', a bonobo-controlled robot with a video camera that the apes can remotely drive through adjacent laboratory areas and around the facility's grounds (Dolins et al. 2017). Such novel ACI technology may hold a great promise for older and less mobile zoo, sanctuary and research animals.

Fig. 9 Orangutan using a voluntary electrocardiogram device. Photo: Zoo Atlanta

3 Conclusions

The purpose of built facilities and built-in features is to support the aims, activities and management programs suggested in the other chapters, contributing to the long-term optimal welfare, the actual thriving, of wild animals in human care and especially those disadvantaged by physical and mental disabilities including the elderly. These types of facilities not only support good animal wellbeing, but also allow animal care staff to care for the animals in a variety of ways, contributing to job satisfaction and good human wellbeing (Brando and Coe 2022). Good physical design may prevent or delay the onset of some disabilities such as degenerative joints in big cats, foot and arthritic difficulties in elephants, feather plucking in birds, hair loss in primates and skeletal deformation in snakes, while providing opportunities for self-care of skin, fur and feathers and encouraging physical exercise, mental stimulation and social engagement. Separate 'retirement' facilities have been mentioned. However, social species should be kept in appropriate social groups to the extent possible. Both groups and individual animals should be empowered to satisfy their own unique needs and preferences 24/7 in conditions benefiting lifelong wellness as described by Brando and Buchanan-Smith (2018) and Maple (2019).

References

Ang M, Shender M, Ross S (2017) Assessment of behavior and space use before and after forelimb amputation in a zoo-housed chimpanzee (pan troglodytes). J Zoo Biol 36(1):5–10. https://doi.org/10.1002/zoo.21345

Ausleisure Newsletter (2019, January 22) Melbourne zoo builds hydro gym for reptiles. Australasian leisure management https://www.ausleisure.com.au/news/melbourne-zoo-builds-hydro-gym-for-reptiles/ Accessed 29 Oct 2022

Baines FM, Chattell J, Garrick D, Gill I (2016) How much UVB does my reptile need? The UV-tool, a guide to the selection of UV lighting for reptiles and amphibians in captivity. J Zoo Aquar Res 4(1):42–63. https://doi.org/10.19227/jzar.v4i1.150

Bergamin M, Zanuso S, Alvar BA, Ermolao A, Zaccaria M (2012) Is water-based exercise training sufficient to improve physical fitness in the elderly? Eur Rev Aging Phys Act 9:129–141. https://doi.org/10.1007/s11556-012-0097-1

Brando S, Buchanan-Smith HM (2018) The 24/7 approach to promoting optimal welfare for captive wild animals. Behav Process 156:83–95. https://doi.org/10.1016/j.beproc.2017.09.010

Brando S, Coe J (2022) Confronting back-of-house traditions: primates as a case study. J Zool Bot Gard 3(3):366–397. https://doi.org/10.3390/jzbg3030029

Browning H, Maple TL (2019) Developing a metric of usable space for zoo exhibits. Front Psychol 10:791. https://doi.org/10.3389/fpsyg.2019.00791

Centre for Excellence in Universal Design (CEUD). https://universaldesign.ie/ Accessed 29 Oct 2022. https://universaldesign.ie/About-Us/ Accessed 20 Oct 2022

Chamove A, Anderson J, Morgan-Jones S, Jones S (1982) Deep woodchip litter: hygiene, feeding, and behavioral enhancement in eight primate species. Int J Study Anim Prob 3(4):308–318

Chebyniak M (2019) Design thinking process overview. https://m-cheba.medium.com/design-thinking-process-overview-7d13c607c0dc Accessed 29 Oct 2022

Coe JC (2004) Mixed species rotation exhibits. Paper presented at Zoo and Aquarium Association 2004 Annual Conference, 7–11 March 2004, Christchurch, NZ

Coe JC (2014) Next generation rotation exhibits: networks and space to explore. Paper presented at Zoo and Aquarium Association Annual Conference 25–28 March 2014, Auckland, NZ. https://www.joncoedesign.com/pub/technical.htm. Accessed 29 Oct 2022

Coe JC, Hoy J (2020) Choice, control, and computers: empowering wildlife in human care. In Animal Centred computing: enriching the lives of animals. Morrison, A., Bollard, B. (Eds). Multimodal Technology Interaction, 4(4), 92. https://doi.org/10.3390/mti4040092

Coe JC, Scott D, Lukas KE (2009) Facility design for bachelor gorilla groups. Zoo Biol 28(2):144–162. https://doi.org/10.1002/zoo.20233

Colahan H, Breder C (2003) Primate training at Disney's animal kingdom. Appl Anim Welf Sci 6(2):235–246. https://doi.org/10.1207/S15327604JAWS0603_08

Dallaire JA, Field N, Mason GJ (2012) Activity and enrichment use in disabled Asiatic black bears (Ursus thibetanus) rescued from bile farms. Anim Welf 21:167–176. https://doi.org/10.7120/09627286.21.2.167

Dam RF, Siang TY (2020). https://www.interaction-design.org/literature/article/what-is-design-thinking-and-why-is-it-so-popular. Accessed 29 Oct 2022

Dolins FL, Schweller K, Milne S (2017) Technology advancing the study of animal cognition: using virtual reality to present virtually simulated environments to investigate nonhuman primate spatial cognition. Current Zool 63(1):97–108. https://doi.org/10.1093/cz/zow121

Earth.com. https://www.earth.com/earthpedia-articles/more-dirt-more-clean-what-is-a-dust-bath/ Accessed 29 Oct 2022

Erwin J, Hof P, Ely J, Perl D (2004) One gerontology: Advancing understanding of aging through studies of great apes and other primates https://doi.org/10.1159/000061455
Fiby M (2020) Mulch in indoor exhibits. ZooLex Design Organization Publications. https://zoolex.org/slist/publications. Accessed 30 Oct 2022
French F, Mancini C, Sharp H (2020) More than human aesthetics: Interactive enrichment for elephants. DIS '20: Proceedings of the 2020 ACM Designing Interactive Systems Conference, ACM 1661–1672. Eindhoven, Netherlands https://doi.org/10.1145/3357236.3395445
George A (2017) Influencing factors on age-friendly interiors. Forensic science and addiction. Research 1(4):39–45. https://doi.org/10.31031/FSAR.2017.01.000517
Hediger H (1950) Wild animals in captivity. Butterworth, London
Hosey G, Melfi V, Pankhurst S (2009) Zoo animals: behaviour, management and welfare. Oxford University Press, Oxford
International Council on Active Aging (ICAA). https://www.icaa.cc/business/Wellness-model-.htm. Accessed 29 October 2022
KardiaMobile EKG Monitor. https://www.alivetec.com/pages/alivecor-heart-monitor
Kelling N, Gaalema D, Kelling A (2012) Elephant in the break room: the use of modified operational sequence diagrams for the determination of zoo exhibit inefficiencies. Proceedings of the Human Factors and Ergonomics Society 56th Annual Meeting Los Angeles CA, pp 1519–1523
King M (2016) Principles and application of hydrotherapy for equine athletes. The veterinary clinics of North America. Equine Practice 32(1):115–126. https://doi.org/10.1016/j.cveq.2015.12.008
Kramer A, Erickson K, Colcombe S (2006) Exercise, cognition, and the aging brain. J Appl Physiol 101(4):1237–1242. https://doi.org/10.1152/japplphysiol.00500.2006
Krebs B, Marrin D, Phelps A, Krol L, Watters J (2018) Managing aged animals in zoos to promote positive welfare: a review and future directions. Animals 8(7):116. https://doi.org/10.3390/ani8070116
Maple TL (2019) Beyond animal welfare, the art and science of wellness. Palmetto Publishing Group, North Charleston. https://doi.org/10.1159/000061455
Mayntz M (2020) Why do birds take dust baths? The spruce. https://www.thespruce.com/how-and-why-birds-take-dust-baths-386438 Accessed 29 October 2022
National Disability Association (NDA). http://universaldesign.ie/What-is-Universal-Design/#:~:text=Universal%20Design%20is%20the%20design,%2C%20size%2C%20ability%20or%20disability. Accessed 29 Oct 2022
Panagiotou M, Michel S, Meijer JH, Deboer T (2021) The aging brain: sleep, the circadian clock and exercise. Biochem Pharmacol 191:114563. https://doi.org/10.1016/j.bcp.2021.114563
Ross S, Schapiro SJ, Hau J, Lukas K (2009) Space use as an indicator of enclosure appropriateness: a novel measure of captive animal welfare. Applied animal. Behav Sci 121:42–50. https://doi.org/10.1016/j.applanim.2009.08.007
Sherwin S, Hemsworth L, Beausoleil N, Embury A, Mellor D (2018) An animal welfare risk assessment process for zoos. Animals 8(8):130. https://doi.org/10.3390/ani8080130
Snowdon CT, Savage A (1989) Psychological Well-being of captive primates: general considerations and examples from callitrichids. In: Segal EF (ed) Housing, care and psychological Well-being of captive and laboratory primates. Noyes Publications, Park Ridge, NY
Watters JV, Margulis SW, Atsalis S (2009) Behavioral monitoring in zoos and aquariums: a tool for guiding husbandry and directing research. J Zoo Biol 28(1):35–48. https://doi.org/10.1002/zoo.20207
Westerlaken M, Gualeni S (2016) Becoming with: towards the inclusion of animals as participants in design processes. ACI '16. Proceedings of the Third International Conference on Animal-Computer Interaction 1:1–10. https://doi.org/10.1145/2995257.2995392
Whitham JC, Miller LJ (2016) Using technology to monitor and improve zoo animal welfare. Anim Welf 25:395–409. https://doi.org/10.7120/09627286.25.4.395

Environmental Enrichment for Ageing Animals in Zoos

Julian Chapman

J. Chapman (✉)
Chapman Zoo Consultancy, Birmingham, UK

S. Brando, S. Chapman (eds.), *Optimal Wellbeing of Ageing Wild Animals in Human Care*, https://doi.org/10.1007/978-3-031-30659-4_5

Rupert. Rupert was a male yellow-throated marten (*Martes flavigula*) who lived at Twycross Zoo, in the United Kingdom, for 12 years. He was very slow to give his trust but was very sweet once he got to know you. This made training him quite challenging. Rupert was very intelligent and got the better of several trainers when he did exactly what he was asked but just not in the way they were planning. He never successfully bred with either of his mates Steffi or Lilly but had lots of good interactions with them. Rupert was always a little more shy than Steffi, so sometimes she would come to collect the food from the front of the enclosure and take it back to him. It seemed this favour was often returned in kind when he allowed her to curl up directly on top of him when napping during the day. His favourite past time was to curl up so close to his mate Steffi so you could hardly tell which parts belonged to who! Favourite enrichment strategies included giving small carcass feeds (usually rabbit) tied to a post around the enclosure, which really brought out their predatory side and showed how strong they were. Rupert was always very good at puzzle feeders and enjoyed manipulating them to get the good stuff making their 'egg day' great fun to watch for both visitors and keepers. Steffi would often bury the eggs to snack on later, but Rupert preferred to enjoy them straight away. A bone cancer (Osteosarcoma) affecting his left forelimb was found during a health assessment to investigate lameness and he was euthanised in July 2022 with his caregivers by his side.

Abstract

The use of environmental enrichment is commonplace in zoos, aquariums, sanctuaries and other facilities and aims to enhance the lives of animals under human care by promoting natural behaviours. There are various types of enrichment used including food-based, sensory, manipulative and novel. Enrichment programmes can be created by listing the options for each type of enrichment and, using a 4-week timetable, can be planned by the caregivers to ensure that enrichment is not overused but is always provided. As animals age, the enrichment items offered may require adaptation to ensure safe use and lack of frustration or engagement by the individual. This chapter will describe the types of enrichment available and the use of the timetable system and end with practical ways to adapt enrichment for the older animals.

Keywords

Enrichment · Safety · Monitoring · Ageing · Zoo

1 Introduction

Care of animals in zoos, aquariums, sanctuaries and other facilities (henceforth zoos) involves a wide range of aspects including habitat design and management, social management, nutrition and veterinary care in order to promote good animal wellbeing. One core component of an animal care programme is providing environmental enrichment, which can be in a variety of forms with specific behavioural goals. This chapter outlines the types of enrichment that can be used, examples of methods and devices available and how these can be adapted for ageing animals.

2 Environmental Enrichment

It is generally accepted within the zoological community that environmental enrichment, in one form or another, should be provided for all animals who are in captivity (Young 2006). There have been various definitions of enrichment over the years with one referring to modifications that act to enhance the level of physical and social stimulation provided by the captive environment (Würbel et al. 1998). Enrichment should continuously be designed to stimulate the animals both physically and mentally. It is important to link habitat management and a well-designed environment together with an active environmental enrichment programme, including observations for evaluation and adjustment (Alligood and Leighty 2015) in order for programmes to remain dynamic for the animals. As animals in captivity are generally living longer lives than their wild counterparts, we have to consider age-appropriate activities and opportunities when choosing the types of enrichment we provide and how these are implemented.

3 Types of Enrichment

An overview of how different types of enrichment can be divided is given below. There are a variety of ways to do this, and it is often a case of personal preference. The preference here is to initially keep it as simple as possible with the view that it is easy enough to add extra categories for the enrichment, but it is not so easy to dispose the category after it has been used for a while. With that in mind, it is recommended to use three categories of enrichment to begin with. If your facility already uses more categories than this, then continue with the system that is already in place. The categories that are used in this chapter are food-based, sensory and manipulatory.

3.1 Food-Based Enrichments

The food-based enrichments involve the animal working in one way or another to gain a food reward (Roe and Cleave 2005; Rozek et al. 2010).

It is important that any food-based enrichment should use food that is taken from the animals' daily, or other frequency, ration (how much food is allocated to the animal) and is not given as an extra treat to ensure attention to healthy weights and body composition. In the case of providing a seasonal food that they do not get regularly, for example, pumpkins, you will have to remove an equivalent part of their daily ration to compensate for the use of food in the enrichment. Additional feed during enrichment has been seen to cause weight gain; hence, there is a need to ensure that the dietary intake is managed appropriately.

3.2 Sensory Enrichments

Sensory enrichments are designed to stimulate one or more of the animal's senses (Martínez-Macipe et al. 2015; Wells and Irwin 2008). Wherever possible, they should do this without providing the animal with a food reward as this would then become a mixed enrichment item. This category includes such examples as temperature changes, scents, noisemakers, shiny or colourful objects and tastes.

3.3 Manipulatory Enrichments

Manipulatory enrichments are usually items that are put into an enclosure for the animal to investigate, play or interact with, or they might involve new or temporary changes to the layout of the enclosure.

The fourth category is novelty. Any time a new enrichment item is introduced to an animal for the first time, it is going to be a novel experience for that animal. However, it is very difficult to continuously provide enough new items for the animals for this to be a viable option for a permanent category; however, a combination of many individual items and continued attention to changes and new ideas can assist in satisfying this more challenging category. Studying the reaction of the animals to novel items can be a very interesting subject, and more work is recommended to be carried out in this area. Facilities caring for animals are encouraged to provide a monetary budget for enrichment, to have time available to engage in enriching activities for both animals and staff and to follow the process from planning to readjusting (Alligood and Leighty 2015).

4 Enrichment Items

When looking at a new enrichment programme for an animal, the first step is to list everything that is already provided either regularly or occasionally. Then, ask the animal caretakers to research other enrichment items that have been provided for that species by other facilities. This is best done either via online searches such as Pinterest boards, Facebook groups or Google Scholar for peer-viewed enrichment or by contacting other care staff of that species directly. They can also take the opportunity to develop their own ideas for enrichment as well. At the end of this chapter, a list of enrichment items will be provided that can be made at little or no cost. Commercially made enrichments are also available if you have the budget to purchase these items.

The next step is to separate all the items into one of the three categories. The chances are that there will be far more food-based enrichments than either of the other two categories. This is normal, as these tend to be the easiest enrichments, and the ones that appear to have a more beneficial effect on the animals, as they will almost always interact with them in a positive way. Later, in this chapter, the use of enrichment rotas or timetables will be outlined to prevent this from potentially becoming a problem and to ensure that any enrichment is not overused. Familiarity may no longer make it enriching for the animals (Young 2006). One enrichment item may well be suitable for use in different categories with just a minor alteration. A commercially available item such as a KONG™ would initially be in the manipulatory item as many animals just like the random way move around if they are knocked or dropped. Placing the KONG in a freezer for a while would mean it could be used in the sensory category, and finally, if it is stuffed with food, it can be put into the food category.

5 Enrichment Methods and Devices

Listing enrichment items should be carried out with all the species within the facility and combined with the enrichment rotas is a good starting point for any programme. Inevitably there are going to be some older animals within certain groups, and we must take them into consideration and adapt the programmes according to their needs and preferences. One way of thinking about these older animals is the analogy of what a young person would do if they saw a nice juicy apple growing on a tree. The chances are that if they wanted it, they would think nothing of quickly climbing up that tree to pick it. Compare that with an older person who is perhaps middle-aged and is starting to feel the effects of an active life on their body. The chances are they would prefer to wait for the fruit to fall, or more likely they would go to the shops to buy an apple!

A tiger example showcases how we can relate this to the animals in our care. Most people caring for tigers will have provided them with an enrichment that is commonly named 'the tiger spring'. This involves a strong spring that is attached to a wire or rope and has a method of fixing a piece of meat or maybe a hessian/burlap bag on the end. A pole or tree within the enclosure is then used so that the device can be raised to the required height above the ground. The idea is that the animal will run at the tree or pole with enough speed for it to climb up and reach the reward. The tiger will then hold the meat or item in his jaws and leap to the ground, landing on all its four feet as cats do.

This is a great enrichment device, and we are continuously trying to devise ways of making it harder for the animals, so they do not adapt to much and grow stronger and fitter. For more details on the benefits of the climbing pole, see the chapter 'The Longevity Legacy: The Challenges of Old Animals in Zoos' in this volume. We will make them have to climb higher or we will make the pole wobble (Law and Kitchener 2020) or move slightly while they are trying to climb it. These are all good and will certainly get the animals to use muscles and strengthen bones and tendons that they may otherwise not use on a regular basis. The animal will also have to think about how they are going to get to the reward. No one can fail to acknowledge that the sight of one of these magnificent animals carrying out this manoeuvre is always thrilling for the visitors and animal carers alike. Check out the video on YouTube to watch the feeding pole in the Tiger Territory in action at ZSL London Zoo.

Now let us compare the tigers and their burlap bag to the humans and their apples. The young tiger will happily climb the pole to get his reward and probably not think anything of it. However, as that animal ages, we do not always change the enrichment to suit. This means that the middle-aged tiger, who might well have mild arthritis or other age-related problems, has no choice but to climb the pole to get the food that it needs. Unfortunately, what we cannot see by observing is the pain that the animal may be suffering as it lands because it can no longer absorb the impact the way that it could when it was younger. In addition, they will lose the ability to turn in mid-air, and once again, we may not realise until the one time that they land awkwardly and injure themselves. Although a review of the use of tiger feeding poles in 2019 found no incidents of trauma (Law and Kitchener 2020), there have been anecdotal reports of big cats breaking their backs with this sort of enrichment and animals might all together stop climbing the pole and not feeding. It is important not to immediately stop this kind of enrichment but to adapt the enrichment to suit the animals as they age.

In addition, in environmental enrichment programmes, close attention needs to be paid to animal behaviour and implementation of preventive and adapted care at a geriatric age. The chapter 'Veterinary Care of Ageing Zoo Animals: Description of a Proactive Approach' in this book will lay out ways to work out when your animals are starting to fall into the ageing animal category and when it is encouraged to put in place protocols adapted to the next phase of life. By planning for and implementing changes at this stage, the aim is to prevent the enrichment from aggravating and accelerating any problems that may be developing.

Using the tiger feeding pole as an example, the following examples are ways of adapting the enrichment.

1. Start to lower the maximum height that the meat or bag is hung so that the animals do not have to climb as high. Do not simply extend the length of the wire that is hanging down as this will potentially cause problems of its own such as entanglement or hanging. It is better to put in a new ring or pulley at a lower level to prevent this.
2. A deep layer of a soft substrate such as bark or wood chips beneath the pole will provide a softer landing for the animals.
3. Do not fix the meat or bag to the wire as tightly; there is a good possibility that their teeth are not going to be in as good condition as they used to be, and we do not want to increase the chances of any damage leading to the need for dental work to be carried out.
4. As time progresses, the meat or bag can be lowered so that they can reach it by simply stretching up on their hind legs.
5. Finally, with a very aged animal, consider changing the device from vertical to horizontal. To do this, put the spring inside a log (real or artificial) or pipe so that they do not have to climb but can still interact with the food for longer. By placing this in front of a viewing window, you can still create an interesting experience for the visitors.

Obviously, this is just one enrichment item for one species, but it does demonstrate the need to think imaginatively or 'out-of-the-box' and be aware of the changes that are happening to the animals as they age.

As with carnivores, we see a large number of older primates in captivity, which due to their social nature present another set of challenges. The biggest challenge is that many primate species live in large groups that consist of animals of all different ages, possibly resulting in having to provide the same enrichment in several different ways to suit the needs of different individuals.

One of the easiest ways to enrich the whole group is identifying the lowest common denominator and only to provide items that are suitable for the older animals. This is certainly an option and can be used for a limited number of times, but the stimulation for the younger animals will likely be reduced. This means that for a certain percentage of the time you will have to either put in a mix of enrichments, so animals have a choice, or provide separate enrichments to certain animals while the others are occupied with something else.

A problem that often occurs with a group-housed older animal is bullying. With some species in the wild, an animal that is no longer perceived as useful to the group is not tolerated and pushed out. Even with more tolerant species, older animals will still be at the bottom of the pecking order, and we must be aware of this. One of the first things is ensuring that there are enough enrichment items to go around. This is particularly the case when we are carrying out a food-based enrichment that comprises a part of the groups' daily ration. If you have 20 baboons and you put in 19 items, some animals will be left without. Be prepared that the most dominant animal will often take more than one item. We need to consider that one animal can carry or monopolize multiple enrichments, so it is best to provide many more than the actual group numbers, provided in multiple locations and levels.

My experience has taught me to always be observant and open to modification, illustrated by the following example with baboons. A group was given a termite mound feeder with enough feeding outlets for the whole of the group to feed at the same time. Unfortunately, it was noted that the dominant male always wanted to see what was happening at the other feeding points in case the output was better. This had a knock-on effect down the groups' hierarchy and increased the aggression levels. It was decided to decrease drastically the number of feeding points and see the result. What occurred was that the male would go to the feeder and stay there until he had enough or was bored and then the next in line would have a go. This basically mimicked what would happen in the wild when a resource is limited and had the desired result of reducing the aggression as hoped. This has also been seen in other research where baboon behaviour was managed using a variety of feeding enrichment methods (Huber 2009).

The example above shows us two things: 1) The outcome that we are expecting from any enrichment that we use is not always the one that we will get, and 2) it is very important that we observe and study our animals before, during and after enrichment to ensure that they are achieving the planned goal. If we do not observe we could miss that, for example, enrichments provided to the animals can be totally ignored, both physically and mentally. In one example, the aim of the enrichment was to reduce pacing in an animal, and it was thought that although they did not engage physically with the item, it did engage them mentally and they were therefore distracted from pacing.

When using an enrichment from the sensory and manipulatory groups, the number of items is not as critical. This is because they tend to last longer in the environment whether this is because they are a durable item such as a KONG or a scent that will take a long time to fade away. It is important to also think of enrichment as a tool in some situations. If there is an ageing animal in the group who may require some extra feeding to maintain his or her weight or medication for a condition, then enrichment can be used to distract the rest of the group while the older individual is given these extra rations or medication. Or if each animal can be given their enrichment individually, provide another way of giving extra rations or medication. Another challenge with older animals is that they are often a lot slower at eating their food due to the condition (or lack) of their teeth. Obviously, if a concern is noticed, it is advised to have a health assessment carried out with particular attention paid to the mouth. If there are no problems found, then all efforts should be made to ensure that they do not miss out when enrichments are given. Once again, we will normally be concentrating on the food-based enrichments.

Scatter feeding is one of the ways of foraging, which can be seen as enriching for older animals and can also be used for the whole group. This may be provided on a regular, or even a daily, basis. If this is the case, it should be removed from any enrichment rota that you have and instead be put down as part of the animal's daily husbandry routine. The thing to do here is to have another look at scatter feeds and see if there is anything that you can do to adapt them and possibly make them more interesting for the animals.

The principal reason a scatter feed is used is to increase foraging time (Troxell-Smith et al. 2017). This will help the older animals by bringing their cage mates feeding rates closer to their own. The food should be spread out as much as possible in their enclosure(s) as possible. This type of enrichment can be used with almost every species in some form or other.

The following points should also be considered:

1. If your scatter feed consists of a chopped fruit, it is important to know that it loses quality as soon as it is cut (Plowman et al. 2006). A scatter feed should therefore be fed immediately or made up of small whole items such as grapes and blueberries. Dependent on the species and if possible, vegetables are probably a better option than fruits due to a higher sugar content in the diets.
2. One study with primates reported that it was more beneficial for the subordinate animals to feed the group whole food items. It was found that when the fruits and vegetables were cut into small pieces, the dominant animals could pick them up and swallow them immediately. However, when whole foods where fed it took them longer to process the pieces, thus allowing the other animals to get a better share (Huber 2009). Also in primates, the dominate animals could only hold a couple of pieces in their mouths and one in each hand before retiring to eat them.

3. When whole food is put on the roof of the enclosure, the lower ranking animals will get a portion of this as some pieces will fall to the floor, and this could be supplemented with some smaller pieces thrown in once the dominant animals are engrossed in the food on the roof. Many animals prefer to work, also called contra free loading (Inglis et al. 1997), for food, so dominant animals may stay up high even though there is a more easily accessed food supply on the floor.
4. Another consideration is the use of intake and out-take studies where you measure how much food is being taken by the animals through enrichment. This involves weighing the food that is offered and weighing food that has not been consumed. Weighing the animals and performing body condition scoring of the animals are also valuable to monitor their nutritional status to ensure adequate amounts of food are being taken and that animals are not becoming overweight.

6 Enrichment Rotas or Timetables

Enrichment timetables are commonly used in facilities to ensure that all the animals get enrichment regularly, allow the staff to plan and prepare activities and enrichment (devices) in advance and promote a wide range of types of enrichments provided.

Separating the enrichment items into different categories is the first step in creating the timetables. At this point, it is recommended to name and catalogue all the enrichments that will be used. This can be done either on a computer or in a folder, as long as all the animal staff can have access to the information. It is preferred to enter the data into a data spreadsheet for analysis and review, to make evidence-based decisions. When cataloguing an item, ensure the items required to make it, and instructions on how to construct it are recorded. Also state which animals the enrichment can be used for, along with any changes that would need to be made. A photograph of the finished item and any other relevant points can be very useful. State what the aim/goal of the enrichment is and finally any safety checks that need to be done before it is used.

Once this step is completed, a 4-week calendar with columns for the day, date, enrichment category and enrichment used should be created, depending on team and facility resources, needs and preferences. Commence with the food-based category, then manipulatory, back to food-based and then sensory before starting the list over again. The reason that more food-based enrichments will be used is because there are probably more options on the assembled list than the other types. The day after each of these enrichments, insert the word refresh or repeat. This means that your first five enrichments will cover 10 days. See Fig. 1 below for an example.

In my experience, animals still used an enrichment item enough on the second day of use therefore it was worthwhile putting the item in again. After the second day, the animals' interest had been reduced significantly, or they had worked out

TYPE OF ENRICHMENT	DAY	DATE	WHAT WAS GIVEN
FOOD	SAT	01-10-22	½ basket feeder
REFRESH	SUN	02-10-22	½ basket feeder
SENSORY	MON	03-10-22	Scented toilet rolls
REFRESH	TUE	04-10-22	Scented toilet rolls
FOOD	WED	05-10-22	Seeded bark
REFRESH	THU	06-10-22	Seeded bark
MANIPULATORY	FRI	07-10-22	Paper sacks
REFRESH	SAT	08-10-22	Paper sacks
NONE	SUN	09-10-22	
FOOD	MON	10-10-22	Cage with whole food
REFRESH	TUE	11-10-22	Cage with whole food
SENSORY	WED	12-10-22	Sun catchers
REFRESH	THU	13-10-22	Sun catchers
FOOD	FRI	14-10-22	Whole food kebabs
REFRESH	SAT	15-10-22	Whole food kebabs
MANIPULATORY	SUN	16-10-22	Egg boxes or trays
REFRESH	MON	17-10-22	Egg boxes or trays
FOOD	TUE	18-10-22	Whole feed on roof
REFRESH	WED	19-10-22	Whole feed on roof
SENSORY	THU	20-10-22	Scented wood chips
REFRESH	FRI	21-10-22	Scented wood chips
FOOD	SAT	22-10-22	Pinatas
REFRESH	SUN	23-10-22	Pinatas
MANIPULATORY	MON	24-10-22	Hanging sacks
REFRESH	TUE	25-10-22	Hanging sacks
FOOD	WED	26-10-22	Bamboo feeder
REFRESH	THU	27-10-22	Bamboo feeder
NONE	FRI	28-10-22	

Fig. 1 Example of an enrichment timetable

how to use the enrichment very quickly. When an item is refreshed, it is advised to move the item to a different area or at least change the location slightly if possible. This repeat day makes it easier for the care staff and also means that by just using 13 items enough enrichment can be provided for 4 weeks.

Because it does not mathematically work out that 13 items will cover every day over the 4 weeks, any spare days can be used either to give the animals a day without enrichment or to use an item that does not fit into any of aforementioned categories, or this spare day gives the care staff a chance to try out new/novel enrichments to see if they are suitable.

It is recommended to start with creating a timetable for one species that already has a lot of enrichments supplied. Over time, other species can be introduced, and extending beyond the original 4-week calendar will be possible. This means that the animals can go longer between seeing the same enrichment, which will mean that they are more likely to interact with it to a greater extent.

To ensure that the animal care staff are not tempted to use their favourite enrichments more often than the less popular ones, there should a separate list of all the enrichments placed into the categories. This will have columns for the items, the date used and any comments about the enrichment. Every time an enrichment is used for a species/group/enclosure, the date is recorded, and the staff are not allowed to use that enrichment again until all the other ones in that category have also been used. This will prevent repetition and will ensure that diverse items and activities are used.

Having these lists and timetables already prepared means that when the care staff have time available, they can prepare many of the upcoming enrichments in advance. This is particularly useful with items such as ice blocks or when time for a scent to permeate into something is required, such as a piece of wood, before it is used. When possible, a facility is encouraged to have volunteers or groups of people to help with enrichment preparations and activities; they can also be used to help with the making of things such as piñatas or feeders (which are described below).

6.1 Enrichments

Below is a list of a few examples of enrichments that either are already suitable for use by ageing animals or can be easily adapted for them.

1. *Buried food.* Even animals that would not normally dig will often at least scratch at the ground with hands, hooves or horns to try and uncover food items. When putting fresh substrate into an enclosure (such as bark chippings), do not be tempted to spread them out in a lovely, neat, flat surface. Instead leave them in a heap with a little of the animals' daily food ration hidden inside and let them dig into to find them. Alternatively, when regularly burying food in the enclosure, ensure adapting and changing the location as the animals age. Choose a softer substrate (bark or sand for

example) and do not bury it as deep. For example, a younger elephant will happily dig down several feet looking for a treat. However, an older elephant would not find it so easy to get down on their knees to dig and would certainly struggle to stand up again.

2. *Food balls*. These are often commercially bought feeders that come in a variety of sizes. Larger ones can be filled with part of the animals' daily ration of pellets which fall out as the item is moved around by the animal. Smaller ones can be used with animals such as meerkats and can contain mealworms that slowly crawl out over an extended period. With older animals, enlarging the holes may be helpful so that the food falls out more easily to prevent frustration. With the mealworm feeders, try hanging them above the enclosure so that they just drop randomly to the floor.
3. *Ice blocks*. These can either be simply plain water or have flavours or scents added or have whole pieces of the daily diet in them. Most of these would be suitable for the older animals, and as with other enrichment, observing them is highly encouraged. If they are becoming more reluctant to interact with them, it might be an indication of a sensitive tooth that would need to be assessed. Ice blocks, depending on size, can leave an area or indoor habitat wet and damp, which can affect wellbeing negatively, so location and weather conditions must be considered.
4. *Scents*. This is suitable for animals of all ages. Applied all over an enclosure and the only change suggested for an ageing animal would be not to encourage them to any part of the enclosure where they might injure themselves by falling or becoming trapped. As animals age, they will have probably started to avoid these areas naturally, but an interesting scent might just make them slightly more daring. Scents, like other enrichments, may also be overwhelming or making a certain area less attractive; therefore, it is advised to think about where and what sorts of scents (or other enrichments) are placed.
5. *Hidden food*. It is another enrichment that is suitable for all ages. Simply hide enough items that any animal that is slower has a chance to find them, and make sure that some of them are in easily accessible places. This enrichment is often combined with scent trails that the animal can follow to the hidden item. To mimic a natural situation, not every trail ends with finding a reward, like not every hunt or search in the wild will result in a meal.
6. *Puzzle feeders*. A good enrichment for slowing down the feeding rate of the animals and for keeping them occupied mentally as well are puzzle feeders of all kinds. The only adaptations that will need to be made are to ensure that, for example, older primates can get their fingers into the devices (see Fig. 2). If they are unable to manipulate the puzzle feeders correctly, it may well lead to frustration, which we are trying to avoid; challenges can be difficult, but over time solutions must be available to find.

Puzzle feeders are not just for primates but can also be adapted for use with many species including birds, reptiles, fish and other mammals. Observing or researching feeding techniques and thinking about how a puzzle feeder can be adapted allows for new devices to be developed for different species. Puzzle feeders are not just for animals with fingers, manipulating a piece of food along tubes until it falls through a hole for them to access. Mammals such as giraffes that have a prehensile tongue can also be fed hay in a puzzle feeder. To create a puzzle feeder for giraffes is simple: 1) Take a large, thick-walled plastic container, 2) drill several holes around the sides and base large enough for the animal to put the tongue through without any danger of it becoming stuck or damaged, 3) smooth the edges of the holes to prevent the tongue being scratched, and 4) cut several short lengths of planed wood to different lengths so that they easily fit inside the barrel. Once completed, you can place some of their daily ration of hay/lucerne/alfalfa into the barrel with the bits of wood mixed in. They will have to work to pull the forage out through the wood. The difficulty can be altered by, for example, altering the amount of wood in the barrel or by bolting metal rods securely through the wood. The idea is mimicking the animals trying to access the leaves of Acacia trees through the thorns that protect them.

7. *Papier-mâché*. This is a very versatile enrichment as the piñatas that are made from the papier-mâché can be constructed in various sizes and strengths depending on the animals they are intended for. While piñatas tend to be used predominately with primates, there is no reason that they cannot be used with other species such as elephants or mice. Piñatas are constructed from pieces of a clean paper (unprinted newspaper is ideal); these are dipped into a paste made of flour and water before being shaped around a mould such as a balloon—we recommend considering the environment when using materials. The mould is removed once the paper has dried and the piñata can be filled as required (see Fig. 3). Only make piñatas as thick as they need to be in case an animal starts to eat them in excessive amounts. To help prevent this, try to use only dry materials inside them, and never use blood as the animals will certainly eat the paper as well. The animals most prone to problems with papier-mâché are birds. The materials can sometimes collect in their crops and cause a blockage. If any animal is observed eating a lot of the piñatas, then stop giving them to them immediately.
8. *KONG*™. These are readily available commercial enrichment items that can be used for a variety of species in different ways (see example in Fig. 4). It is important that enough KONGs are provided, in a group setting, to ensure that each animal has access to prevent any potential aggression. Also, ensure that you get the right size KONG for your animal, and regular checks should be done to ensure that they are still in a good condition.

9. *Whole food*. This is an enrichment that may or may not work depending on the situation. With primates, if it appears that the younger dominant animals are managing to get the bulk of the food at feeding times, the general solution is to chop the food up in order to spread it out more, providing access for all animals. When whole food is provided, they can carry less items in their cheek pouches and hands before they have to stop and eat them. This allows the other animals a chance to access some of the items themselves. The main drawback of using wholefoods is determining whether elderly animals have a good dentition that can deal with a larger and harder piece of food.
10. *Mealworm dispensers*. These come in a variety of styles but mainly consist of either a) the ones that are placed inside the enclosure for the animals to have access too or b) the ones that are suspended above the enclosure to allow a slow but continual supply of mealworms to fall out. With dispensers that are placed inside the enclosure, they need to be small containers with a hole drilled in them big enough for the mealworms to crawl out. These are simply filled with the insects and placed in the enclosure. Ensure that there is at least one container per animal to prevent aggression. This means that even if the older animal is moved on there will still be a container for them, and it is impossible for any animal to guess where the next mealworm will appear. For animals such as meerkats, 35-mm film canisters (see Fig. 5) can be used; however, in this digital age, these might be harder to get such items; therefore, any plastic container of a similar size can be used instead. Ensure to increase the strength of the container as the size of the animal increases. It may be sensible to glue the lids onto the containers for safety, with a non-toxic glue brand, and typically for animals bigger than meerkats, these small canisters are discouraged to avoid ingestion. For a more naturalistic style, use coconuts or lengthy bamboos. With suspended feeders, you can use a piece of drainpipe with mesh over one end (this is the bottom of the feeder) and fill with straw or other substrates. Put the mealworms in the top end and place a cover to prevent wild species stealing the mealworms and suspend it over the enclosure. Share the mealworms between several of these to avoid aggression. The density of the substrate will vary the speed in which the mealworms emerge.
11. *Scatter feed*. As mentioned with whole feeds (9), scatter feeding can cause problems depending on species and individuals. For ungulates, spreading their pelleted food over a large area will certainly slow down feeding and allow all animals to eat. However, do ensure that it is over a large enough area to allow all animals to eat peacefully. There will always be some dominance shown from one animal over another at feeding time, but if the number of feeding areas is adequate, then this will be minimal. This is also a good method when feeding large flocks of birds.

12. *High feeders*. This is aimed at primates but will also work with other climbing species as well. If part of the animals' enclosure consists of a caged area with a meshed or other partly open material–based roof, this can then be used to feed animals with whole food items on top. The younger and fitter individuals will climb up to feed, working for their food as they would in the wild. Once they have managed to get a piece of the food, they will eat a small amount and then discard it in favour of the next piece. These discarded pieces can then be accessed by the less agile individuals at lower levels to all the way on the floor. It is also possible to supplement the food on the ground, while some of the animals are occupied above.
13. *Horizontal feeding poles*. These are a variation of the tiger feeding poles that were mentioned earlier. They consist of a horizontal tube or pipe placed either inside the enclosure or just outside with one end facing into the enclosure. They need to be robust enough to resist the interaction with the animals and firmly anchored to the ground. Inside the tube is a spring that is of a suitable size and strength for the animals in question. One end is fixed at one end of the pipe, and the other is attached to a chain, wire or rope that feeds out of a small hole in the other pipe end. The enrichment item is then fixed onto this. The idea is that the animal can still pull the item which will 'fight back' because of the spring. However, if the animal becomes tired or is unable to remove the food from the device, the animal is still able to eat the food on location. These can vary from vehicle suspension springs for large carnivores to elasticated rope for mongoose. They can also be camouflaged to resemble fallen logs if required. Ensure that all metalwork is protected by fire hose or similar, to prevent damage to the animals' teeth.
14. *Scratching posts*. These are much appreciated by animals of all ages and can be either natural, for example, logs, branches and rocks, or artificial such as broom heads, road sweeper or car wash brushes. Ensure that they are placed in such a way that an animal can access them with all parts of their body including the back and belly. With the road sweeper brushes, ensure that they do not contain any metal bristles and check regularly to remove loose bristles to avoid ingestion.
15. *Feeding trees*. These are branches that are placed either into sockets in the ground or into a more temporary socket such as a cable reel. These are suitable for most species of animal (see Fig. 6). Once they are secure, simply place pieces of the animal's daily ration onto the branches and give the animals access. As the animal's mobility decreases with age, simply make allowances by making the pieces more accessible. However, some exercise and stretching may be beneficial so it may be best to discuss this with your veterinarian and other behavioural staff which options are suited for the individual.

16. *Browse in enclosure*. While some zoo enclosures are of a suitable size to provide natural browse, this may not be the case for all animals, so we must provide browse for our animals if it is a normal part of their diet. Hanging browse from a pole is a common way of doing this, but be aware of the limits of your older animals. Importantly, ensure that the branches are tied in such a way that a noose is not left in the rope if the browse falls out. For an older animal, it might be best to attach metal piping to walls or just outside the enclosure (depending on the fencing) at an angle low to the ground. Branches can then be placed in these, and if the angle is correct, they will not have to stretch too much to access the leaves.
17. *Various size perching*. Using the metal tubes mentioned above in the floor of an aviary allows changing some of the perching on a regular basis. Put the tubes in the ground vertically, and place a suitable branch into it. Make sure that the branch is edible as many birds will chew the wood or eat the leaves and change it regularly. In addition, providing the stimulus of the leaves, it also allows them to use perches of different sizes, allowing them to exercise and move their feet in different positions. This helps to prevent problems caused due to placing weight and pressure on one part of the animal's foot all the time. The size of the bird and the aviary will determine the size of the sockets. The same principle can be used for smaller primates and other animals. When the sockets are not in use, they should be covered, or have a short branch in them to prevent any animals becoming stuck.
18. *Seasonal*. These are the enrichments that are only available at certain times of the year and can often be provided in large amounts. Because of this, there is usually enough for all the animals to gain access to them. The main potential problem with the older animals could be gastrointestinal upsets. With enrichments such as pumpkins or Christmas trees, try to limit the amount, introducing them gradually and removing some of the normal daily ration to compensate for what they eat (see Fig. 7).
19. *Substrate*. Try to provide older animals with a variety of substrates as this can help with any sore or painful joints that they may have, in conjunction with medication for these problems if necessary and in collaboration with the veterinarian. When working with a species that enjoys digging, provide them with a substrate that is easier for them to carry out this behaviour.
20. *Bubbles and Compact Discs (CDs)*. These are enrichments for any age of animal and are usually used as a distraction for them. A bubble machine will blow bubbles across an enclosure that many animals are curious about and will follow. In addition, CDs threaded onto a rope and suspended outside an enclosure will reflect sunlight and shine light across a darker corner if positioned carefully.

Fig. 2 Puzzle feeder being used for Sulawesi crested macaques (*Macaca nigra*). Photo J. Chapman

Fig. 3 Piñata being investigated by a variegated spider monkey (*Ateles hybridus*). Photo J. Chapman

Fig. 4 A KONG™ being manipulated by a Sulawesi crested macaques (*Macaca nigra*). Photo J. Chapman

Fig. 5 A mealworm dispenser made from an old film canister and suspended on a thin rope. Photo J. Chapman

7 Conclusions

With any enrichment, it is important to research, plan, observe, record and evaluate how the animals interact with the items. Items must be safe and not cause any harm to the animals. As animals age, enrichment should be adapted, and changes implemented should enable older animals to continue to be engaged and be enriched in different life stages. The use of enrichment timetables is a time-saving method of providing continued, varied enrichment to all species, especially the elderly animals.

Fig. 6 A feeding tree being enjoyed by an anoa (*Bubalus depressicornis*). Photo J. Chapman

Fig. 7 Christmas trees being provided for meerkats (*Suricata suricatta*). Photo J. Chapman

References

Alligood C, Leighty K (2015) Putting the "E" in SPIDER: evolving trends in the evaluation of environmental enrichment efficacy in zoological settings. Anim Behav Cogn 2(3):200–217

Huber HF (2009) Environmental enrichment for gummivorous primates. Unpublished thesis. Texas State University-San Marcos, San Marcos, Texas

Inglis IR, Forkman B, Lazarus J (1997) Free food or earned food? A review and fuzzy model of contrafreeloading. Anim Behav 53(6):1171–1191

Law G, Kitchener AC (2020) Twenty years of the tiger feeding pole: review and recommendations. Int Zoo Yearb 54(1):174–190

Martínez-Macipe M, Lafont-Lecuelle C, Manteca X, Pageat P, Cozzi A (2015) Evaluation of an innovative approach for sensory enrichment in zoos: semiochemical stimulation for captive lions (Panthera leo). Anim Welf 24(4):455–461

Plowman A, Green K, Taylor L (2006) Should zoo food be chopped? Animal nutrition III. Filander Verlag, Germany

Roe S, Cleave R (2005) Are we just feeding carnivores or are we providing enrichment as well?. In Seventh international conference on environmental enrichment, p 248

Rozek JC, Danner LM, Stucky PA, Millam JR (2010) Over-sized pellets naturalize foraging time of captive Orange-winged Amazon parrots (Amazona amazonica). Appl Anim Behav Sci 125(1–2):80–87

Troxell-Smith SM, Whelan CJ, Magle SB, Brown JS (2017) Zoo foraging ecology: development and assessment of a welfare tool for captive animals. Anim Welf 26(3):265–275. UFAW

Wells DL, Irwin RM (2008) Auditory stimulation as enrichment for zoo-housed Asian elephants (Elephas maximus). Anim Welf 17(4):335–340

Würbel H, Chapman R, Rutland C (1998) Effect of feed and environmental enrichment on development of stereotypic wire-gnawing in laboratory mice. Appl Anim Behav Sci 60(1):69–81

Young RJ (2006) Environmental enrichment for captive animals. UFAW animal welfare series. Blackwell Publishing, p 240

The Role of Learning and Training for Ageing Animal Care and Wellbeing

Sabrina Brando and Debra Marrin

S. Brando (✉)
AnimalConcepts, Teulada, Spain
e-mail: sbrando@animalconcepts.eu

D. Marrin
San Francisco Zoo, San Francisco, CA, USA
e-mail: DebbieM@sfzoo.org

S. Brando, S. Chapman (eds.), *Optimal Wellbeing of Ageing Wild Animals in Human Care*, https://doi.org/10.1007/978-3-031-30659-4_6

Stella. Stella was a variegated spider monkey (*Ateles hybridus*) who lived at Twycross Zoo in the United Kingdom. She lived at the zoo for 15 years until she passed away at the age of 26 years 8 months in September 2019. She had regular elderly animal health check-ups during her later years as part of the zoo's proactive ageing animal care programme. During her assessments, she was diagnosed with diabetes and was given oral treatment for this condition along with on-going close monitoring. She was also found to be suffering from ovarian cysts, and these were removed by laparoscopic surgery. Stella was beautiful, and had a mind of her own. Training her was a very slow process, and her personality was great. She was always there to give a greeting to her caregivers when they walked into their back-of-house areas. If the caregivers were doing something she did not want them to do, she would definitely let them know.

Abstract

Understanding animal learning and training can greatly support animal care and good animal wellbeing through all stages of life, including when animals are ageing and often require special care. Thinking and planning for the needs and integrating the preferences of an individual animal are fundamental in short- and long-term wellbeing programmes, including animal training programmes. Good animal care and wellbeing programmes pay attention to the types of training methods specifically and learning in general, with a focus on positive reinforcement and good human–animal interactions. Through proactive planning and monitoring, age-related health concerns can potentially be detected early, and species-specific concerns can be anticipated. Animals will experience better wellbeing and potentially an extended longevity due in part to a lifelong preventive care and wellbeing programme when this includes voluntary medical behaviour training. In addition, (training staff for end-of-life decisions and training animals for euthanasia requires organization, leadership and team support. Designing for age-specific and adapting environments will facilitate care and treatments, making it easier and more comfortable for animals to participate in training sessions. Commitment and dedication of the animal care teams help provide animals with the best care possible at all stages of life.

Keywords

Training · Learning · Preventive care · Proactive care · Habitat adaptations · Human–animal interaction · Positive reinforcement

1 Introduction

Understanding animal learning and training can greatly support animal care and good animal wellbeing through all stages of life, including when they are ageing when they often require special care. Animal welfare considers an animal's life 24/7 across lifespan (Brando and Buchanan-Smith 2018), starting from before an animal is born, or transferred from another facility, to the day they die. Animals are considered elderly when they have reached or passed 75% of their lifespan. Planning, record-keeping and monitoring are all part of a progressive and proactive programme (see the chapter 'Holistic Approaches for Promoting Good Wellbeing for Ageing Wild Animals' for more information on a holistic animal wellbeing approach). Thinking and planning for the needs and integrating the preferences of an individual animal are fundamental in short- and long-term wellbeing programmes, including animal training programmes. Animal training programmes are nestled and highly connected to overarching animal care and wellbeing programmes and should be considered in their own right. While learning and training are often used interchangeably, in this chapter, we will distinguish them by using learning from a more general perspective and a bird's eye view. What are the necessities of an animal during his or her life, for example, what is there to learn while living in human care, effects of routines, choice and control in general or specific needs at different ages? Training—sometimes also referred to as shaping—is used when discussing individual and group training sessions and activities that have specific goals, for example, taking medication, mouth inspection, weighing, shifting (moving an animal from one part of the enclosure to another) or injection training. Understanding the effects of animal learning and training can be challenging due to the many factors that influence behaviour, as well as intricacies and debates about the learning theory. However, some rules and principles such as learning how to use positive reinforcement and how to write a training plan can be fairly easily understood and implemented. In this chapter, we will use jargon-free descriptions and focus on the use of positive reinforcement and available best practices and add publications to develop and maintain animal training programmes for elderly animals.

Animal wellbeing comprises the psychological and physical experiences of an individual animal, as perceived by them. Such experiences may vary widely in their valence, from very positive to very negative. A variety of inputs may affect animal wellbeing. These may act together and may be synergistic rather than simply additive in their consequences. Therefore, the evaluation of wellbeing should consider all such measurable factors and their interactions. An animal with good welfare has the ability to exert agency over their life to a meaningful extent and ideally to fulfil all their needs and to satisfy most of their preferences. To promote good welfare, the focus should be on promoting agency, choice and control and predominantly positive wellbeing, and the monitoring and assessment of animal welfare should include a focus on likely experiences (Brando 2022). While many other definitions can be found, contemporary animal welfare encompasses environmental, behavioural and psychological aspects, promoting positive welfare states (Mellor et al. 2020) and necessarily subjective consideration of the animals' feelings (Veasey 2017). Professional zoos and sanctuaries globally are evolving welfare standards from surviving (early modern zoos) to coping (better present zoos) to flourishing, promoting optimal animal wellbeing (Brando and Herrelko 2021). Good animal care and wellbeing programmes pay attention to the types of training methods specifically and learning in general. Learning is defined as a change in an organism's behaviour or thought resulting from experience (Chance 1988). In animal training sessions, we look for changes in behaviour towards the desired outcome. For example, in training, we try to set the animal up for success by breaking down our goals into smaller steps and we look for changes to reward these small steps towards the desired goal. However, we also aim to be sensitive to changes in behaviour due to illness, social context or other influences that may support, halt or make training activities successful (as in getting to the desired goal or step). Learning through experience can include coming towards the caregiver when keys are jingled as this is associated with feeding or moving into a shelter when it rains or strong winds; these are all examples of changes in the environment, which caused animals to learn something. Teaching animals to collaborate in their care with positive reinforcement as one of the basic procedures described by Skinner (1953), meaning adding something to the animal such as a piece of food and a beloved toy or scratching after successful completion of the desired behaviour. For an in-depth review of learning theory, see Dorey (2020).

1.1 Skills and Resources When Caring for Ageing Animals

Caring for ageing animals can require a considerable amount of time as they benefit from adapted environments, special diets, appropriate enrichment and specific behaviours that facilitate care. Organizations, teams and individuals should identify what time and resources are necessary to promote excellence in animal care, provide support and reduce the strain on staff and the care of other animals and programmes. Highlighting the opportunities such as sharing stories of how ageing animals are cared for, including through animal training, can be rewarding and educational activities. Providing staff with continuous personal development on a wide variety of topics, including how to train animals, supports job satisfaction (Brando et al. 2023). In this chapter, animal care staff and trainers are used interchangeably to highlight the importance of this skill for all staff directly working with animals. Having a good understanding of the theoretical knowledge and being skilled in the theory, practical applications and methods of learning and training—with a focus on positive reinforcement, is important for animal caregivers in general. These are especially important when faced with caring for an animal who is or might be experiencing age-related challenges overtime, such as high blood pressure or painful joints, in need of daily/regular care and medications. Staff need to be prepared to assist ageing animals, including through formal training programmes.

2 Training Ageing Animals

Positive reinforcement training is an essential component in a holistic approach to animal management. It is a powerful tool to help promote positive welfare for animals at every stage of life. Formally training animals should start as soon as possible, after arrival in their new home, or as soon after the birth as safe for mother and baby. What animals are learning, including while still in the womb, should be considered to set the stage for a lifelong and positive animal care and training programme. When animals are young and agile, able to use all their senses and body as expected, they will require less effort and time compared to when they are older; when they may have challenges to move around and use limbs; and potentially experience and/or suffer from reduced eyesight, hearing loss or cognitive decline. A solid training foundation will assist in the transition of, for example, visual stimuli to touch or hearing, or vice versa for auditory stimuli. A cue that was the first visual is transferred to an auditory tone or tactile cue, preferably before impairment is present, and animals can be trained for behaviours that will be useful and important in the last phases of their life. For ageing animals, the benefits of positive reinforcement training increase when the animals experience age-related health issues.

Caregivers and veterinarians sometimes do preventive health check-ups on a yearly or regular basis, and this is of particularly interest when an animal passed the 75% of longevity mark. Some parts of preventive health check-ups can be conducted while the animal is awake such as training for an open-mouth behaviour to check teeth and gums, presenting for palpitations, taking radiographs, blood draws and urine or faeces collections. More thorough examinations may have to be conducted under general anaesthesia, such as dental operations, or a full physical examination. This can still be possible to do with the collaboration of the animal for the anaesthesia if this can be done safely for animal and staff. An ageing animal trained to participate in blood draws or vaccines or treatment of asthma or other respiratory afflictions may also learn to collaborate safely with the administration of oral or injectable anaesthetics. Once the animal is sedated, the care staff can go inside to safely move the animal to a stretcher for transfer to the operation or examination room, or the procedure may be conducted within the animal's facility. Full and thorough physical examinations are more likely to yield a wide variety of information on the health status of the animals (see the chapter 'Veterinary Care of Ageing Zoo Animals: Description of a Proactive Approach'), which turn informs the training programme and priorities. After a diagnosis, the animal care staff may want to find ways to support the animal in his or her treatment to, for example, ease pain of arthritic joints and spondylosis; treat kidney disease; and monitor and treat diabetes, vision loss and other ailments associated with ageing (Dennis 2013). Caregivers may also need to address dental disease or overgrown claws and hooves, urine soiled fur and other results from an animal's inability to groom properly. In order to diagnose and treat age-related issues in the safest and least stressful manner, a variety of voluntary medical and husbandry behaviours are helpful and provide the animals with the highest level of choice and control. Many voluntary behaviours that are useful throughout all stages of life, such as voluntary hand injection, weighing (see Fig. 1), hoof care, giving treatment (see Fig. 2), ultrasound (see Fig. 3) and blood draws, are also very valuable in the old age for close monitoring and proactive care.

2.1 Age-Related Behaviour Changes

When training ageing animals, the scientific principles and techniques we use for training do not change. How we assess behaviour and develop behaviour plans remains largely the same while adapting it to the needs and preferences of the

Fig. 1 Larry, the Sumatran tiger (*Panthera tigris sumatrae*), being rewarded for standing on the scales

Fig. 2 Snow leopard (*Panthera uncia*) receiving subcutaneous antibiotics

Fig. 3 Leanne, the Sumatran tiger (*Panthera tigris sumatrae*), participating in ultrasonography

individual's life stage as the animal's physical and mental abilities change. Many issues may prevent the animal from participating in training; for example, some behaviours may cause so much pain so they refuse them because the avoidance of pain is a far greater motivator than the reinforcers we may offer. The behaviours the animals could easily perform in the past may become behaviours they are physically incapable of performing, for example, climbing to a sleeping hammock or holding on to a target above their head or to the sides. This in turn can affect the types of behaviours that can be trained. It can also inform the type of progression of behaviours that can be performed during, for example, physiotherapy (chapter "Pain: Physiology, Recognition, and Management in Zoo Animals"). Mental faculties may also diminish with age, making it more difficult for the animals to learn new behaviours and to concentrate, or they may have increased feelings of insecurity or inactivity. Participation in established or new behaviours may not occur during periods of unresponsiveness to their environment caused by age-related cognitive decline. Medications prescribed for pain management may also affect mental capacities and physical abilities, but these would not be used usually only towards the end of life when no new behaviours will likely be trained. A lack of mental alertness may prevent training from occurring at certain times of a day or when the animal is doing poorly that day. The responsibility of trainers is to make interacting easy for the ageing animal whenever possible and to plan sessions according to behavioural observation to increase ease and success for the animal. Trainers can also observe how medications affect the animal's behaviour at different times of a day and plan sessions according to those observations too. Asking the animal to participate when he or she is physically comfortable, capable and mentally alert will create the most optimal conditions to make the interaction successful and collaborating positively easy. For example, an animal receiving medication to treat arthritis should be given the medication, and time for the medication to take effect should then be considered before the caregiver asks the animal to shift or participate in other behaviours. Some medications may take a longer time to circulate and be effective. Understanding the effects of medication on behaviours, and also the effect of, for example, anaesthesia potentially causing nausea, can all affect the animal's experiences and subsequent desire to collaborate again or not. A close collaboration of the animal caregivers with the veterinary staff is key to support a long-term and effective ageing animal care and training programme tailored for each animal. See also the chapter 'Pain Physiology, Recognition, and Management in Zoo Animals' regarding the physiology, recognition and management of pain in zoo animals.

2.2 Each Animal is an Individual

All animals age in different ways. As each animal grows older, he or she will experience a unique combination of (progressive) age-related changes. Changes in physical and cognitive abilities often lead to needed changes in behaviours, which in turn change how we should manage each individual. Each animal should be assessed and managed with a customized plan based on his or her physical and psychological needs, including health challenges. Behaviour goals should be set with input from multiple team members including caregivers, curators, veterinary staff and staff specialized in behaviour (Watters et al. 2015). On-going welfare assessments, including health, will pinpoint diagnostic and treatment needs and guide the training goals for that individual. As the animal's needs change, caregivers and/or veterinary staff may train new behaviours to optimize the care and monitoring routines. Staff should be encouraged to think widely and train novel behaviours in addition to the traditional repertoire, so animals are more likely to be flexible and open to new and different activities and behaviours when needed. Using research and predicting the types of concerns could be expected for each species and individual assists in being prepared ahead of time. Being open-minded and trying new ideas are how we progress the quality of care. Many treatments, diagnostic procedures or therapies can be candidates for training, such as physical therapy, blood pressure readings, ultrasonography, acceptance of subcutaneous fluids, acupuncture, laser therapy and other non-traditional treatments. Appropriate diagnosis, treatment and management of age-related issues are best accomplished through teaching animals to be active participants in their care. Every animal's training and care needs should be considered on a case-by-case basis and include aspects such as personality, friends and family, preference in caregiver and more. Nonetheless, there are general factors to consider when deciding when, how and what or which behaviour to train to help care for the specific needs of ageing animals. Safety concerns, characteristics of the animal, the diagnostics or treatments needed, the behaviour required to succeed at the diagnostics or treatment and the human–animal relationship will all help determine the best approach for each individual.

2.3 Animal Safety

Voluntary participation on the animal's part improves safety for both the animal and the staff. This is true at all stages of life, but as animals age, risks may increase for the animal in terms of both physical and psychological aspects. Older animals may not handle anaesthesia as well (Carpenter et al. 2005; Hughes 2008) or may be more easily injured during restraint due to decreased bone densities or other age-related changes (Navaratnarajah and Jackson 2013). Therefore, the voluntary behaviours that increase diagnosis and treatment options also increase safety. Voluntary participation in health care also reduces or eliminates the negative stress an animal may experience when forced treatments take away all choice and control (Herron and Shreyer 2014; Brando 2012; Yin 2009). Negative stress may have a significant negative impact on an aged animal's overall wellbeing, as they may not bounce back as quickly from these experiences (Kirkland et al. 2016) as a younger and more resilient animal might. With a probable increase in medical issues due to ageing, staff can prevent animals from experiencing negative experiences through the training of behaviours through positive reinforcement. Keeping interactions positive between animals and humans will also reduce fight, freeze and/or flight behaviours that can cause unsafe situations for all. Fine-tuning medication dosages to reach the optimal amount are sometimes only possible through the use of voluntary medical behaviours. Behaviours such as blood pressure monitoring and blood draw are important to evaluate the effectiveness of medications. For instance, anaesthesia can significantly alter animals' blood pressure or haematological parameters (Boström et al. 2002), making it difficult to determine whether a medication is having the desired effect without voluntary participation of the animal under treatment. Negative stress can also increase the necessary dose to anaesthetize an animal, which in turn can increase the risk for the animal. Veterinarians can better determine whether the medication dose is having the desired effect when the animal is not anesthetized during testing. Anaesthesia comes with risks and should be used only when necessary, making it not an optimal choice for regular repeated diagnostics.

2.4 Whole Life Training

Early training in an animal's life establishes a foundation of behaviours that help them to participate voluntarily in their own health care. It provides veterinary and animal care staff options for diagnosis and treatments that would not be possible otherwise. Already-established behaviours provide instant options for veterinarians and animal staff caring for ageing animals as new challenges arise. When new behaviours are needed, animals with an established learning history will learn new behaviours more quickly, making novel diagnostics or behaviours achievable sooner. New behaviours may be mastered in a much shorter time than if the animal has no training history, and foundational behaviours already established provide a shorter path to a new behaviour goal. For example, a big cat or a great ape previously trained for voluntary blood draws may learn a new behaviour to allow blood pressure readings within days or weeks instead of months. By prioritizing training throughout animals' lives, we can set the stage for optimal care and wellbeing as they age and successfully monitor animals closely for more effective preventive and proactive care.

2.5 Planning Considerations

Veterinary staff should also play an important role in setting behaviour goals. They can advise the animal care and training team on which behaviours would be the most beneficial for the individual animal to learn, across time. Once the goals are set, the veterinary staff should also inform the trainers what the animal will be required to do, for how long and under what conditions. In order for a procedure to be successful, the trainers and animals need to practice as closely as possible what they will need to do when the behaviour is utilized. Trainers need to know what time of the day it will occur, who will be doing the procedure, how many people will be involved, what equipment will be used or if the equipment makes noise, vibrates or has flashing lights. Care staff will also need to understand how long the veterinarian needs the animal to remain in position, what type of preparation may be needed like shaving hair or fur, applying alcohol or gel or whether the animal can eat before and during the procedure. Frequency of training sessions will likely be informed by knowing how often the behaviour will be needed (e.g. daily or monthly). In a best-case scenario, the people who will be involved on the day of the procedure will be part of the training process as early and often as possible to reduce fear for unknown staff or participants. This may necessitate including veterinary staff into training sessions regularly. Introducing animals to new people and teaching them to be flexible throughout their life will benefit them when teaching and utilizing medical and other behaviours. The time and effort invested in positive relationships and voluntary behaviours of all kinds will make it easier for staff to support positive wellbeing for ageing animals.

2.5.1 Long-Term Planning

Daily behavioural observations, or any other meaningful timeframe, through the use of ethograms as well as monitoring and assessing processes using animal-based indicators are at the core of long-term planning and inform an animal training program overtime. Training a novel behaviour is only part of what the staff need to consider. Behaviours need to be maintained to remain reliable for future use, and behavioural observations in and outside the training sessions are also an important part of the programme. Part of the discussion among team members should include how frequently the behaviour will need to be utilized. This may be informed by the individual's need for veterinary care, pain management or other care considerations. Once the care staff know how frequently a behaviour will be needed, they can set a training schedule that will maintain it. The training and maintenance of every behaviour is a 'study of one' as every animal is unique and will behave slightly differently during training. The training and maintenance plans should therefore be customized for the individual. A behaviour that is needed frequently, for example, every 3 days, may dictate a need to increase the time dedicated to training and maintenance in order to keep behaviour reliable, especially for behaviours that may cause pain or discomfort to the animal. For example, a Sumatran tiger (*Panthera tigris sumatrae*) diagnosed with cancer needed weekly injections and monthly blood draws. In order to keep the behaviours positive, the animal was trained daily so the number of sessions without a needle insertion remained higher than the sessions with a needle insertion. This strategy kept participation in the behaviour reliable and positive for the tiger throughout the treatment.

Diets may also need to be adjusted so that high-value reinforcers are used strategically for procedure days. A nutritionist can be helpful in setting healthy diets that support training goals. In some cases, the animal may receive more food reinforcers than is optimal within their diet, but the training benefits for a determined period of time, in case of a treatment, may outweigh the diet concerns. In other cases, it may just be a matter of rearranging when the preferred diet items are used. All these need to be monitored to make an informed decision.

2.6 Human–Animal Relationships

An important component of training at every age is developing a positive trusting relationship between caregivers and the animals (Pryor 2002; Zeligs 2014; Marrin et al. 2019). The stronger the relationship between human and animal, the more likely the animal will participate voluntarily, particularly when food is not motivating or if an animal feels ill (Ashley and Wingfield 2012). Relationship strength may also be the difference between success and failure when new stimuli are added to the environment. A good trainer will have developed a positive relationship with the animals and will ensure that new stimuli and people are associated with positive outcomes. A good human–animal relationship is based on a two-way communication principle where animals and people are attuned to each other and where animal care staff provides options for choice and control and supporting animals in being agents of their own life. The trainer's ability to imagine and see the environment from the animal's perspective coupled with empathy and compassion will help them to be a flexible and open-minded teacher and a student, where both the human and non-human animal are teacher and student. It is the caregiver's responsibility to observe their animal's behaviour and adjust antecedents, that is, the events, actions and circumstances that occur immediately before a behaviour, and adjust behavioural criteria accordingly so it is easy for the animal to engage in the desired behaviours. Behaviour is not stagnant, and as the animals age, behaviour changes may be due to changes in physical and mental health issues that make previous reliable behaviours more difficult to perform (Chiu and Bodley 2010; Brando and Watters 2016; Krebs et al. 2018). Competing motivators may be the cause when there is a change in compliance with once reliable cues. The animal may refuse a specific behaviour because the motivation to avoid pain associated with doing the behaviour may be stronger than the motivation to acquire food or other reinforcers the trainer may offer (LaGraize et al. 2004). Another possibility is a change in cognitive function that affects the animal's behaviour (Azkona et al. 2009; Bellows et al. 2016; Peters et al. 1996).

Animals living in human care, in an environment with their physical, emotional and medical needs are taken care of, may survive far longer than their wild counterparts. These extended lifespans may lead to negative health issues unique to older animals, including issues that would likely lead to death in animals not in human care. Zoos with a cradle-to-grave approach therefore find the need for training and maintaining voluntary medical and husbandry behaviours, an essential component to maintaining positive welfare. Adding new behaviours to training programmes that eventually become standard practice helps to raise the bar for best practices globally. Managers need to set the staff up to succeed to provide the animals the best care and quality of life possible. This can be accomplished by considering what possible age-related factors may affect the animal's life including behaviour. A physical or mental change in the animal may lead to behaviour changes during training. A refusal to perform a behaviour is information to the trainer, veterinarian or care staff. Animals who have participated reliably for years in voluntary care and medical behaviours may suddenly refuse to respond to cues that previously elicited their participation. If we can determine the cause of the change, we can adjust the environment and our expectations to achieve good wellbeing for the animals and training goals. Sometimes care staff blame the animal or use descriptors such as lazy, grumpy, vicious or stubborn, labelling the animal and not the animal's behaviour. Labelling the animal is not beneficial to the evaluation and planning process. Instead, we need to do a functional assessment that considers the antecedents, observable behaviour and the consequences that follow (Friedman 2007) and use observation with animal-based indicators to gain more information into the experiences and states of wellbeing of the animal. Different software and/or technologies can be used for behavioural observations and recording over days and time such as the Care and Welfare Module in ZIMS by Species360 or ZOOMONITOR by Lincoln Park Zoo. Active and proactive monitoring of ageing animals is part of an animal training programme as well as overall animal wellbeing assessments. Regular behaviour observations as part of a monitoring programme are a key in spotting subtle differences in behaviour which can give insights into the animal's experience, as well as how to progress in training programmes.

2.7 Record-Keeping and Environmental Changes

Meaningful training and record-keeping is a key to successfully train ageing animals. Sometimes to change the behaviour, change the environment; for example, a softer area to sit can increase the likelihood the animal wants to sit for a longer duration for a particular behaviour. Declining strength or mobility is common in older animals, and simple modifications to their training space can support positive training outcomes such as providing cushioning mats and a better overview of the environment by additional lighting. Many common age-related changes in animal's bodies can result in painful or limited ability to navigate a once-accessible enclosure. Simple modifications such as additional handholds for primates can help them navigate their environment without or with less pain. Tiered platforms, steps and ramps may help animals to be reliable with behaviours such as recall and securing, just like the modification made in other indoor and outdoor habitats. Voluntary medical behaviours rely on specific environmental conditions to make the goals possible. The diagnosis or treatment goals often rely on the ability to access specific body parts. Some animals may be trained in a free contact situation making it easy to access any body part for tests, treatments or physical therapies. Animals that are trained in a (semi)-protected contact scenario pose a greater challenge as we may not be able to access part or all of the animal's body or body parts. Facility modifications may be needed such as blood sleeves, tail ports (Fig. 4), ultrasound windows (Fig. 5), slots for blood draws (Fig. 6) or training benches. Facility modifications that are made for training should focus on animal comfort and human safety (Marrin et al. 2018). When new facilities are designed, providing appropriate training spaces should be a priority, including the opportunity to modify it fairly and easily according to the lifespan needs. Compiling a list of all of the behaviour goals for the animals that will live in the environment assists in the design process. Older facilities not designed or built with training in mind can be successfully modified to allow for training; however, modifications after the build may be more costly or worse, never added. Facility design or modifications will vary greatly depending on the species. A positive learning environment will set up the animal and trainer to

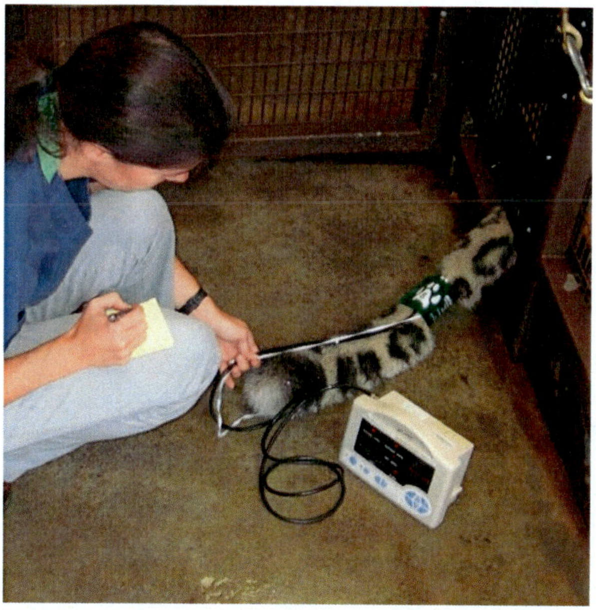

Fig. 4 Enclosure modification with a space for passing the tail of a snow leopard through for measuring blood pressure

Fig. 5 Using an auditory stimuli for shifting with elderly sheep, Hazel

Fig. 6 Opening in the fence to take a blood sample from the tail

succeed, and in general, across all species, the environment needs to be positive for the animal (see the chapter 'Environmental Enrichment for Ageing Zoo Animals' for more details on environmental design).

Wild animals in human care can be challenging from a clinical perspective as they often hide symptoms of disease until the disease has progressed significantly. Therefore, behaviour can be a helpful indicator of different aspects of wellbeing. Animal staff may become aware of age-related health issues during regular training sessions or when observing behaviour between sessions. Both scenarios provide information that may lead to changes in trained behaviour expectations and needs. Observable changes in animal behaviour in and out of training sessions are information that aids in the evaluation of an animal's state of wellbeing. When evaluating conditioned behaviours, we look at the animal's response to cues. Are their responses timely or latent? Do they completely refuse some behaviours? When an animal engages a requested behaviour, are their movements stiff and slow? Does the animal exhibit aggression or pain indicators in response to normally reliable behaviour cues? Is there an appearance of confusion or lack of understanding? These observable behaviours tell us about not only the animal's welfare state, but also where there may be a need to modify the environment, behaviour criteria or medications. Behaviours outside of training sessions that indicate poor health may include changes in sleep patterns, weight shifting, resting in different positions, changing positions often, heavy open-mouth breathing, over-grooming specific body parts and a loss of interest in conspecific relationships, discussed in the first seminar on caring for elderly animals organized by AnimalConcepts with the support of the San Francisco Zoo in 2016 and reviewed in Krebs et al. (2018). Cognitive dysfunction may also occur and manifest itself in observable behaviour changes, such as acting disoriented, experiencing a lack of control of urination or defecation, new stereotypies or unresponsiveness to environmental stimuli. As wild animals normally mask signs of pain and discomfort, familiarity with possible age-related changes is important to evaluate accurately the cause of a lack of engagement for established behaviours or to learn new ones. Is the function of the behaviour to avoid something negative or to gain something positive? The value of a reinforcer may change over time and is dependent on multiple variables. Latency in performance or refusal to engage in a behaviour may be due to, for example, pain avoidance, inability or mental confusion. Lack of motivation may also be the result of satiation, pairing the reinforcer with a negative event, a reduction in the value or rate of reinforcement and a lack of social support from within the group. Animals are unable to communicate verbally or communicate clearly with their behaviour, body language, sounds and facial expressions so if we are paying attention to details, the messages can become clear. Care staff should modify their expectations and the environment when needed based on the animal's behaviour, social needs or others. If caregivers can recognize changes in responses to cues early on, they may be able to address the health-related causes before a total breakdown of the behaviour occurs. Diagnosing and treating medical issues early on can also provide better health outcomes promoting good wellbeing. We cannot simply ask our animals 'How do you feel?' or 'Where does it hurt?'; thus, their quantitative and qualitative behavioural expressions, including sounds and smells, are the best ways to assess their overall wellbeing. All conditioned and other behaviours can help create a non-verbal dialogue between us and the animals in our care. Behaviour speaks at every age, and behavioural changes in ageing animals are important to recognize. Knowing what normal and historic behavioural responses by each individual gives us important insights that help us solve wellbeing concerns and behavioural issues and evaluate wellbeing (Watters et al. 2015; Marrin et al. 2018). An animal's behaviour also helps us know when our changes to medication doses and environmental conditions are effective. An observable behaviour combined with diagnostic tests and samples obtained through positive reinforcement training allow us to utilize the most positive and least intrusive methods to care for ageing animals.

3 Practical Examples and Case Studies

Voluntary behaviours including health-related behaviours such as weighing that were first developed in the film industry and marine mammal community in the 1970s and 1980s have influenced the zoo community (Brando 2010). Animal training has become common and part of the best practices in zoos and beyond. As animal caregivers, veterinarians and managers see the practical benefits of training, the resources provided often increase, including habitat and back-of-house modification, as well as special equipment and more time to observe and train. Staff continually develop new training goals that directly benefit animal wellbeing throughout the animal's lives, including behavioural assessments and monitoring. Below are some examples and case studies of individual animals at the San Francisco Zoo, which complement some of the other individual stories living in other organizations mentioned across this book.

4 Polar Bear Mobility Issues and Treatment

Pike, a polar bear that lived her whole life at the same facility, suddenly became unreliable for shifting at the age of 32 years. Her normal routine was to 'shift and secure' to receive her first feed of the day. Keepers originally thought that the issue was related to food motivators. The reinforcement magnitude and value were increased for shifting. When her behaviour to shift did not improve, a behavioural assessment was conducted, which led to a different conclusion. During the observations, the bear sometimes did not respond at all when called. Other times she would start to move towards her den and then sit back down for long periods of time. Her movements were slow and stiff, and she was often doing open-mouth breathing. At that time, it was theorized that the behaviour issue was caused by pain and that the bear was motivated to avoid pain than to acquire the reinforcers the care staff were offering. At that time, the bear was on medication to treat arthritis that she received once a day in the morning. It was thought that perhaps her pain levels peaked just prior to receiving her medication in the morning and that the avoidance of pain and the desire to acquire food were likely competing motivators. In order to test the theory and find a solution, when and how the bear's medication was given was changed, and when she was asked to shift was also changed. Care staff would go up on the roof of her habitat first thing every morning to toss her the fish containing the medication. Like this she was receiving the fishes wherever she was lying down at the time. After the medication had been consumed, the care staff would wait a minimum of 1.5 hours before returning and asking her to shift. This small adjustment solved the behaviour problem and led to reliable shifting on a daily basis.

5 Medical Monitoring of a Snow Leopard

Rigel, a 17-year-old snow leopard (*Panthera uncia*), was diagnosed with high blood pressure during a routine life stage and his medication was started. The veterinary staff explained that the only way to know if he was on the correct dose was to monitor his awake blood pressure on a regular basis. The animal already had a training history of voluntary medical behaviours such as weights, injections and blood draw from his tail. With this solid foundation, the staff were able to train regular blood pressure readings using a cuff on his tail. The new behaviour was conditioned in approximately 1 month as his previous experience with training helped train the new behaviour in the short time frame. All of his husbandry behaviours combined contributed to his increased longevity and good quality of life. Without training, most of the options utilized for diagnosis and treatment would not have been available. When he passed at his 20 years, he was the oldest living snow leopard in an Association of Zoos and Aquarium (AZA)-accredited institution in the United States.

6 Kidney Disease and Treatment in a Dolphin

Stormy, an ageing female Bottlenose dolphin (*Tursiops truncatus*) suffering from kidney disease, needed freshwater multiple times a day. Dolphins do not drink water naturally as they acquire fluids from the fishes they consume, so there was a need to train her to accept fluids from her care team voluntarily. Already-established behaviours such as gastric (stomach) sample collection provided the foundation to teach her to receive freshwater multiple times a day. Large amounts of water were given to her by attaching a funnel to the gastric tube once it was passed to her stomach. Smaller amounts of water using thin aquarium tubing attached to a 100-ml syringe were also given to her. The tube was passed a short distance past the back of her tongue, and water was given multiple times. In addition to the water, she learned to eat unflavoured Jell-O that was added to her hydration and over time became an additional reinforcer. Because dolphins swallow their food whole, the behaviours were not irritating and were easy to use on a daily basis, and she was more reliable in medication compliance and diet goals.

7 Loss of Eyesight and Use of Auditory Target Training in Domesticated Sheep

Hazel, an elderly female sheep (*Ovis aries*), lost her eyesight and was unable to navigate her environment without difficulty. When shifting her to different locations, she would regularly run into fencing. Shifting was necessary for daily husbandry and for keeping her with her social group. To help her navigate, she was taught to follow jingle bells sounds using classical conditioning. Once this behaviour was acquired, the bells were attached to a food bowl, and she was taught to follow the moving sound for short distances. Through successive approximations, the distance was slowly increased before receiving reinforcement. This very simple training allowed her to safely navigate and shift without negative incidents. Her confidence and wellbeing improved in a short amount of time with an easily achievable behaviour.

8 Voluntary Sedation in a Sulphur-Crested Cockatoo

Cici, a 46-year-old sulphur-crested cockatoo (*Cacatua galerita*), was diagnosed with an idiopathic epilepsy, experiencing seizures at times when startled or stressed. He was due for a routine health examination that would include radiographs, blood work and a recheck of a mass on his abdomen, but restraining him to administer anaesthetics was not an option as the stress was likely to trigger a seizure. Care staff decided that the best course of action would be to train him for a voluntary sedation via intranasal administration of a combination of midazolam and butorphanol. The training plan included a voluntary passive restraint with a towel and the simultaneous insertion of a syringe into both the nostrils. The towel behaviour was a safety precaution in case the bird flipped backwards or shook his head. The trainers draped a towel over the bird's back and applied gentle pressure with their hands to support him. The syringes would then administer the sedation drugs. During the training, the bird was conditioned to hold for both air and moisture coming through the syringes. On the day of his procedure, staff dispensed 0.25 ml of liquid medication into each nostril successfully (see Fig. 7). The towel was removed; the bird crated

Fig. 7 Cici, the sulphur-crested cockatoo (*Cacatua galerita*), participating in a voluntary sedation via intranasal administration of a combination of midazolam and butorphanol.

voluntarily and was transported to the hospital. At approximately 7 min after the administration of the midazolam and butorphanol, the bird was sedated and recumbent. His full examination and transport were conducted without any negative incidents or stress.

9 Transitions and End-of-Life Decisions

Good animal care and wellbeing programmes can anticipate many of the age-related changes and concerns. As mentioned earlier in this chapter, preparing and modifying to suit the animal's life stage is key. Environmental changes, changing cues to suit different sensory systems, considering social needs, the impact of human–animal relationship and understanding the quality of life all revolve around transitions from one phase or one moment to another. Transitions also include knowing when it is time to say goodbye to an animal, when good quality of life can no longer be supported and promoted. Using behavioural observations, record-keeping, decision trees and other tools and processes can assist in knowing and deciding together for end-of-life decisions (see the chapter 'Euthanasia of Geriatric Zoo Animals: Decision-Making and Procedure' for a team-based approach on end-of-life decisions). The last phase and time with animals in our care may include training animals for behaviours necessary for euthanasia, including weighing, longer duration stationing, voluntary injection or inhalation. These tend to be difficult times and moments as care staff prepare to say goodbye and may feel conflicted about training euthanasia-specific behaviours. Organization, leadership and team support are fundamental to support staff training for behaviours, which will assist in a good death for the animal in their care. Utilizing trained behaviours can reduce stress on the animals and staff who care for them. Training animals, the final activities and time together and opportunities to enrich or say goodbye can all be facilitated but are not limited to animal training. The last moments together may be more positive supporting the human–animal relationships built over the animal's life. When the end comes, natural or induced, every person should be able to decide how this moment should be lived for them. A good death for the animal and a personal way to find closure for people must be considered. See the chapter 'Caring for Elderly Wild Animals: The Human Experience' for more details on caring for elderly animals and the human experience.

10 Conclusion

Understanding the effects of animal learning and training can be challenging as there are many factors and influences. However, learning how to observe animals and use positive reinforcement can be achieved by those who are committed to learning and practising to train animals to participate in their care. Commitment and dedication of the animal care teams help provide animals with the best care possible at all stages of life. Focusing on positive reinforcement training and considering special needs at different stages of life are fundamental to optimal wellbeing. Utilizing science- and ethics-based best practices to train animals to participate in their own health and other care is considered the best practice in accredited zoos and aquariums. Through proactive planning and monitoring, age-related health concerns can potentially be detected early, and species-specific concerns can be anticipated. Designing for age-specific and adapting environments will facilitate care and treatments, making it easier and more comfortable for animals to participate in training sessions. Individuals and organizations are encouraged to conduct more research on the benefits of animal training and specifically in ageing animals and to publish their findings in grey or preferably peer-reviewed literature for a more evidence-based approach to animal care and wellbeing. Training staff for end-of-life decisions and training animals for euthanasia require organization, leadership and team support. Animals will experience better wellbeing and potentially an extended longevity due in part to a lifelong preventive care and wellbeing programme when this includes voluntary medical behaviour training.

References

Ashley NT, Wingfield JC (2012) Sickness behavior in vertebrates: Allostasis, life history modulation, and hormonal regulation. In: Demas GE, Nelson RJ (eds) Ecoimmunology. Oxford University Press, New York

Azkona G, García-Belenguer S, Chacón G, Rosado B, León M, Palacio J (2009) Prevalence and risk factors of behavioural changes associated with age-related cognitive impairment in geriatric dogs. J Small Anim Pract 50(2):87–91

Bellows J, Center S, Daristotle L, Estrada AH, Flickinger EA, Horwitz DF et al (2016) Evaluating aging in cats: How to determine what is healthy and what is disease. J Feline Med Surg 18(7):551–570

Boström I, Nyman G, Hoppe A, Lord P (2002) Effects of meloxicam on renal function in healthy beagle dogs with low blood pressure during anaesthesia. Doctoral dissertation, thesis: Effects of non-steroidal anti-inflammatory drugs and acepromazine on renal function in Dogs with Anaesthesia-Induced Low Blood Pressure, Swedish University of Agricultural Science, Uppsala

Brando SI (2010) Advances in husbandry training in marine mammal care programs. Int J Comp Psychol 23(4)

Brando SI (2012) Animal learning and training: implications for animal welfare. Vet Clin Exot Anim Pract 15(3):387–398

Brando S (2022) Marine mammals. In: Knight A, Phillips CJC, Sparks P (eds) Routledge Handbook of Animal Welfare. Routledge, London, pp 293–308

Brando S, Buchanan-Smith HM (2018) The 24/7 approach to promoting optimal welfare for captive wild animals. Behav Processes 156:83–95

Brando S, Herrelko ES (2021) Wild animals in the city: Considering and connecting with animals in zoos and aquariums. In: Bovenkerk B, Keulartz J (eds) Animals in Our Midst: The Challenges of Co-existing with Animals in the Anthropocene. Springer, Cham, pp 341–360

Brando S, Watters J (2016) Caring for elderly animals. In: 1st International symposium of caring for elderly animals. Hosted by San Francisco Zoo, 6–7 December 2016, San Francisco, USA

Brando S, Rachinas-Lopes P, Lima D, Rodrigues Goulart V, Hart LA (2023). Understanding job satisfaction and occupational stressors of distinctive roles in zoos and aquariums. Animals 13(12):2018

Carpenter RE, Pettifer GR, Tranquilli WJ (2005) Anesthesia for geriatric patients. Vet Clin Small Anim Pract *35*(3):571–580

Chance P (1988) Learning and behavior, 2e. Wadworth Inc, Belmont, CA

Chiu E, Bodley K (2010) A psychogeriatrician's home visit to the zoo: a case report. Int Psychogeriatr 22(4):671–673

Dennis PM (2013) The golden years - managing the health of geriatric animals. In: Paper presented at North American Veterinary Community Conference, Cleveland Metroparks Zoo and Ohio State University Cleveland, OH

Dorey NR (2020) Learning theory. In: Melfi VA, Dorey NR, Ward SJ (eds) Zoo Animal Learning and Training. Wiley, Newark, pp 3–13. Kindle Edition

Friedman SG (2007) A framework for solving behavior problems: functional assessment and intervention planning. J Exot Pet Med 16(1):6–10

Herron ME, Shreyer T (2014) The pet-friendly veterinary practice: a guide for practitioners. Vet Clin Small Anim Pract 44(3):451–481

Hughes JML (2008) Anaesthesia for the geriatric dog and cat. Ir Vet J 61(6):1–8

Kirkland JL, Stout MB, Sierra F (2016) Resilience in aging mice. J Gerontol A Biol Sci Med Sci 71(11):1407–1414

Krebs BL, Marrin D, Phelps A, Krol L, Watters JV (2018) Managing aged animals in zoos to promote positive welfare: A review and future directions. Animals 8(7):116

LaGraize SC, Borzan J, Rinker MM, Kopp JL, Fuchs PN (2004) Behavioral evidence for competing motivational drives of nociception and hunger. Neurosci Lett 372(1–2):30–34

Marrin D, Krebs B, Watters J (2018) An approach to assessing and supporting the behavioural wellness of ageing zoo animals. ABMA conference

Marrin D, Poole A, Krebs B, McInture K (2019) A detailed behavioral approach to rebuilding confidence, trust and a positive relationship with a Sumatran tiger (Panthera tigris sumatrae). ABMA conference

Mellor DJ, Beausoleil NJ, Littlewood KE, McLean AN, McGreevy PD, Jones B, Wilkins C (2020) The 2020 five domains model: Including human–animal interactions in assessments of animal welfare. Animals 10(10):1870

Navaratnarajah A, Jackson SH (2013) The physiology of ageing. Medicine 41(1):5–8

Peters A, Rosene DL, Moss MB, Kemper TL, Abraham CR, Tigges J, Albert MS (1996) Neurobiological bases of age-related cognitive decline in the rhesus monkey. J Neuropathol Exp Neurol 55(8):861–874

Pryor K (2002) Getting started: Clicker training for dogs. Sunshine Books, Waltham, MA, p 100

Skinner BF (1953) Some contributions of an experimental analysis of behavior to psychology as a whole. Am Psychol 8(2):69

Veasey JS (2017) In pursuit of peak animal welfare; the need to prioritize the meaningful over the measurable. Zoo Biol 36(6):413–425

Watters J, Sulzner K, Marrin D, Huang S, MacDonald C, Ostapak S et al (2015) Assessing quality of life in geriatric zoo animals. Markus Gusset1 & Gerald Dick2 24:37

Yin SA (2009) Low stress handling, restraint and behavior modification of dogs & cats: techniques for developing patients who love their visits. CattleDog Pub, Davis, CA

Zeligs JA (2014) Animal training 101: the complete and practical guide to the art and science of behavior modification. Hillcrest Publishing Group, Minneapolis, MN

Supporting Geriatric Zoo Animal Welfare Through Nutrition

Francis Cabana and Amy Plowman

F. Cabana (✉)
PetCubes, Singapore, Singapore
e-mail: francis@petcubes.com

A. Plowman
Bumblebee Conservation Trust, Paignton, UK

S. Brando, S. Chapman (eds.), *Optimal Wellbeing of Ageing Wild Animals in Human Care*, https://doi.org/10.1007/978-3-031-30659-4_7

Nox. Nox, a female agouti (*Dasyprocta azarae*), is very curious and is always looking for things to eat, especially nice treats. She likes to explore her habitat, together with her companion Dourada, turning over all nooks and crannies every day. When new things are introduced or changes are made in the habitat, they both are the first to check it all out. If the care staff are easy-going and calm, she will come over to take treats from their hands, the delicacies attracting her to interact. If it takes too long to clean the indoor and/or outdoor areas, she can get a little impatient, making little grunts and stamping her feet. In the past, when she became too impatient, she would throw over her water bowl, so the care staff made a special bowl holder for her to avoid such happening. The agoutis share their habitat with a couple of common marmosets who definitely have the upper hand when it comes to new enrichment and food, but Nox patiently waits her turn. One of their favourite items are coconuts, which they love to climb on and work at until breaking. As behoves a real agouti, her favourite food are nuts, walnuts in particular, and avocado. She is always busy hiding pieces of her food in different places around the habitat. Nox makes herself a little nest in the hay as a preferred place for sleeping.

Abstract

Aging wild animals undergo many physiological changes that are not visible to their keepers. Changes in their digestive system render them less efficient at absorbing nutrients, changes in their brain reduce mental acuity and inflammation builds up throughout the body, especially the joints. Eventually symptoms will be observed such as muscle loss, decreased activity and longer reaction times, thereby impacting their welfare and eventually their quality of life. Some of these changes may be prevented or delayed by meaningful changes to the diet when an animal reaches 70% of their life span in captivity. Protein, vitamins and omega-3 fatty acids have shown high potential in the maintenance and improvement of the welfare and quality of life of geriatric wild animals.

Keywords

Bone mass · Brain function · Dietary supplements · Gut function · Osteoarthritis · Prebiotics

1 Introduction

As described by the authors throughout this book, many wild animals in human care, especially those in modern zoos with high standards of husbandry and veterinary care, far outlive the typical life expectancy of their free-living conspecifics. As a result, these animals become subject to age-related changes in their metabolism and physiology due to accumulated deoxyribonucleic acid (DNA) damage, oxidative stress, etc. Some of these changes may be caused or accelerated by or alternatively prevented or delayed by the aspects of diet throughout the animal's life. Some of these age-related changes may themselves have an impact on the animal's nutritional status, through changes in nutrient requirements; ability to find, consume and digest food; and ability to absorb and assimilate nutrients from the diet.

Free-living wild animals rarely live long enough to show any degenerative signs of ageing; therefore, very little is known about the ageing process for most species of wild animals held in zoos and aquariums. Production animals similarly do not generally live to old age, but there is some literature available on the nutrition of companion animal species, especially cats, dogs and horses, but not others, for instance, popular hobbyist fish species. Research animals are another source of information on ageing, especially primate models involved in studies of human ageing conditions. Together these studies do provide us with some indicators about the potential relationship between aspects of ageing and nutritional factors in a range of zoo-housed species. From this available evidence, we have developed and tested the ability of aspects of dietary management to prevent, delay or mitigate age-related conditions in zoo animals.

2 Nutrition-Relevant Age-Related Changes

Physiological changes in geriatric animals are many, and more links to nutrition are being discovered each year. Here, we briefly describe three commonly described symptoms in numerous geriatric species found in zoos, namely declining brain function, changes in bone mass and degenerative joint disease and gastrointestinal tract (GIT) changes.

2.1 Gastrointestinal Tract Changes

Changes to the gastrointestinal tract, which includes the mouth, teeth, oesophagus, crop, stomach and intestines, will have an impact on how an animal can consume the diet and assimilate nutrients from it. The most obvious change to affect feeding is missing or broken teeth, which is one of the most common conditions seen in geriatric domestic dogs (>50%, Srikala et al. 2020). However, older animals show a range of other GIT changes including decreased production of saliva and gastric juices, increased gastric emptying time, shortening of the absorptive surfaces, slower gut cell regeneration and a reduced ability to move digestate along the GIT. Constipation (39%) or diarrhoea (31%) are also common conditions seen in elderly dogs (Srikala et al. 2020), indicating that gut changes do have an impact on digestion likely leading to a decrease in nutrient assimilation (e.g. Patil et al. 2004: Siciliano 2002). Therefore, relative to their body weight, ageing animals may need an increased amount of food or higher nutrient concentrations in their diet; in particular, higher protein concentrations may be required. Prebiotics in the diet help to support the numbers and activity of the gut microbiota, which can have significant beneficial effects on the ability of animals to absorb nutrients from their diet. Fermentable soluble fibres such as gum arabic, pectin and β-glucans (found in oats) are efficient prebiotics that can be added to the diets of geriatric animals to support gut function.

2.2 Declining Brain Function

Less obviously associated with nutrition, declining brain function is synonymous with ageing in the majority of the chordate species. Much recent research has shown a preventative link with nutrition. The process of cognitive decline includes but is not limited to a declining sense of smell and taste, deficits in memory and learning, disorientation, decreased social interaction, decreased activity and sleep disturbance (Head et al. 2008). Amongst other changes in the brain over time is an increase in deposition of the amyloid-β peptide. This is proposed to be an early factor in the pathogenesis of Alzheimer's disease and has also been shown to be deposited significantly more in geriatric dog's brains than in younger dog's brains. Amyloid-β decreases the ability of every affected brain cell to absorb nutrients (Selkoe 1996). It is thought to be deposited throughout the animal's life at a rate similar to the oxidative stress experienced by the brain. The dogs' (and other chordates') brains have protective mechanisms to prevent amyloid-β damage, such as the antioxidant superoxide dismutase that is produced endogenously within the body. These protective mechanisms appear to decrease in effectiveness as animals age. The brain is highly concentrated in unsaturated lipids that make it particularly vulnerable to oxidative stress (Rofina et al. 2006) because of their susceptibility to lipid peroxidation. Antioxidant support provided in unprocessed, whole foods such as fruits and vegetables showed significant positive effects in increasing brain antioxidant concentration and slowing the development of cognitive decline in both younger and older dogs (Head et al. 2008).

2.3 Changes in Bone Mass and Degenerative Joint Disease

Age-related changes in bone mass and degenerative joint disease have been recorded in many animal species, for example, non-human primates (National Research Council 2003), horses (Siciliano 2002) and dogs (Bui and Bierer 2003). Calcium and vitamin D are important for maintaining bone mass and so must be provided in adequate quantities for geriatric animals. Glucosamine, chondroitin and methylsulfonylmethane (MSM) are supplements that have been commonly used for many years to prevent and relieve the symptoms of arthritic pain through anti-inflammatory action and some cartilage-protecting effects. More recently, Omega 3 (or n-3) fatty acid supplements have been used to support osteoarthritis in dogs and horses (e.g. Bui and Bierer 2003: Manhart et al. 2009). Unsaturated fatty acids are essential in the diets of all vertebrates, but it is not only the total amounts that are important but also the ratio of the different types to each other. In particular, the ratio of n-6 to n-3 fatty acids is crucial and should be between 4:1 and 5:1. Most foods available to humans and animals in human care have a much higher concentration of n-6 fatty acids than n-3 fatty acids, and overall diet ratios of over 20:1 are common. Most readily available vegetable oils, nuts and seeds are very high in n-6 fatty acids and very low in n-3 fatty acids. Linseeds and linseed oil are good sources of plant-based n-3 fatty acids in the form of α-linolenic acid (ALA). However, this is not the active form of n-3 needed by vertebrates, and different species have varying abilities to convert ALA to the active forms they need, eicosapentaenoic acid (EPA) and docosahexaenoic acid (DHA). Most carnivorous species have little to no ability to convert ALA into EPA and DHA as they would never normally consume ALA in a wild diet. These species therefore require n-3 supplements containing EPA and DHA such as in the forms of fish oil (salmon, cod liver, etc.), mussels, krill oil, algae extract and even micro-distilled from various sources. These may be more effective sources of omega-3 fatty acids for most animals irrespective of their ability to convert ALA into EPA and DHA.

3 Nutrition as Part of Geriatric Animal Care Programmes at Singapore Zoo and Toronto Zoo

At Singapore Zoo in Singapore and Toronto Zoo in Canada, any animal that has reached 75% of its published in-care lifespan (see Chapter "Veterinary Care of Ageing Zoo Animals: Description of a Proactive Approach" for more information on this method) is included in the geriatric care plan. Individuals in the plan are classified as green (no signs of age-related conditions), amber (some symptoms of age-related conditions) or red (obvious old-age conditions and deteriorating health), and their plan was managed according to these categories. Designation of one which category is chosen for which animal is based on the symptoms presented and decided by the veterinary team. In all categories, the care plan may involve changes to the normal adult diet for that species, in terms of ingredients, supplements, quantities and/or presentation, in order to enhance individual welfare. Nutrition adaptations and supplements are incorporated at the green stage based on taxonomic group. In most cases, by the time an animal reaches the amber or red stage, nutrition will have a smaller impact on their health and symptoms, and the main emphasis of the care plan will be handled by the veterinarians on pain relief and other medication to manage clinical conditions and animal care staff on daily care needs. The real value of the nutritional aspects of the care programme, therefore, lies in the prevention or delay of onset of conditions while the animals are still in the green phase.

In a pilot study conducted at Singapore Zoo on nutritional supplementation for geriatric animals, we focused on arthritic conditions because of their ease in visual scoring and because they were unfortunately relatively common in the older animals. These conditions can also be very painful and cause restricted movement that can affect other aspects of animal welfare and impact visitor appreciation. Using literature from a range of domestic species in human care, we developed bespoke supplement mixes and other dietary adjustments designed to reduce arthritic symptoms in many species of carnivorous birds, carnivores, Artiodactyls, Perissodactyls

Table 1 Implementation of dietary changes for geriatric animals to reduce or delay the onset of symptoms of arthritis

	Carnivorous birds	Carnivores	Artiodactyls	Perissodactyls	Primates
Supplements	Calcium $n-3$ fatty acids Antioxidants L-carnitine Taurine	Calcium $n-3$ fatty acids Vitamin E	Calcium $n-3$ fatty acids	Calcium $n-3$ fatty acids Pectin Vitamin C Vitamin B complex	Calcium $n-3$ fatty acids Pectin Carotenoids Vitamin C Vitamin E
Other changes		20% red meat replaced with chicken	Added fermentable fibre	Added fermentable fibre (beet pulp)	Added oats and egg whites
Species	3	10	5	4	2
Individuals	3	22	6	10	2

and primates (Table 1). All the supplements contained calcium and n-3 fatty acids as there is consistent evidence across taxa that these can be effective in arthritic conditions (primates: Cordingley and Cornish 2022: cats: Johnson et al. 2020: dogs: Roush et al. 2010; horses: Vandeweerd et al. 2012).

We monitored the effects of these supplements in a total of 43 'amber' individuals across 24 species in these groups (Table 1) using the method of Roush et al. (2010). This method involved two parallel scoring systems over 12 weeks from the first use of the supplements on Indian vultures (*Gyps indicus*), banteng (*Bos javanicus*), gaur (*Bos gaurus*), warthogs (*Phacochoerus africanus*), zebras (*Equus zebra hartmannae*), lion-tailed macaques (*Macaca Silenus*), Sumatran orangutans (*Pongo abelii*), Asiatic lions (*Panthera leo leo*), fishing cats (*Prionailurus viverrinus*), leopard cats (*Prionailurus bengalensis*) and green-winged macaws (*Ara chloropterus*). The first, scored by the veterinary team, rated lameness, weight-bearing, body condition and pelage quality on a scale of 1 (ideal) to 5 (very bad) every 4 weeks. The second, scored every 2 weeks by the keeper team, rated a range of behaviours as 1 (better than the last time), 2 (same as the last time) or 3 (worse than the last time) to give an indicator of the progression of a condition. The behaviours scored were overall activity level, aggression, climbing, jumping, lagging behind, limping, walking, running, stiffness and rising from rest.

There was an improvement in the animals' conditions, across all the taxonomic groups. Lameness score decreased from an average of 2.4 in Week 0 to 0.5 in Week 12, indicating a very substantial improvement. Six species also had an improvement in the score for weight-bearing but other vet-scored measures (body condition and pelage condition) did not change. The average keeper scores across all species for activity, limping, rising from rest, stiffness and walking were all below 2 (1.6–1.8), indicating a consistent improvement in the previous assessment over the 12 weeks. There was variation amongst species, with many having a steady score of 2 for most keeper-rated behaviours and overall activity (all 10 species), walking, stiffness and rising from rest (9 species only) being the most likely to have an average score below 2. Over 12 weeks, none of the vet-scored or keeper-scored measures indicated a worsening of arthritic symptoms.

4 Geriatric Diet Modifications

Through trials at both the zoological institutions, the following diet adaptations gave positive initial results, although a larger sample size and time length are needed to truly measure any impact, positive, null or negative.

1. Reduction in energy of the diet to reflect any observed decrease in energy/weight gain but maintenance of protein/fat/carbohydrate concentrations within the diet and, in some cases, slight increases in protein.
2. Addition of creatine in the diet to help slow down muscle wastage (sarcopenia).
3. Addition of whole foods high in water (e.g. leafy vegetables, melons and frozen coconut water enrichment) to encourage a higher water intake.
4. Addition of whole food or targeted supplements high in a variety of antioxidants including carotenoids, flavonoids and polyphenols and if appropriate based on nutrient profile provided: vitamin C, vitamin E or selenium (e.g. almonds, sunflower seeds, capsicum, sweet potato, kale, broccoli and mushrooms).
5. Provision of a bioavailable n-3 fatty acids supplement such as fish oil.
6. Glucosamine, chondroitin, MSM and calcium supplements.
7. Prebiotics such as gum arabic, pectin, β-glucans (found in oats) and other fermentable soluble fibres.

Future trials will be necessary on the effects of species-specific probiotics and gut health in ageing animals to maintain or improve nutrient absorption, hormonal cycle and disease prevention. A long-term focus on animal ageing would also be mitochondrial factors and how to maintain the cell's mitophagy ability to clean up stressed mitochondria to prevent this significant source of oxidation.

References

Bui LM, Bierer TL (2003) Influence of green lipped mussels (Perna canaliculus) in alleviating signs of arthritis in dogs. Vet Ther 4(4):397–407. PMID: 15136981

Cordingley DM, Cornish SM (2022) Omega-3 fatty acids for the management of osteoarthritis: a narrative review. Nutrients 14(16):3362. https://doi.org/10.3390/nu14163362

Head E, Rofina J, Zicher S (2008) Oxidative stress, aging, and central nervous system disease in the canine model of human brain aging. Vet Clin Small Anim Pract 2008:167–178. https://doi.org/10.1016/j.cvsm.2007.10.002

Johnson KA, Lee AH, Swanson KS (2020) Nutrition and nutraceuticals in the changing management of osteoarthritis for dogs and cats. J Am Vet Med Assoc 256(12):1335–1341. https://doi.org/10.2460/javma.256.12.1335

Manhart DR, Scott BD, Gibbs PG, Coverdale JA, Eller EM, Honnas CM, Hood DM (2009) Markers of inflammation in arthritic horses fed omega-3 fatty acids. Prof Anim Sci 25(2):155–160. https://doi.org/10.15232/S1080-7446(15)30702-6

National Research Council (2003) Nutrient requirements of nonhuman primates. Second Revised Edition. The National Academies Press, Washington, DC. https://doi.org/10.17226/9826

Patil A, Cupp C, Perez-Camargo G (2004) Incidence of impaired nutrient digestibility in aging cats. Compend Contin Educ Pract Vet 26(Suppl 2A):72

Rofina JE, van Ederen AM, Toussaint MJ, Secreve M, van der Spek A, van der Meer I, van Eerdenburg FJCM, Gruys E (2006) Cognitive disturbances in old dogs suffering from the canine counterpart of Alzheimer's disease. Brain Res 1069(1):216–226. https://doi.org/10.1016/j.brainres.2005.11.021

Roush JK, Dodd CE, Fritsch DA, Allen TA, Jewell DE, Schoenherr WD, Richardson DC, Leventhal PS, Hahn KA (2010) Multicenter veterinary practice assessment of the effects of omega-3 fatty acids on osteoarthritis in dogs. J Am Vet Med Assoc 1:59–66. https://doi.org/10.2460/javma.236.1.59

Selkoe DJ (1996) Amyloid beta-protein and the genetics of Alzheimer's disease. J Biol Chem 271:18295–18298. https://doi.org/10.1074/jbc.271.31.18295

Siciliano PD (2002) Nutrition and feeding of the geriatric horse. Vet Clin Equine Pract 18:491–508. https://doi.org/10.1016/s0749-0739(02)00028-7

Srikala D, Kamran A, Kumari NK, Ramesh PT, Rao S, Yathiraj S (2020) A study on different clinical manifestations of geriatric dogs. Pharma Innov J 9(10):206–210. https://doi.org/10.22271/tpi.2020.v9.i9Se.5263

Vandeweerd JM, Coisnon C, Clegg P, Cambier C, Pierson A, Hontoir F, Buczinski S (2012) Systematic review of efficacy of nutraceuticals to alleviate clinical signs of osteoarthritis. J Vet Intern Med 26(3):448–456. https://doi.org/10.1111/j.1939-1676.2012.00901.x

Veterinary Care of Ageing Animals in Zoos: Description of a Proactive Approach

Sarah Chapman and Phillipa Dobbs

S. Chapman (✉)
Chapman Zoo Consultancy, Birmingham, UK

P. Dobbs
Twycross Zoo, Birmingham, UK

S. Brando, S. Chapman (eds.), *Optimal Wellbeing of Ageing Wild Animals in Human Care*, https://doi.org/10.1007/978-3-031-30659-4_8

Irma. Irma is a 14-year-old female snow leopard. She has lived at Twycross Zoo since March 2010 and has successfully reared six cubs. They have gone on to rear a further 22 individuals all around the world. She can often be seen sitting on a rock in the main exhibit watching visitors watching her. She is quite lazy when it comes to using her senses to find food hidden in her paddock, probably because she has become accustomed to many things being done for her. Irma has a sassy personality. When she meets new caregivers, she stares at them as if she is judging them. She has an elderly health check-up each year under anaesthesia and is showing only mild ageing changes so far. She really enjoys positive reinforcement training and is trained for hand injection of vaccinations and anaesthesia and blood sampling, although she feels the vets need more training in how to successfully obtain blood from her!

Abstract

This chapter describes a proactive ageing animal care programme that was implemented in a zoo setting, across all taxa, with individuals identified as aged if they were 75% through their expected natural longevity. Animals were prioritised, based on their history and known health conditions, and then assessed which included a full review of the animal's history and health assessment through 'hands-off' methods, for example, positive reinforcement training and behavioural observations, followed by 'hands-on' clinical assessment. Subsequently, treatment plans were instigated where needed, and cases were closely monitored, through a combination of subjective observations and follow-up hands-on health assessments, to see whether the animal's condition improved with the medication or management changes. Criteria for euthanasia were used where appropriate to aid in decision-making around timely euthanasia. The criteria included response to treatment, ability to deliver medication consistently, safety of the treatment and overall impact on the individual's behaviour in some cases. This proactive approach led to enhanced animal welfare and improved teamwork across the veterinary and animal teams.

Keywords

Welfare · Zoo · Ageing · Health · Treatment · Euthanasia · Case studies

1 Introduction

Many species kept in zoos, aquariums and sanctuaries (henceforth zoos) live longer than they would do naturally in the wild (Tidière et al. 2016). Older animals can be described using a variety of terminology including ageing, elderly or geriatric and will present with a variety of clinical conditions that cause negative welfare and require treatment and management (Longley 2011; Paré and Lentini 2010; Baine 2012), including pain (see more in the chapter 'Pain Physiology, Recognition, and Management in Zoo Animals'). These clinical conditions include joint, heart and/or dental disease. These conditions can be hard to detect in non-domestic animals that hide clinical signs of pain well but can cause significant negative welfare to the individual. This chapter outlines the model of an ageing zoo animal care programme that can be applied across all taxa. The aim of this programme was to identify geriatric animals, identify welfare concerns, diagnose any contributory medical conditions and implement a treatment/management/reassessment plan. Information from this health care programme was also used to aid in the creation of criteria for timely euthanasia to prevent suffering.

The main principles used throughout this programme included the following:

1. The application of basic veterinary principles to every species to allow early detection and treatment of clinical conditions
2. The importance of gathering information prior to hands-on assessments
3. Careful preparation for procedures
4. Appropriate clinical follow-up
5. High standards of record-keeping
6. Good communication between the veterinary and animal-keeping teams
7. Careful and considered decision-making around euthanasia and thorough post-mortem examinations

Following a number of animals presenting acutely but with chronic conditions, an ageing zoo animal care programme was created at a medium-sized zoo, in the United Kingdom, to include the assessment of both behaviour and health. The information from these assessments was then fed into an overall individual welfare assessment. Ageing animals were identified once they were 75% through their longevity. Longevity being the duration of an individual's life. Longevity data from the wild was used, where possible, but where this information was not available, data pertaining to the longevity of that species in captivity was used. The data on longevity was taken from various publications and online sources such as AnAge Database. Once individuals were identified as ageing, the individual's behavioural, reproductive and clinical history was reviewed. A priority list was created after discussion with the animal care team, and health assessments were planned. An individual was given higher priority if there was an obvious concern about a particular disease or condition, for example, stiff movement possibly due to joint disease or difficulty eating due to dental issues. Two methods of assessment were used, hands-off and hands-on, and are described in more detail below. Clinical conditions such as arthritis, dental disease and kidney disease were identified, and treatment plans were formulated with input from animal care staff. These treatment plans would also include medication options, and ideal time frames for follow-up assessments were then discussed as a team. The results of assessments combined provided an overall welfare assessment of the individual animal. In cases where clinical conditions were considered to significantly reduce welfare, for example, severe joint disease affecting multiple joints or where treatment was refractory, that is, no response to treatment, euthanasia was recommended. If treatment was instigated, a reassessment plan was created including expected treatment goals, criteria for reassessment of welfare and criteria for euthanasia with the veterinary and animal care teams if this was needed. The computer software programme Zoological Information Management System (ZIMS, Species 360, MN, USA) was used to record data including training, behaviour and clinical information within the various modules of Medical ZIMS.

2 Health Assessment

Prior to any health assessment, the animal's behaviour was discussed with the animal care staff and any welfare concerns identified, for example, stiff movement or poor appetite. Training needs were discussed, for example, entering a crate, weigh scales and injection training, and training plans were put into place. A combination of hands-off and hands-on health assessments was used in the majority of cases. In some cases, the hands-on assessment could be carried out using manual restraint, for example, a small reptile species, whereas chemical restraint (i.e. general anaesthesia) was used in animals that were more dangerous and/or were required to remain still for diagnostics such as radiography. Training animals to participate regardless of species was preferable over manual restraint where safe and possible.

2.1 Hands-off Methods of Health Assessment

This included the following methods:

1. Case history review
2. Behavioural observations
3. The use of positive reinforcement training
4. Body condition scoring
5. Thermal imaging

2.1.1 Case History Review

A complete review of the animal's history was undertaken. This included any medical history and treatment, behavioural records and notes from animal care staff.

2.1.2 Behavioural Observations

The next stage of assessment utilised observations described in daily records including subjective assessments (e.g. demeanour, mobility and interaction with care staff) and more objective assessments (appetite, thirst, urination, defecation, enclosure use and sleeping patterns). Additional observations, carried out by the veterinary team, were also used, for example, assessing expected activity level, response to stimulation, vision, resting respiratory rate and gait analysis. Basic behavioural ethograms were used to assess various aspects including social interactions, enclosure use (split into sections including area and height) and gait analysis. Ethogram data was usually collected by observing the individuals or groups for 30-min periods at various times during the day, which allowed a complete daytime budget of behaviours to be created. Reassessments were based on repeating the ethograms after predetermined periods of time, which varied depending on the individual case. For example, a chimpanzee (*Pan troglodytes*) was given medication for pain relief for arthritis, and ethograms were repeated weekly for the first month of treatment; then, based on the satisfactory response being seen, these reassessments were reduced to every 2 to 3 months. Video recordings were a useful part of data collection, especially for gait analysis and behavioural data overnight; for example, night-vision closed-circuit television (CCTV) within the enclosure of an arthritic elephant was useful to monitor sleep patterns and mobility when getting up from sleeping. Zoo volunteers were used to assist in collating the video data that was then reviewed by the veterinary team.

2.1.3 The Use of Positive Reinforcement Training

Positive reinforcement training was utilised to allow safe access to the animals and gather data wherever possible. It was important that the animals were comfortable with the training prior to any procedures and that this was a positive experience for them with appropriate food or drink rewards. The rewards would also help when providing individual medication to the animals. One of the main goals of training was to have the ability to weigh the animals routinely. This was achievable for most species either with the use of built-in scales within the enclosure or by placing scales within the enclosure only during designated training sessions. This non-invasive method of health monitoring was invaluable for monitoring clinical cases and ensuring accurate dosing of medications. Close inspection and, in some cases, examination of the animals were achieved through training in a protected contact setting, for example, elephants, great apes and small-/medium-sized monkeys. The collection of biological samples (e.g. blood and urine) formed part of the training plans for certain individuals depending on the facilities available and need for assessment. Examples included blood sample collection in an ageing Asian elephant (*Elephas maximus*) and urine sample collection in Bornean orangutans (*Pongo pygmaeus*).

Training was also used to ensure individual dosing of medications without the need to separate individuals from their social group at the time of treatment. An example was using stick-feeding of medicated meat to an ageing dhole (*Cuon alpinus*). Diagnostic tests were performed through training in some individuals; examples include foot radiography in an elephant, electrocardiography in chimpanzees and abdominal ultrasonography in a red-faced spider monkey (*Ateles paniscus*) to investigate pregnancy. Clinical follow-up was also achieved through training in many cases, for example, oral assessments following a tooth extraction in a chimpanzee.

2.1.4 Body Condition Scoring

Visual body condition scoring was used in many species, in conjunction with body weights, to monitor health; however, this was not possible in species with thick fur or feathers, for example, snow leopards and birds. Published body condition scoring systems were used where available for certain species for example, elephants (Schiffmann et al. 2017). Some individuals were scored using basic principles from domestic animal scoring systems when species-specific scoring systems were not available. When using body condition scoring, the whole animal was assessed focusing on areas such as the neck, shoulders, chest, belly or abdomen, hips and pelvic area. These areas can highlight if an animal has a low body condition score, for example with prominent bones seen, or a high body condition score, for example having a rounded or bulbous body shape. These scores are subjective, and it was important that the same people carried out the scoring on a regular basis, for example, monthly to ensure that trends in body condition could be monitored. This method was used alongside weighing the animal wherever possible. Photographs and video of the animals were also taken to help with data collection and review of the body condition over time.

2.1.5 Thermal Imaging

Thermal imaging was a useful adjunct diagnostic tool and is well described in other publications (Hilsberg-Merz 2008). This technology uses an infrared camera to take images of the animal, which show the surface temperature of the animal in the imaged area. This can allow the identification of body areas with excess heat that can be related, for instance, to underlying inflammatory processes such as acute osteoarthritis (bone inflammation). Thermal imaging was used as a tool to monitor progress of conditions and response to treatment. It was essential that the imaging was performed consistently, for example, in same enclosure area and at the same time of day, to eliminate factors such as environmental temperature on the readings. This was especially noticeable in elephants if readings were taken after bathing or periods of being outside in high environmental temperatures. Daily weather was also recorded in keeper daily records so that this information could be used to assist with the interpretation of the data.

2.2 Next Steps

Once the hands-off assessment had been completed, enough information had been gained in some cases to proceed with a treatment plan. With these individuals, the options of treatment, feasibility, cost, availability and safety of medications were then discussed with the animal care team. Examples included primates with obvious mobility issues and stiffness that were given pain medication for suspected joint disease to see whether an improvement would be seen. These were often individuals where common conditions such as spondylosis (bone spurs on the spine) had been found in others of the same species. However, in some cases, there was not enough conclusive information available to make evidence-based decisions. For these individuals, a hands-on method of health assessment was also conducted.

2.3 Hands-on Methods of Health Assessment

With any hands-on health assessment, the physical safety of animals and staff and the facilities available were considered during the planning and execution stages. The handling of individuals was discussed with senior animal care staff to reduce potential stress or trauma to the animal and to ensure staff safety. With animals living in large social groups, additional mitigations were put in place to minimise the chance of reintegration issues after the procedure. Maximising the chance of a successful procedure required detailed planning and effective communication with the teams involved. This included describing clear objectives, contingency planning for each stage of the procedure and ensuring that all staff were aware of their specific roles and responsibilities, objectives and their role in the procedure. Effective preparation ensured safe, efficient procedures that achieved the objective(s) while minimising risks to patient(s) and staff. In some species, handling was possible without the need for chemical restraint, but in others, general anaesthesia was required to allow safe handling. Chemical restraint involving general anaesthesia carries multiple inherent risks for patients and staff, which was also considered as part of the process (Lamont and Grimm 2014).

A hands-on health assessment included the following:

1. Weight and body condition score
2. Clinical examination including detailed dental examination
3. Collection of biological samples (e.g. blood, urine and faeces)
4. Diagnostic imaging (e.g. radiography and ultrasonography)
5. Treatment as appropriate (e.g. dental extractions)

Blood samples were sent to an external veterinary laboratory for haematology and biochemistry and a range of other biochemical markers dependent on species and individual risk assessments with additional tests as per individual needs.

In-house biochemistry and total white cell counts were performed on acutely sick patients or if a quicker result was required with the understanding that these may carry greater variability in results than the external laboratories.

Good record-keeping during these procedures was essential with checklists and labelled photographs being used. These photographs were stored digitally on the zoo's computer system in an individual animal's folder. A dedicated note-taker was often in place during more complex procedures to ensure that no information was lost or forgotten.

Following health assessments, the clinical findings and results of tests were discussed with the appropriate animal managers and treatment options were outlined. This information was fed into an overall welfare assessment that was carried out in conjunction with veterinary and animal care staff. At this point, timelines for clinical follow-up were agreed, and criteria for euthanasia were developed in some cases. Training needs were also discussed as were any enclosure modifications or dietary alterations that could be beneficial. Individuals placed on long-term medication became part of monthly 'rounds' that involved veterinary and animal care staff visual assessment and discussion of the wellbeing of the individual animal. These rounds ensured that no individual went without veterinary assessment for long periods of time, and any subtle changes were identified which may not have been captured by the daily observation records. More in-depth meetings took place, involving relevant team members, if individual cases required more discussion or if changes to treatment plans were put forward for more frequent assessment.

2.4 Euthanasia

Creation of clear, written criteria for euthanasia was carried out where appropriate. When discussing each individual, the following were considered:

1. The impact(s) of treatment on the individual including time away from the social group and side effects of medication
2. The ability of the animal care team to give the treatment including time and safety
3. The impact(s) on other group members
4. The clinical response(s) to treatment
5. Deterioration of the condition(s)

When individual-specific criteria were met, the decision was then taken together to euthanise the animal. For instance, in cases of renal disease, a set amount of weight loss and reduction in body condition would have been part of the criteria. Another example was the lack of response to analgesia for cases of osteoarthritis within a set time frame. Common situations often involved a combination of events such as deterioration of the condition, lack of response to medication and a negative impact on the individual of proposed treatment, leading to the decision to euthanise the individual.

2.5 Post-Mortem Examination

A thorough post-mortem examination was carried out on all individuals by an experienced zoo veterinary surgeon, and animal care staff were given the opportunity to be involved in the examination. Body and limb radiographs were taken prior to the examination in the majority of cases, along with body weight and measurements. In some cases, pathology was worse than expected, including additional concurrent diseases that had been undiagnosed. Tissue samples from all organs were collected into formalin for histopathology and research projects. Fresh-frozen samples were taken and stored in case of additional testing; for example, if bacteriology/virology was required. Histopathological diagnosis often confirmed clinical diagnoses, for

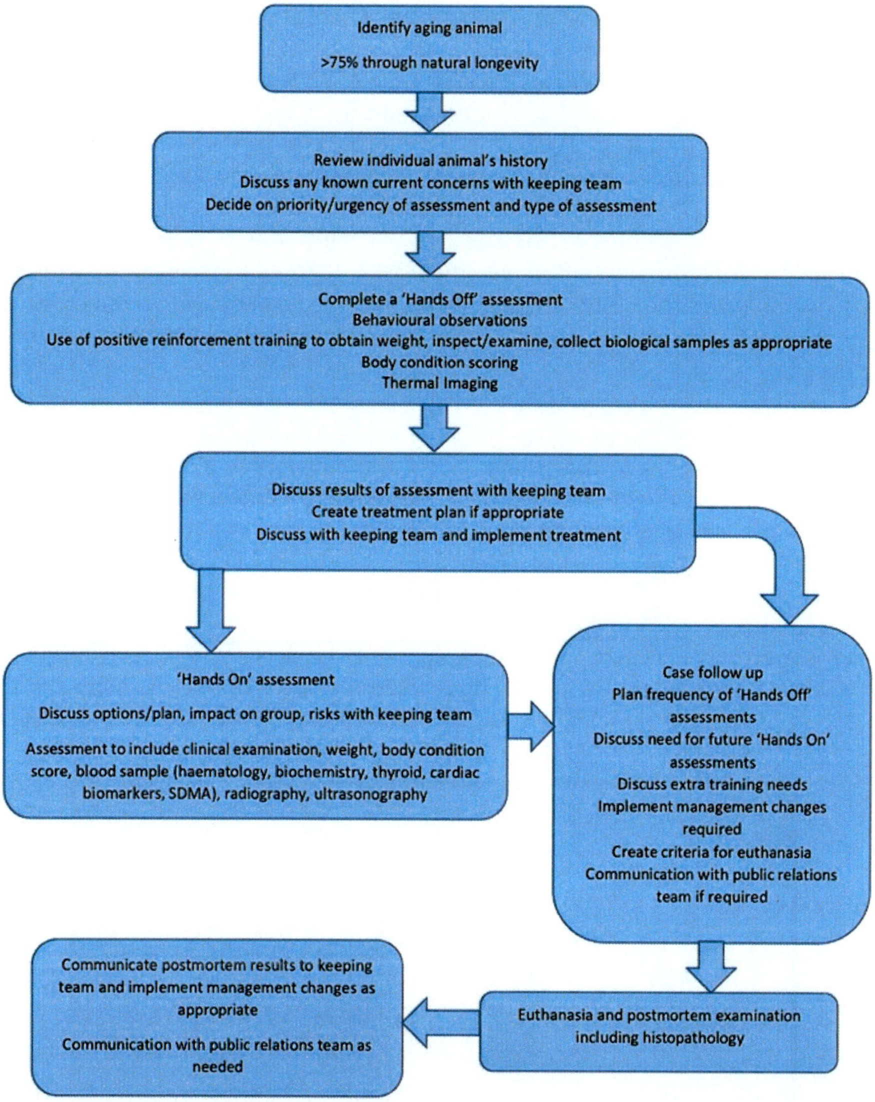

Fig. 1 Ageing zoo animal care flow chart

example, renal disease. Some carcasses were kept for educational purposes or sent to museums for further studies. Any research project undertaken was subject to ethical review by the zoo's ethics committee.

Reviews were conducted on a regular basis to ensure that the health protocols in place were still adequate for any remaining individuals. Collation of results and retrospective reviews allow adjustments to decision-making in future similar cases.

Figure 1 is a flow chart that outlines the processes involved in this programme.

2.6 Examples of Health Assessment Outcomes

A variety of clinical conditions were identified following health assessments. Common conditions included dental, osteoarthritis, spondylosis and renal disease with appropriate treatment plans put in place for each case. Many conditions found had not been obvious prior to examination from visual/behavioural assessment of the cases. During 1 year of the programme, 42% of mammals assessed had a new condition identified. Of the 47 primates assessed in that year, there were seven cases of dental disease, four cases of spondylosis, (see Fig. 3 for an example) two cases of osteoarthritis and two cases of renal disease identified and treated accordingly. As the programme progressed over the years, the number of new conditions identified reduced.

See Fig. 2 for an example of an infected canine tooth in a grey gibbon (*Hylobates muelleri*). This gibbon was eating well with no abnormalities noted externally.

Individuals benefited from analgesia for conditions such as osteoarthritis and teeth were extracted where appropriate. Poor body condition was sometimes not appreciated visually, and renal disease was found to be the cause in some cases. This allowed for management changes, extra individual feeding and regular weighing to be instigated if the individuals were not clinically sick. Figure 4 shows the extent of osteoarthritis in a Sulawesi crested black macaque (*Macaca nigra*). This macaque had showed only slight changes to his gait prior to his death.

These examinations also identified subclinical diseases such as cardiomyopathy (disease of the heart muscle) and hepatic haemosiderosis (excessive iron deposited

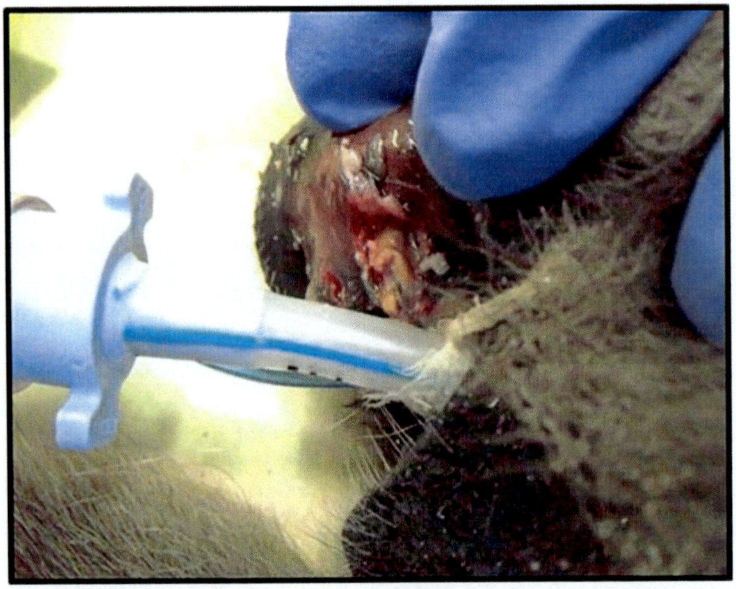

Fig. 2 The fractured and infected canine of a grey gibbon (*Hylobates muelleri*). Photo courtesy of Dr Sarah Chapman

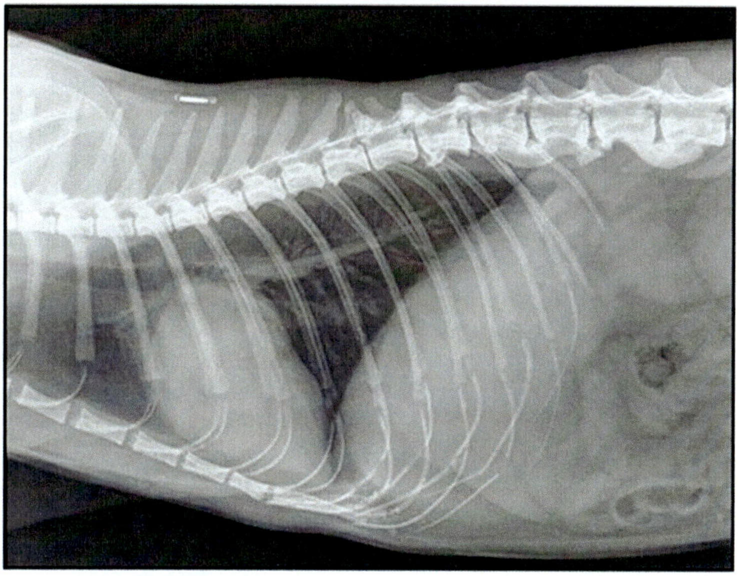

Fig. 3 A radiograph of a black howler (*Alouatta caraya*) showing significant spondylosis of the spine. Photo courtesy of Dr Sarah Chapman

within liver tissue) in primates. A female ageing bush dog (*Speothos venaticus*) was previously diagnosed with bilateral stifle osteoarthritis during an elderly health assessment, which was being managed with appropriate pain relief. Her annual health assessments were otherwise unremarkable until she was found to have changes on her blood biochemistry indicative of kidney disease. Due to her worsening condition, she was euthanised 2 months later. Figure 5 shows the kidneys of this individual photographed during post-mortem examination showing extensive changes.

Figure 6 shows the abnormal heart of an ageing chimpanzee on post-mortem examination, which was confirmed as idiopathic myocardial fibrosis (thickening of the heart muscle) on histopathology. This chimpanzee was being treated for cardiac disease for many years prior to her death. She sustained a blood clot to one of her main arteries (secondary to heart disease) leading to her euthanasia.

Inappropriate diet and excess body weight are important risk factors for multiple diseases including type 2 diabetes mellitus, osteoarthritis and dental disease. An ageing black howler had a routine health assessment including blood and urine sample collection. The results of this assessment revealed a diagnosis of previously undetected type 2 diabetes mellitus. Dietary changes were made to reduce calorific intake (in an attempt to reduce body condition score), and oral medication was administered to reduce blood sugar levels.

Additional examples of conditions found during these health assessments included a Humboldt penguin (*Spheniscus humboldti*) and a monitor lizard (*Varanus salvator*) both with osteoarthritis. Both individuals were managed with oral joint supplementation and pain relief. The penguin also responded well to Class III laser

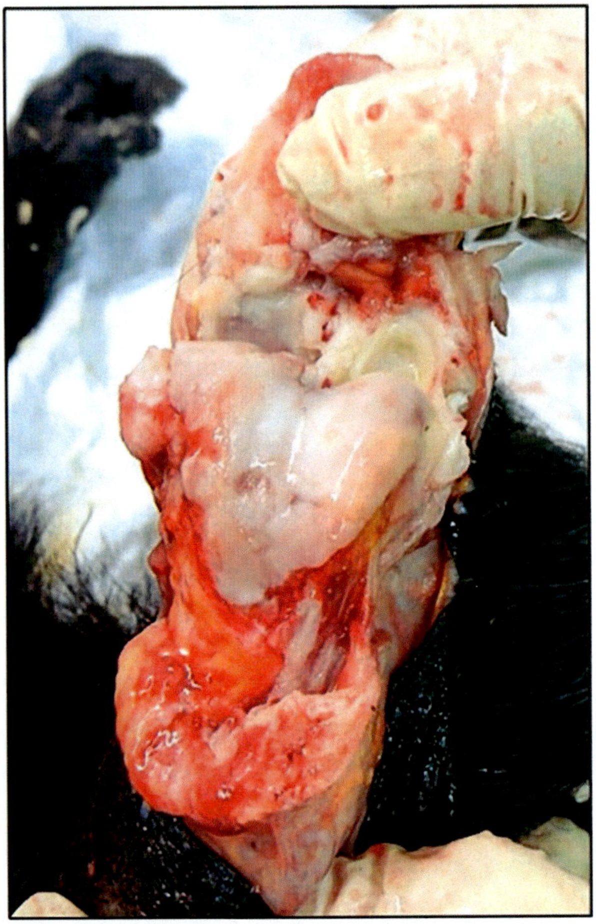

Fig. 4 The stifle of a Sulawesi crested black macaque (*Macaca nigra*) showing considerable cartilage erosion and new bone formation indicative of osteomyelitis. Photo courtesy of Dr Sarah Chapman

therapy, and monthly videos of her gait were useful as a monitoring tool for her condition.

3 Conclusion

Wild animals can be especially challenging from a clinical perspective as they often hide symptoms of disease until the disease has progressed significantly. Various methods of ageing zoo animal care have previously been used practically and reported in the literature (Föllmi et al. 2007; Whitham and Wielebnowski 2009; Vogelnest and Talbot 2018). The programme of ageing zoo animal care described

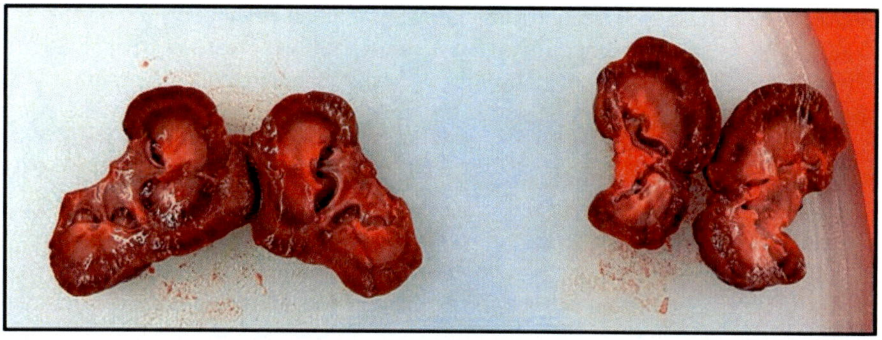

Fig. 5 Kidneys of a bush dog (*Speothos venaticus*) showing severe renal changes—confirmed as severe, chronic multifocal interstitial nephritis. Photo courtesy of Phillipa Dobbs

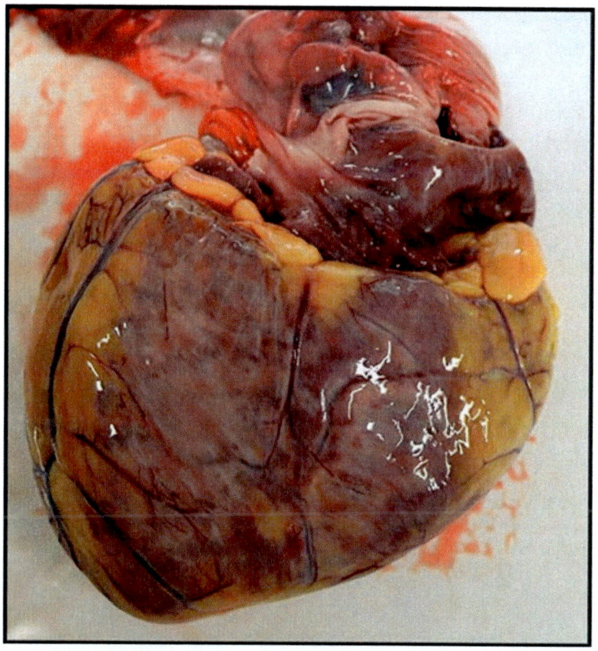

Fig. 6 The heart of a chimpanzee (*Pan troglodytes*) showing fibrosis of the myocardium indicative of idiopathic myocardial fibrosis. Photo courtesy of Phillipa Dobbs

here demonstrated that previously unknown clinical conditions can be frequently found when undertaking proactive health assessments. As suggested by the published literature, conditions such as osteoarthritis and spondylosis were common findings, especially in primates (Kershaw et al. 2020; Nunn et al. 2007; Rothschild and Woods 1992; Rothschild and Woods 1989). Treatment and management of these conditions led to improvement in welfare, and cases were followed up appropriately to ensure that these improvements were maintained. In certain cases, enclosure changes were effective mitigations for animals with reduced mobility, for example, by adding different climbing structures or resting areas (Waitt et al. 2010). The use of positive reinforcement training was a key element to the programme. The zoo was very supportive of the use of training to support husbandry and veterinary care and thus provided the keepers with the time to spend training the animals. This positive approach to training had a positive impact on welfare as animals were assessed, weighed, medicated and monitored appropriately without the need for frequent handling or anaesthesia.

Euthanasia can be challenging and emotional. Creating a decision-making process with clear and written criteria for euthanasia was extremely useful. Having a team approach to welfare assessment and development of euthanasia criteria also promoted a supportive environment for all involved.

The euthanasia procedure itself required careful planning and communication between the veterinary and keeping teams. Time was given to explain the actual euthanasia process and discuss any concerns. All euthanasia events were carried out with a calm and quiet approach to reduce stress as much as possible. Other considerations included conspecifics, the plan for handling the body and how the post-mortem examination would be performed. For certain species, members of the social group were given access to the body following euthanasia. This included elephants, apes and primates. This was thought to be an important part of the process despite the risk of a delay to undertaking a post-mortem examination if the group stayed with the body for a long period of time.

A thorough post-mortem examination was vital to ensure the diagnosis of all clinical conditions and to identify any subclinical disease present. Survey radiographs taken also enabled the creation of a large database for comparison with other clinical cases or for use in research projects. It was important to gain as much information as possible from all individuals. Knowledge of the presence of disease and the underlying causes can only benefit the future care of ageing individuals. The involvement of animal care staff in post-mortem examinations was voluntary, and many accepted and appreciated the opportunity to learn about the animals they cared for. This also enhanced their understanding of the processes involved and ultimately improved teamwork between the animal and veterinary teams.

The use of a proactive ageing animal care programme in a zoo vastly improved the welfare of ageing animals by providing treatment and husbandry options for clinical conditions.

Animal carers' feedback was that their initial hesitance and concerns were allayed by the outcomes of the health assessments including no anaesthesia-related deaths, positive welfare improvements in response to treatment and increased knowledge for them obtained when attending health assessments and post-mortem examinations. Many started to ask when animals were due their ageing animal health assessment as they had observed changes in some animals, which had been similar to others on medication for conditions such as arthritis.

Overall, this approach led to the implementation of a positive and proactive programme for assessing the welfare of ageing animals on a regular basis. This model can be applied to all taxa and settings where animals are kept in captivity, for example, zoos, sanctuaries and aquariums. Early detection of clinical conditions found in ageing animals enabled treatment and management changes to be implemented. An appropriate follow-up ensured close monitoring, adjustment to treatment and changes in husbandry options to promote optimal welfare, and creation of criteria for euthanasia assists in decision-making for the timely planning of euthanasia. While challenges were encountered, teamwork and good communication were vital in implementing a successful ageing animal care programme.

References

Baine K (2012) Management of the geriatric psittacine patient. J Exot Pet Med 21:140–148

Föllmi J, Steiger A, Walzer C, Robert N, Geissbühler U, Doherr M, Wenker C (2007) A scoring system to evaluate physical condition and quality of life in geriatric zoo mammals. Anim Welf 16(3):309–318

Hilsberg-Merz S (2008) Infrared thermography in zoo and wild animals. In: Fowler M, Miller E (eds) Zoo and wild animal medicine current therapy, vol 6. Saunders/Elsevier, pp 20–32

Kershaw T, Hal LEJ, Dobbs P, Liptovszky M, Strong V (2020) An exploration of the value of elective health checks in UK zoo-housed gibbons. Animals (Basel) 10(12):2307

Lamont LA, Grimm KA (2014) Clinical pharmacology. In: West G, Heard D, Caulkett N (eds) Zoo animal and wildlife immobilisation and anaesthesia. Wiley, pp 5–41

Longley L (2011) A review of ageing studies in captive felids. Int Zoo Yearb 45:91–98

Nunn CL, Rothschild B, Gittleman JL (2007) Why are some species more commonly afflicted by arthritis than others? A comparative study of spondyloarthropathy in primates and carnivores. J Evol Biol 20:460–470

Paré JA, Lentini AM (2010) Reptile geriatrics. Vet Clin North Am Exot Anim Pract 13:15–25

Rothschild BM, Woods RJ (1989) Spondyloarthropathy in gorillas. In: Seminars in arthritis and rheumatism. Elsevier, Philadelphia, PA, pp 267–276

Rothschild BM, Woods RJ (1992) Erosive arthritis and spondyloarthropathy in Old World primates. Am J Phys Anthropol 88:389–400

Schiffmann C, Clauss M, Hoby S, Hatt JM (2017) Visual body condition scoring in zoo animals – composite, algorithm and overview approaches in captive Asian and African elephants. J Aquar Res 5(1):1–10

Species360 Zoological Information Management System (ZIMS) (2021). zims.species360.org

Tidière M, Gaillard J-M, Berger V, Müller DW, Lackey LB, Gimenez O, Clauss M, Lemaître J-F (2016) Comparative analyses of longevity and senescence reveal variable survival benefits of living in zoos across mammals. Sci Rep 6:36361

Vogelnest L, Talbot JJ (2018) Quality of life assessment and end of life planning for geriatric zoo animals. In: Miller E, Fowler M (eds) Zoo and wild animal current therapy, vol 9. Elsevier Health Sciences, pp 83–90

Waitt C, Bushmitz M, Honess PE (2010) Designing environments for aged primates. Lab Prim Newsl 49:5–9

Whitham JC, Wielebnowski N (2009) Animal-based welfare monitoring: using keeper ratings as an assessment tool. Zoo Biol 28:545–560

Pain: Physiology, Recognition, and Management in Zoo Animals

Heather Bacon

H. Bacon (✉)
University of Central Lancashire, Preston, Lancashire, England, UK
e-mail: HBacon@uclan.ac.uk

S. Brando, S. Chapman (eds.), *Optimal Wellbeing of Ageing Wild Animals in Human Care*, https://doi.org/10.1007/978-3-031-30659-4_9

Magellan. Magellan, a male California sea lion (*Zalophus californianus*), was born on 11th June 1993 in England, and he arrived in France at Zoo d'Amneville 10 years later. He sired eight pups during his breeding years. He is a mellow animal, who likes contact with his trainers, and enjoys touch and chest rubs. In recent years, he has been retired from shows, but still comes out to meet the public for a quick introduction occasionally just before a scheduled showtime. Magellan is trained for a variety of voluntary husbandry behaviours, which allows him to participate stress-free in his own healthcare (e.g. eyedrops, submerged eye ultrasound, radiographs, ultrasound of the entire body, blood draws, taking liquid medications orally). His daily husbandry routine includes going onto scales to take his weight, therapeutic eye salt baths, eyedrops, and oral liquid herbal supplements, to help with joint health and mobility. Magellan's favourite food is Atlantic herring. He enjoys feeder toys a lot, especially a green crocodile that he pushes around his habitats. He also has a hammock, and sometimes shares it with two females: Luna and their daughter Shouka. At the age of almost 30 years, Magellan naps during lunch time, but when he is awake or during sessions he is full of energy, enthusiasm, and commitment. He is definitely a favourite amongst trainers and interns, due to his friendly personality and calm demeanor. Malgosia Kaczmarska, Zoo d'Amneville

Abstract

Pain is a significant welfare concern for zoo animals around the world. As zoos commonly house animals living beyond their lifespans in the wild, pain associated with degenerative diseases of old age, as well as that associated with acute illness or injury, is of increasing concern. Barriers to effective management of pain across the many and varied taxa that zoos house include our current understanding of pain physiology and the recognition of pain states in different species. Additionally, the therapeutic approach to pain management may be a challenge depending on species. This chapter outlines the physiology of pain, types of pain, pain recognition, and the principles of preventing and managing pain experiences in animals.

Keywords

Pain · Welfare · Zoo animals · Analgesia · Pain states · Pain recognition · Pain prevention · Pain management · Degenerative diseases

1 Introduction

Pain is a key contributor to poor welfare in zoo animals and may often be underdiagnosed due to a lack of validated pain assessment methods, or due to misunderstandings in how different species may demonstrate signs of pain. Thus, an understanding of signs of pain in zoo species, pain physiology, and the management of different types of pain is essential to promoting good welfare in zoos. It is not known what percentage of animal species suffer from pain, but poor pain management has been suggested as a significant welfare problem across all species under human care (Rioja-Lang et al. 2020). In order to effectively manage the welfare impacts of pain, we must first understand what the pain experience is, the neurobiology of pain, how to recognise pain in different animal species, and possible therapeutic interventions that are effective in managing pain. This chapter outlines these aspects, focusing primarily on mammals as this is where the majority of pain research and management information is focused, but it will also discuss the applications of these principles in non-mammalian species, and consider a range of pain management strategies and their application in zoo species.

2 Pain Physiology and Recognition

In order to understand pain, it is important that we first understand the processes by which tissue injury and pain are detected and processed by the mammalian nervous system. The mammalian nervous system is divided into two main areas, the central nervous system (brain and spinal cord) and the peripheral nervous system (the nerves that run throughout the body carrying messages to and from the central nervous system (CNS). The nervous system comprises three types of nerves: autonomic, sensory, and motor. Autonomic nerves are responsible for the 'housekeeping' functions of the body such as signalling gastrointestinal movement to aid in digestion or controlling respiratory activity to ensure we are breathing. These activities generally occur without conscious control or thought. Sensory nerves carry information from receptors on the external surfaces and periphery of the body such as the skin, eyes, nose, and mouth to the central nervous system (spinal cord and brain). This sensory information is then processed, and in response motor nerves may be activated, resulting in movement of the body. Movement of the body may also occur in response to reflex activity, this is the involuntary activation of a local loop of sensory and motor nerves (a reflex arc) that occurs without the cognitive processing of that signal. An example of this can be seen in the nociceptive response to unpleasant stimuli.

2.1 Nociception

Pain is sometimes confused with nociception. Nociception is the stimulation of receptors (nociceptors) in the tissues of the body by a noxious (aversive) stimulus such as soft tissue injury, or excessive heat, cold, or pressure. Nociception is the process of nociceptor activation resulting in the transmission of this signal to the central nervous system by the peripheral afferent nerves (A-delta and C nerve fibres). Nociception is responsible for our reflex responses to noxious stimuli, for example if you burn your hand in the oven, you will automatically withdraw your hand quickly from the oven before your brain has had time to consciously process the fact that your hand now hurts because it is burnt (Fig. 1). This process is not driven by a conscious behavioural choice but by a subconscious nociceptive reflex arc.

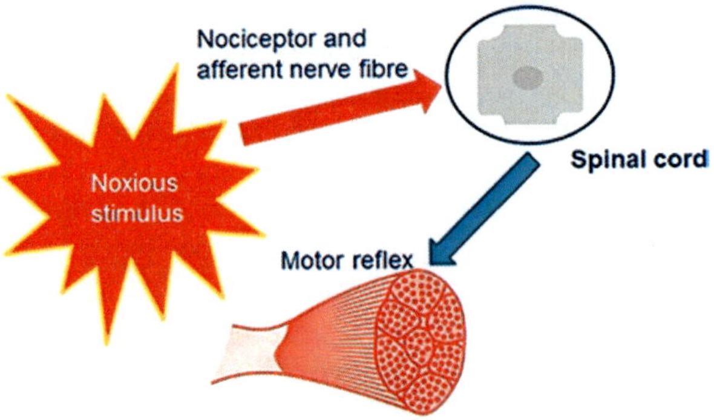

Fig. 1 The nociceptive mechanism (image copyright H. Bacon)

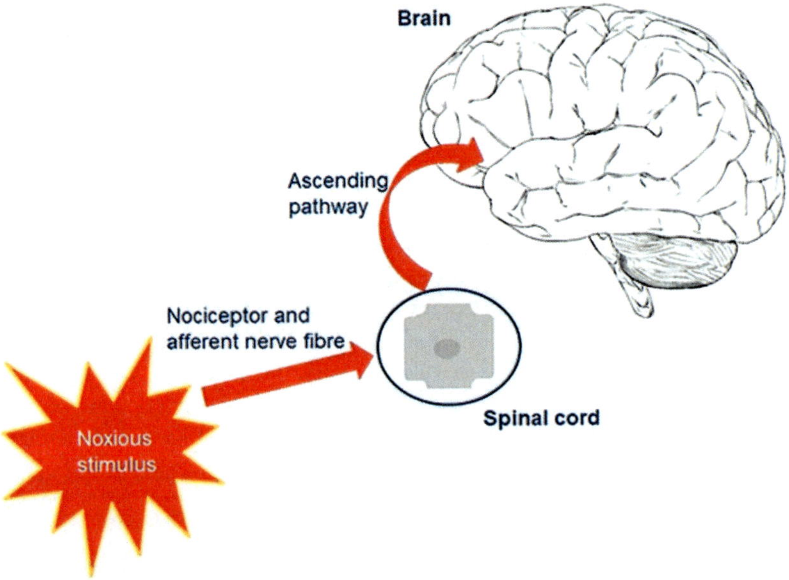

Fig. 2 The pain pathway involves both peripheral detection of noxious stimuli and central processing of this experience (image copyright H. Bacon)

2.2 Pain in Mammalian Species

The pain experience is different to nociception as pain requires the aversive or noxious experience to be processed by the brain. Pain is the experience produced by the brain when it processes the detection of the noxious stimulus; this is often a slower response than nociception (not processed by the brain). Thus, a painful experience is a product of both the *noxious stimulation of tissues* (peripheral mechanisms) and the *unconscious nociceptive processing by the spinal cord and brain* (central mechanisms) (Fig. 2). Whilst the figures in this chapter reflect the mammalian nervous system, the neurophysiological mechanisms are broadly similar across all vertebrate species.

2.3 Pain in Non-mammalian Species

Physiological and behavioural changes as a result of pain have been documented across species including invertebrates commonly used in research. For example, *C. elegans*, a nematode worm, reduces feeding and egg laying behaviours and changes both immediate and long-term behaviours in response to pain (Massingham et al. 2020). As such there is an increasing body of evidence that indicates that whilst neural structures may differ between species, vertebrates species and some invertebrates species are accepted to have (1) the neurologic components (nervous system) to respond to painful stimuli, (2) the endogenous mechanisms to modulate and reduce the pain experience (neurotransmitters) (Machin 2005), and (3) the central processing capacity to demonstrate prolonged memory and make choices to avoid pain (Elwood 2019). Comparisons between insect and mammalian neurological structures suggest functionally similar structures modulated by the same neurotransmitters. This indicates that the circuitries that existed in the common ancestry of invertebrates and vertebrates have been evolutionarily conserved (Fiore et al. 2015). Invertebrates can also demonstrate changed behaviours in response to prior experiences, suggesting some level of emotional capacity. This has been demonstrated in invertebrates including nematodes, arachnids, decapod crustaceans, and gastropods (Mason 2011; Sneddon et al. 2014; Crump et al. 2022). Additionally cognitive bias studies performed in insects suggest states of optimism and pessimism as with mammalian species and operant and classical conditioning tests successful in a range of insects. These studies suggest that invertebrates may have more significant emotional and cognitive capacities than previously thought.

Evidence for the pain experience in vertebrate non-mammals, such as fish, amphibians, reptiles, and birds, is much clearer as the neurological structures and neurotransmitter chemicals, as well as the brain structures, are functionally similar to those found in mammals, and so a similar pain experience can be assumed across the vertebrate taxa. This assumption may sometimes conflict with common husbandry pathologies. For example, when provided with a focal heat source such as heat rocks or ceramic bulb heat lamps, many reptile species will stay close enough to it to suffer thermal burns (Zwart 2001; Gartrell et al. 2020). Such burns generally occur when focal heat sources are provided without associated light. If we consider the evolutionary development and ecology of these species, it makes sense that they would be sensitive to the impacts of diffuse heat in association with light (such as that experienced under the tropical sun) but would have no evolutionary need to detect isolated focal heat sources away from a light source. Despite the lack of behavioural responses to the severe burns associated with the provision of focal heat sources, it should be assumed that the significant tissue damage is a source of pain, and a lack of evolved behavioural response to this artificially induced pathology should not be mistaken for an absence of pain experience. Zoos should consider the potential for experiencing pain in most if not all commonly kept species. This consideration should influence ethical decision-making and the humane treatment of animals regardless of their role in the zoo.

Because the pain experience is modulated centrally and involves emotional as well as sensory components, it is a complex multidimensional experience. The definition of pain used by the International Association for the Study of Pain (ISAP) is "An unpleasant sensory and emotional experience associated with, or resembling that associated with, actual or potential tissue damage" Raja et al. 2020. Pain is both a psychological and physiological experience and therefore may differ between individuals depending on physiological and emotional state. Because pain has emotional as well as sensory components, predicting or measuring pain responses can be challenging; however, there are a number of methods we can use to recognise pain in animals.

3 Pain Recognition

Whilst we cannot measure pain levels directly, and whilst animals cannot self-report their pain experience in the same way that people can, a variety of proxy-indicators can give us useful insights into the pain experience. One of the key indications that an animal may be experiencing pain is a change in behaviour and so pain should be considered when any behavioural change is noticed. Performing regular welfare assessments throughout an animal's life can provide a useful baseline of normal behaviours and physical parameters and means that any changes can be noted and investigated before they become severe (Bacon 2020). Common consequences of pain are outlined below in Table 1.

3.1 Pain Behaviour

> Pain in animals is an aversive sensory and emotional experience, which elicits protective motor and vegetative actions, results in learned avoidance and may modify species-specific traits of behaviour including social behaviour (Zimmerman 1986)

Pain is highly aversive and something that an animal tries to avoid; thus, behaviours related to the pain experience are one of the most useful tools in assessing pain. Because of this, we can use animal behaviours to evaluate whether or not an animal may be in pain, and we can apply this principle across vertebrate species, and to a number of invertebrates species also: '*Unquestionably, the experience of pain must be aversive and unpleasant to motivate a bird or any other animal to change its behavior. In plain language, pain hurts regardless of the species.*' (Paul-Murphy et al. 2004)

Common behavioural responses to pain include avoidant, aggressive, and abnormal behaviours, e.g. self-injurious or stereotypical behaviours. The behavioural signs of pain may be overt or covert and vary according to the species, sex, age, and

Table 1 Examples of the psychological, behavioural, and physiological consequences of pain (Maslak et al. 2013; Page 2013; Frank 2014; Affenzeller et al. 2017; Bacon 2018)

Psychological responses
Anxiety
Restlessness
Mental impairment
Disruption of sleep
Behavioural responses
Aggression
Stereotypy and other abnormal repetitive behaviours
Obsessive behaviours
Self-mutilation
Physiological response
Hypertension and tachycardia, stressing the heart
Impaired gastrointestinal function may lead to paralytic ileus
Decreased respiratory movement leading to pneumonia and hypoxia
Urinary retention, hypokalaemia, and hypertension
Immune system depression
Consequences of unmanaged pain
Impaired wound healing
Reduced immune function/increased susceptibility to disease
Reduced reproductive function
Behavioural abnormalities
Unpleasant emotional experience

previous experience, as well as the current situation in which the animal finds itself. Behavioural responses to pain may include an animal seeking to reduce a painful experience. For example in a case I worked on where welfare concerns for a polar bear were raised, zoo management suggested that her welfare was good as she spent much of her time 'playing under a waterfall' in her enclosure. After reviewing video footage of this behaviour, it was clear that she was not 'playing', but was actually holding her face under the falling cold water (Fig. 3). I immediately recommended a dental assessment which revealed numerous broken and infected teeth. Once these were treated with appropriate extraction and root canal therapy, her behaviour under the waterfall stopped.

The many factors that influence pain behaviours create difficulty in interpreting whether animals are painful. One reliable indicator of pain in animals is a change in their behaviour; thus, any uncharacteristic changes in behaviour should always be investigated. When looking for signs of pain, a single behaviour should never be considered in isolation but always in the context in which it occurs and in conjunction with other behaviours. A comprehensive ethogram can be useful in evaluating an animal's behavioural repertoire and any changes over time. A behavioural response involves not only what an animal does but also when, how, and where he or she exhibits the behaviour.

Biological sex and age have been shown to have an effect on pain behaviours in non-human species. In humans, males and females also differ in their response to

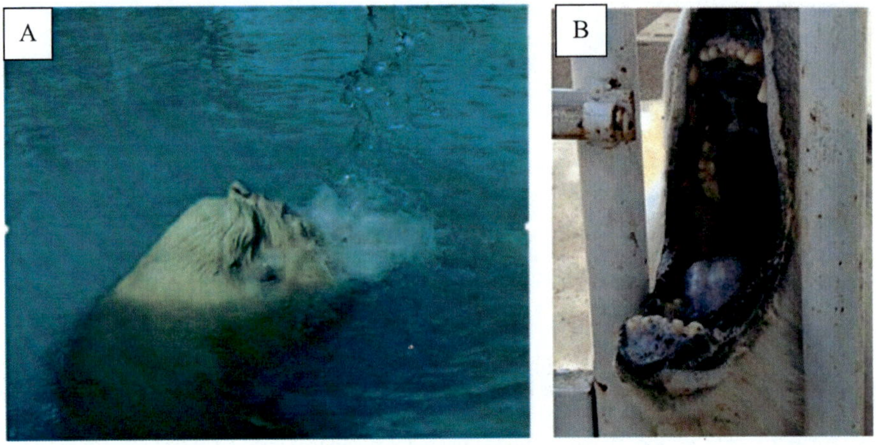

Fig. 3 (**a**) Note the bear holding her muzzle under the falling cold water; the placement of the paw on the face also suggests pain. (**b**) Training an 'open-mouth' behaviour allowed the broken and infected teeth to be visualised and a dental procedure planned (images copyright H. Bacon/ M. Whittaker)

stress and experiences of anxiety (Jones and Zachariae 2004; Witkoś and Hartman-Petrycka 2020), and this may exacerbate psychological as well as physical pain (Arendt-Nielsen et al. 2004; Pester et al. 2022). There are also marked species-specific differences in behaviour and responses to pain (Paul-Murphy et al. 2004). An individual's previous experiences of pain or stress may also influence the behavioural responses exhibited. In particular, pain experienced during the peri-natal period may predispose animals to lower pain thresholds and increased anxiety as they grow up as has been demonstrated in a range of species including but not limited to primates, canids, and ungulates (Meder 1989; McMillan 2017; Lemos de Figueiredo et al. 2021).

The environment in which the animal is assessed for pain will also affect how pain behaviour may be expressed. For example, the presence or absence of others (conspecifics as well as members of other species), familiar or unfamiliar surroundings, the weather, and novel stimuli all play a part in determining not only which behaviours are exhibited but also the duration and frequency of the behaviour. For example, many studies aimed at documenting behavioural signs of pain use remote video technology to avoid the presence of humans inhibiting the display of pain behaviours (Dalla Costa et al. 2014; Mans 2020). As such, the inability to directly observe a change in behaviour should not be taken as reliable evidence of an absence of pain. Any change in an animal's behaviour may be the first indication that an animal is experiencing pain. There are both validated and non-validated measures that can be used for pain assessment in a range of species (Table 2).

In addition to the signs outlined above, an absence of normal behaviour can indicate pain, and as already mentioned, this is why regular objective welfare assessment is a crucial component of effective pain recognition and management. For

Table 2 Pain signs applicable across a range of species

Pain sign	References
Facial expression/grimace in particular ear tension, squinted eyes, and cheek tension Tense body posture Lameness Noise sensitivity/noise phobia Vocalising Licking/nibbling/paying attention to a particular area Feather plucking Overgrooming Hiding Aggression Tooth grinding Repetitive behaviour Increased anxiety behaviour Reduced activity Stereotypy	The Glasgow composite short form pain scale (Reid et al. 2007, 2017) Helsinki Chronic Pain Index (HCPI) (Hielm-Björkman et al. 2011) The UNESP-Botucatu multidimensional composite pain scale (MCPS) (Brondani et al. 2013) The Colorado State Pain Scale (Shipley et al. 2019) AAHA/AAFP pain management guidelines for dogs and cats (Hellyer et al. 2007) Bristol Rabbit Pain Scale (Benato et al. 2021) And a variety of grimace, gait, and other pain scoring methodologies: (Mogil et al. 2020) (Graubner et al. 2011; Dalla Costa et al. 2014; de Grauw and van Loon 2016) (Flower and Weary 2009; Müller et al. 2019; Tschoner 2021) (Kohn et al. 2007, Mans 2020)

example, as animals get older their activity levels may reduce, or the diversity of activities they show may decrease. These subtle changes may go undetected for considerable time or simply attributed to 'old age' but are often symptoms of chronic pain (Bacon 2020). Such cases should always be evaluated for underlying painful conditions as the evidence suggests that often detection of painful conditions in zoo animals is delayed, to the detriment of their welfare (Föllmi et al. 2007; Bacon 2018; Krebs et al. 2018). As such it is important that zoological collections perform regular welfare assessments to ensure baseline activity levels and behaviours are established, and to aid the detection of behavioural changes that may arise as a result of pain (Bacon 2018). An approach to investigating behavioural pathology including identifying pain behaviours is shown below (Fig. 4). Further information on welfare assessments may be found in the EAZA welfare assessment library (European Association of Zoos and Aquaria 2020).

4 Categorising Pain

Pain may be categorised by the duration of the pain experience (acute or chronic) or by the type of injury it is associated with (nociceptive, inflammatory, and neuropathic). Understanding these different types of pain is critical in selecting appropriate pain mitigation strategies.

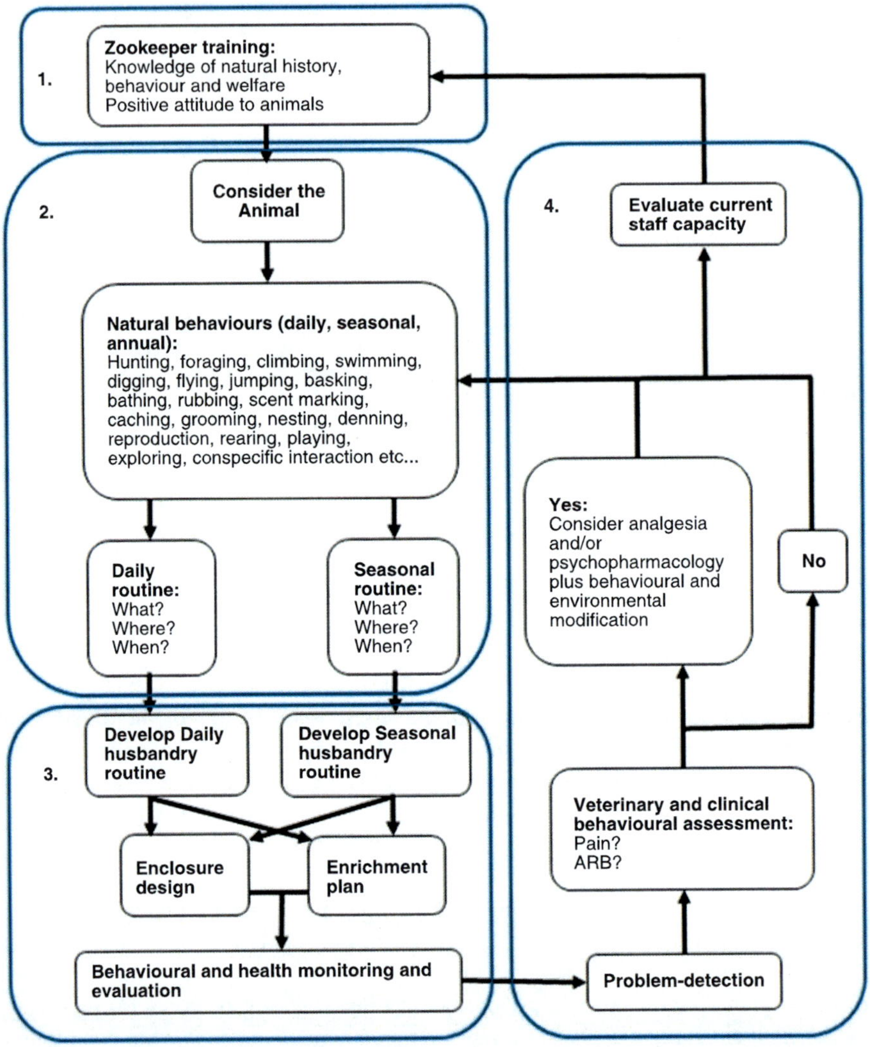

Fig. 4 An algorithm for the holistic four-stage prevention and management of behavioural pathologies based on evidence drawn from animal ecology, welfare, and clinical animal behaviour. (1) Effective and appropriate training of zookeepers to support empathetic attitudes and appropriate species-specific behavioural knowledge. (2) Consideration of the animal, its evolved and species-specific behaviours, its welfare needs, and the temporal and spatial provision of resources to it (what is provided, where is provided, and when is it provided?) (3) The development of appropriate plans and documentation to facilitate monitoring and evaluation. (4) Detection and management of mental, behavioural, or physical health problems (reproduced from Bacon, H. (2018). Behaviour-based husbandry—a holistic approach to the management of abnormal repetitive behaviors. *Animals: an open access journal from MDPI, 8*(7))

4.1 Acute and Chronic Pain

Pain may be acute (short-term) or chronic (long-term). Acute pain has a protective and evolutionary role in that when we experience it, we know that we have an injury and will take steps to protect ourselves; e.g. we will avoid walking on a broken leg. In order for pain to be functional in this way, it has to be associated with the injury and then to subside once that message has been processed by the brain. Because of this, there is a pathway from the brain to spinal cord (the descending pathway), which 'damps down' the pain signalling from the spinal cord to the brain and reduces the pain experience once the brain has processed it. This is why when you injure yourself you may experience an acute pain that gradually subsides (Fig. 5).

However, if for some reason you are not able to avoid continued activation of the peripheral nociceptors (e.g. if you are anaesthetised for surgery), and the pain signals continue to be processed by your nervous system, then a process called 'wind-up' may occur. This is where the injured nociceptors send continuous messages to the central nervous system, and this can result in process where part of the spinal cord called the 'dorsal horn' is activated to send continuous pain messages to the brain and stops responding to the descending pathway's messages to reduce the pain experience (Fig. 6).

This process means that if an animal has surgery without effective pain relief (analgesia) then it will wake up in intense pain, despite being anaesthetised, as the nerves continue to send pain messages to the brain even when the body is anaesthetised. Additionally, because the process of 'wind-up' has occurred and the animal is already in pain, any pain-relieving medication given post-operatively will

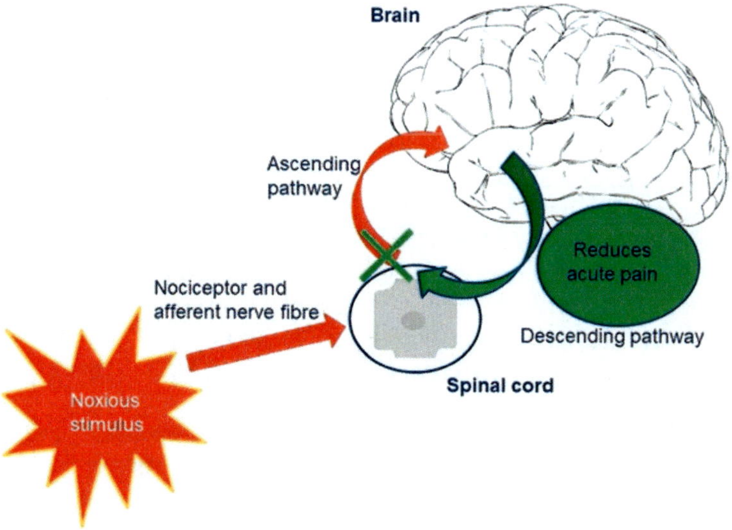

Fig. 5 The processing of painful experiences and the activation of the descending pathway to reduce the pain experience (image copyright H. Bacon)

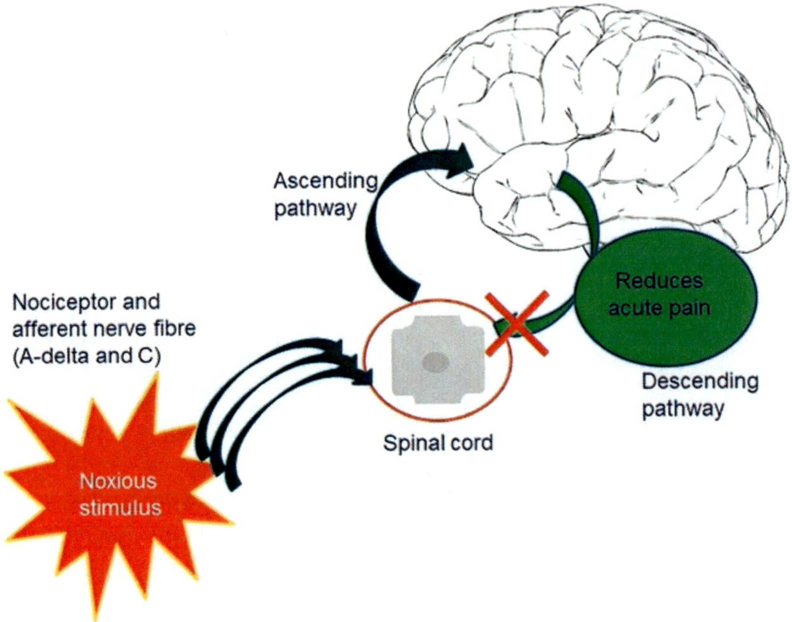

Fig. 6 Wind-up. The dorsal horn of the spinal cord is activated by continuous messages from the peripheral nervous system and stops responding to the descending pathway pain-reduction messages. This results in a more intense pain experience (image copyright H. Bacon)

be less effective. This is why it is so important that animals are given appropriate pain relief before or during surgery, as this can help to stop wind-up from occurring. Administering pain relief after surgery is helpful but is less effective than pre-operative analgesia as 'wind-up' will have already occurred, and the CNS has been sensitised. Surgical pain is 100% predictable and so planned pain management strategies are an integral part of every surgical procedure.

However, there are many cases where pain is not planned—this may be due to a sudden injury or the development of disease processes in the body. Some of these processes may result in chronic (long-term) pain as ongoing nociceptor and peripheral nerve activity leads to chronic dorsal horn activation in the spinal cord and an ineffective descending pathway response. For example, in untreated dental disease or osteoarthritis, the damaged tissue will result in chronic nociceptor activation resulting in chronic dorsal horn activation and a continuous pain experience. Detecting chronically painful conditions can be much more challenging than detecting acute pain, and most animals living with chronic pain will modify their behaviour in order to try and cope with their pain experience. Such animals will reduce their activity, may rest more, and may be more irritable. However, they may also still display highly motivated behaviours (animals with severe dental disease will still eat, dogs with osteoarthritis will still want to chase balls despite pain in their joints) and so it is important to evaluate overall activity and not to assume that the continuation of one or two highly valued activities represents a lack of pain. Perhaps counterintuitively, animals with chronic pain may also increase their activity. For example abnormal repetitive behaviours are associated with chronic pain in bears and other mammals (Swaisgood and Shepherdson 2005; Maslak et al. 2013), and birds may demonstrate feather plucking in response to pain (Malik and Valentine 2018). Chronic pain is pathological and does not serve any protective function in the way that acute pain does. For example, acute pain causes us to rest and immobilise an injury such as not walking on a broken leg and allowing it to heal, but chronic pain such as that experienced in osteoarthritis does not allow for recovery as it is caused by a degenerative and irreversible disease process.

4.2 Types of Pain

In addition to the duration of pain experiences, we can classify pain according to the type of pain it is. Pain may arise due to tissue damage such as injury or surgery (nociceptive pain), tissue inflammation such as swelling associated with infection (inflammatory pain), or damage to the actual nerves in the body (neuropathic pain). Importantly, these different types of pain may occur at the same time. For example, if an animal has dental disease and infection causing inflammation, they will experience inflammatory pain with redness and swelling of the gums, and if they then fracture a tooth, they will experience nociceptive pain, in relation to that injury and exposure of the sensitive tooth pulp. They may also experience neuropathic pain, particularly if infection sets in and the nerves supplying the tooth become chronically irritated. Similarly, an animal with osteoarthritis of the spine may experience nociceptive pain as the joints between rub together abnormally and cause damage to the joint tissues, but also neuropathic pain as the nerves coming from the spine are squeezed by the bony arthritic changes.

These different types of pain are important to understand when choosing medications to reduce pain (analgesics) as different analgesic drugs will have different levels of efficacy for different types of pain and choosing a medication which is not effective for the type of pain you are trying to treat may mean that the treatment is ineffective and the animal's pain experience continues. In particular, many common pain-relieving medications are not effective against neuropathic pain, and so it should be considered if an animal does not appear to be responding to pain-relieving medication.

5 Prevention and Management of Pain

The importance of preparing for planned painful procedures and administering analgesia for predictable pain is discussed above in the acute pain section. A wide range of analgesic medications are available and should be selected and administered in accordance with veterinary advice and local drug regulations. Pain should always be prevented and treated as a priority, to safeguard animal welfare. An animal cannot be experiencing good welfare if it is in pain as the negative consequences of pain are significant. Pain should never be used as a mechanism to attempt to reduce activity or 'rest' an animal. As already outlined in Table 1, there are multiple negative physical and psychological impacts of pain including reduced resting/sleeping and impaired wound healing and so using pain to restrict an animal's activity is both unethical and may create further health and welfare problems. Figure 4 outlines an approach to managing behavioural pathologies including those caused by pain, and provides a useful framework for addressing these symptoms, but more specific information on options for treating pain is given below.

5.1 Pharmacotheraputics

Some examples of medical therapies for different types of mammalian pain are shown in Table 3. In addition to medication, it is important that steps are taken to address any causes of pain; e.g. therapeutic dentistry should be performed if dental pain is suspected and hoof trimming should be performed if it may be contributing to lameness.

Treatment options in other species may be more limited, but analgesic drugs and doses can be found in the literature for a range of amphibians, fish, reptiles, and birds (e.g. Carpenter and Marion 2017).

In all cases veterinary clinicians should aim to use a multimodal approach to analgesia, i.e. using combinations of drugs that work together to more effectively reduce the pain experience whilst minimising side effects (Mathews et al. 2014). Figure 7 shows the neurophysiological targets of some different types of analgesic drugs. By using a combination of drugs, different locations on the pain pathway are targeted and more effective analgesia is achieved.

5.2 Complementary Therapies

In addition to medical therapies, there are a number of complementary approaches to mitigating the animal's pain experience depending on the type of pain and the underlying disease process. For example, one of the best preventative and

Table 3 Types of pain and medications which may be effective as analgesics for different pain types

Nociceptive	Inflammatory	Neuropathic
Non-steroidal anti-inflammatory drugs (NSAIDs)	Non-steroidal anti-inflammatory drugs (NSAIDs)	Gabapentin
Opioids	Steroids	Amitriptyline
Tramadol	Opioids	Nortriptyline
Paracetamol	Paracetamol	Local anaesthetic
Local anaesthetic		Ketamine
Ketamine		

Note efficacy, dose, delivery, and safety of different medications will vary between species. Seek veterinary advice

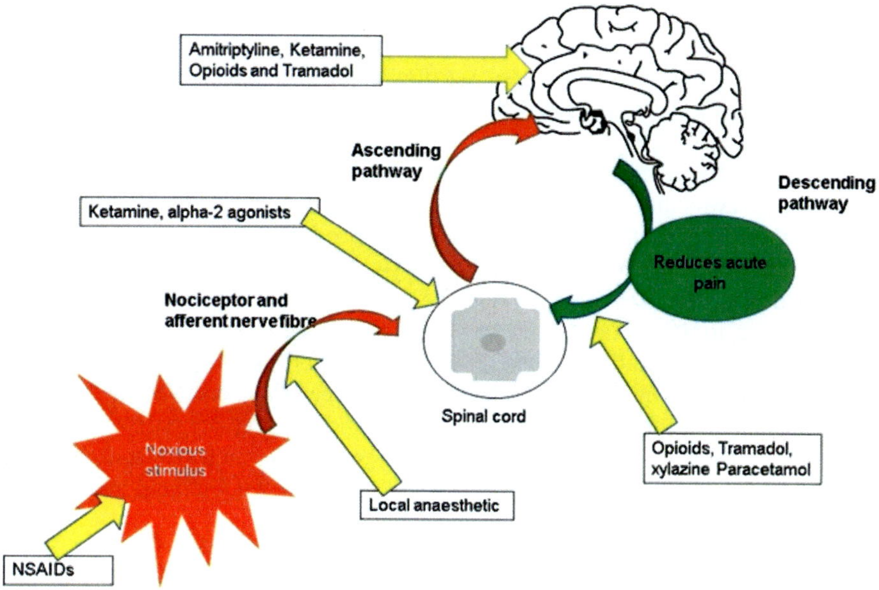

Fig. 7 The different targets of different types of analgesic drugs along the pain pathway illustrating the efficacy of multimodal analgesia at target pain in multiple locations (image copyright H. Bacon)

supportive therapies to prevent the onset of and reduce the pain experience in osteoarthritic animals is weight loss. Excess fat deposits can increase the risks of other diseases such as diabetes or heart disease and reduce lifespan and quality of life. Poor muscle tone and strength also contribute to the ongoing development of joint pathology which may predispose to osteoarthritis. The collection of baseline data is important in detecting degenerative changes and so regular programmes of body condition scoring and/or weighing should be used to monitor zoo animals throughout their lives and ensure they are within health limits for the species and retaining active muscle mass (bearing in mind that some species such as bears should exhibit significant seasonal fluctuations (Maher et al. 2021)). Physical therapy (Wolfe et al. 2015; Goldberg 2019; Nakatani et al. 2019; Shipton and van Herpt 2022) including targeted massage and specific therapeutic activities have been used in both zoo (Komodo dragons, Coatis, Asiatic black bears) and domestic species (dogs, horses) to support recovery from surgery and maintain activity despite chronic illness.

Nutraceuticals or nutritional supplements including glucosamine-chondroitin, green-lipped mussels, and omega-3 fatty acids have been shown to improve mobility in dogs experiencing osteoarthritis (Johnson et al. 2020) and may reduce joint inflammation in horses (Goodrich and Nixon 2006; van de Water et al. 2017) and may also be useful in zoo species. See chapter 'Supporting Geriatric Zoo Animal Welfare Through Nutrition' for more on nutrition.

Low-level laser therapy (Ritzman 2015; Goldberg 2019; Sparrow 2020) has biomodulation properties, and promotes wound healing and cell renewal as well as reduces the pain experience in animals with tissue injury or underlying arthritis. Both traditional and electro-acupuncture are also documented in a range of domestic and zoo species including rabbits, rodents, elephants, tigers, birds, and snakes (Harrison and Churgin 2022) and may be helpful for alleviating painful conditions and improving mobility.

The social and environmental provisions for painful animals should also be considered. Because the pain experience is partially modulated by the emotional response to pain, enhancing an animal's emotional state may mitigate their experience of pain. As such it is important that an animal's comfort is considered in a planned painful procedure. For example, animals recovering from surgery should be recovered in a comfortable environment, where they can adequately rest and recuperate without additional stress. Supporting positive emotional experiences has also been shown to mitigate chronic pain experiences (Vachon et al. 2013; Bacon et al. 2019; Kimura et al. 2022), and so it is important that animals with painful conditions have good social relationships with conspecifics and keepers, and enjoy a range of enrichment activities to enhance their emotional wellbeing and reduce their pain experience.

6 Quality of Life and End-of-Life Decision-Making for Painful Animals

In addition to the significant physical impacts of unmanaged pain, it can have significant behavioural impacts as described above, and these behavioural changes may be disruptive in socially housed animals, and lead to social conflict. It is important that animals experiencing pain are treated to mitigate physical, psychological, and social impacts and to ensure their welfare does not become poor. Even in cases where pain is adequately managed, some animals may still require modified environments, particularly ageing animals who may be experiencing a number of co-morbidities (multiple disease problems). In some cases, the needs of ageing animals might be easily accommodated with environmental modifications to prevent slipping, and ramps to ease access. In cases where this is not possible, where the use of resources to support ageing populations is not viable, or where there are welfare implications of removing animals from social groups, the animal's overall welfare should be evaluated, and alternative approaches to managing chronic pain, including euthanasia, should be considered (also see chapters 'Facility Design for Disabled and Ageing Wild Animals,' 'Environmental Enrichment for Ageing Zoo Animals,' 'Veterinary Care of Ageing Zoo Animals: Description of a Proactive Approach,' 'Physiotherapy and Management of the Musculoskeletal Health of Ageing Wild Animals in Human Care,' and 'Euthanasia of Geriatric Zoo Animals: Decision-Making and Procedure').

Particular attention should be paid to ageing animals as whilst 'long-lived' animals may be publicly celebrated in zoos, longevity is not in itself an indicator of good welfare. Ageing or 'senescence' is a normal physiological process resulting in reduced function of all body systems and eventual death. Old age is not a disease or pathology in itself, but increasing age does predispose to pathological conditions. Many of these diseases may negatively impact upon animal welfare and may occur concurrently with the pain experience, e.g. cognitive decline, reduced physical strength, and increased anxiety.

Euthanasia of animals in pain is a valid approach to alleviating current suffering, but also to preventing suffering from occurring in the first place. People tend to be more comfortable with making decisions about death that they feel are more easily ethically 'justifiable', for example, killing to end suffering rather than killing a healthy animal for population management. However, if we always wait for suffering to occur before making a euthanasia decision, the animal will, by definition, be suffering which is problematic for animal welfare. Rather than waiting for this and making a reactive decision in response to animal suffering, preventative forward-planning of euthanasia decision-making strategies can help to make difficult euthanasia decisions more objective and transparent, and ease the ethical discomfort of animal care staff (Walraven et al. 2018).

Being aware of our own biases and ethical responses to euthanasia decision-making is important. Attitudes towards the acceptability of euthanasia varies by taxa (Powell and Ardaiolo 2016), and people are more empathetic to species more closely related to themselves (e.g. primates) or to those perceived as attractive. These are often the charismatic species where euthanasia decision-making is more difficult, and may be delayed to the detriment of animal welfare (Föllmi et al. 2007). Objective pain management and euthanasia decision-making criteria may help to make decisions based on objective animal welfare indicators rather than on subjective ethical opinions.

7 Summary

Pain is an unpleasant emotional and physiological experience which can potentially affect the majority of, if not all, species housed in zoos. As such we have a duty of care to ensure that planned painful experiences are mitigated and that there is a proactive approach to pain prevention, recognition, and treatment across the animal collection. Pain recognition and reporting is an important part of staff training, and pain management should be holistic and incorporate multiple aspects of the animal's healthcare and husbandry provision, including multimodal analgesia, complementary therapies, enrichment and environmental modifications, and positive social interactions.

References

Affenzeller N, McPeake KJ, McClement J, Zulch H (2017) Human-directed aggressive behaviour as the main presenting sign in dogs subsequently diagnosed with diskospondylitis. Vet Rec Case Rep 5(4)

Arendt-Nielsen L, Bajaj P, Drewes AM (2004) Visceral pain: gender differences in response to experimental and clinical pain. Eur J Pain 8(5):465–472

Bacon H (2018) Behaviour-based husbandry—A holistic approach to the management of abnormal repetitive behaviors. Animals 8(7)

Bacon H (2020) Quality of life assessments. Paper presented at the EAZA annual conference, online. https://www.youtube.com/watch?v=0U82GfqCqg4&t=0s

Bacon H, Walters H, Vancia V, Connelly L, Waran N (2019) Development of a robust canine welfare assessment protocol for use in dog (Canis familiaris) catch-neuter-return (CNR) programmes. Animals 9(8):564

Benato L, Murrell J, Knowles TG, Rooney NJ (2021) Development of the Bristol Rabbit Pain Scale (BRPS): A multidimensional composite pain scale specific to rabbits (Oryctolagus cuniculus). PLoS One 16(6)

Brondani JT, Mama KR, Luna SPL, Wright BD, Niyom S, Ambrosio J, Padovani CR (2013) Validation of the English version of the UNESP-Botucatu multidimensional composite pain scale for assessing postoperative pain in cats. BMC Vet Res 9(1):143

Carpenter JW, Marion C (2017) Exotic animal formulary-E-book: Elsevier Health Sciences

Crump A, Browning H, Schnell A, Burn C, Birch J (2022) Sentience in decapod crustaceans: A general framework and review of the evidence. Anim Sentience 7(32):1

Dalla Costa E, Minero M, Lebelt D, Stucke D, Canali E, Leach MC (2014) Development of the Horse Grimace Scale (HGS) as a pain assessment tool in horses undergoing routine castration. PLoS One 9(3)

de Grauw JC, van Loon JPAM (2016) Systematic pain assessment in horses. Vet J 209:14–22

Elwood RW (2019) Assessing the potential for pain in crustaceans and other invertebrates. In: Carere C, Mather J (eds) The welfare of invertebrate animals. Springer, Cham, pp 147–177

European Association of Zoos and Aquaria (2020) EAZA animal welfare. Retrieved from https://www.eaza.net/about-us/areas-of-activity/animal-welfare/

Fiore VG, Dolan RJ, Strausfeld NJ, Hirth F (2015) Evolutionarily conserved mechanisms for the selection and maintenance of behavioural activity. Philos Trans Roy Soc B Biol Sci 370(1684)

Flower FC, Weary DM (2009) Gait assessment in dairy cattle. Animal 3(1)

Föllmi J, Steiger A, Walzer C, Robert N, Geissbühler U, Doherr M, Wenker C (2007) A scoring system to evaluate physical condition and quality of life in geriatric zoo mammals. Anim Welf 16(3):309–318

Frank D (2014) Recognizing behavioral signs of pain and disease: a guide for practitioners. Vet Clin N Am Small Anim Pract 44(3):507–524

Gartrell BD, Ahn JY, Khude R, Dougherty N, Johnson K, McCutchan J, Hunter S (2020) Thermal burns of the spectacle associated with supplementary heating in native New Zealand geckos. N Z Vet J 68(2):126–133

Goldberg ME (2019) A walk on the wild side: a review of physiotherapy for exotics and zoo animals. Vet Nurs J 34(2):33–47

Goodrich LR, Nixon AJ (2006) Medical treatment of osteoarthritis in the horse—a review. Vet J 171(1):51–69

Graubner C, Gerber V, Doherr M, Spadavecchia C (2011) Clinical application and reliability of a post abdominal surgery pain assessment scale (PASPAS) in horses. Vet J 188(2):178–183

Harrison TM, Churgin SM (2022) Acupuncture and traditional chinese veterinary medicine in zoological and exotic animal medicine: a review and introduction of methods. Vet Sci 9(2):74

Hellyer P, Rodan I, Brunt J, Downing R, Hagedorn JE, Robertson SA (2007) AAHA/AAFP pain management guidelines for dogs and cats. J Feline Med Surg 9(6):466–480

Hielm-Björkman AK, Kapatkin AS, Rita HJ (2011) Reliability and validity of a visual analogue scale used by owners to measure chronic pain attributable to osteoarthritis in their dogs. Am J Vet Res 72(5):601–607

How is pain assessed? (2010) AWIN [Mobile application software]

Johnson KA, Lee AH, Swanson KS (2020) Nutrition and nutraceuticals in the changing management of osteoarthritis for dogs and cats. J Am Vet Med Assoc 256(12):1335–1341

Jones A, Zachariae R (2004) Investigation of the interactive effects of gender and psychological factors on pain response. Br J Health Psychol 9(3):405–418

Kimura LF, Novaes LS, Picolo G, Munhoz CD, Cheung CW, Camarini R (2022) How environmental enrichment balances out neuroinflammation in chronic pain and comorbid depression and anxiety disorders. Br J Pharmacol 179(8):1640–1660

Kohn DF, Martin TE, Foley PL, Morris TH, Swindle MM, Vogler GA, Wixson SK (2007) Guidelines for the assessment and management of pain in rodents and rabbits. J Am Assoc Lab Anim Sci 46(2):97–108

Krebs BL, Marrin D, Phelps A, Krol L, Watters JV (2018) Managing aged animals in zoos to promote positive welfare: a review and future directions. Animals 8(7):116

Lemos de Figueiredo R, Hartley M, Fletcher AW (2021) Assessing the behaviour, welfare and husbandry of mouse deer (Tragulus spp.) in European zoos. Appl Anim Behav Sci 237:105283

Machin KL (2005) Avian pain: physiology and evaluation. Compend Contin Educ Pract Vet 27(2):98–109

Maher CJ, Gibson A, Dixon LM, Bacon H (2021) Developing a reliable welfare assessment tool for captive hibernatory bear species. Animals 11(11):3090

Malik A, Valentine A (2018) Pain in birds: a review for veterinary nurses. Vet Nurs J 33(1):11–25

Mans C (2020) Insights into pain assessment and management in rabbits. Vet Rec 186(18):600–602

Maslak R, Sergiel A, Hill SP (2013) Some aspects of locomotory stereotypies in spectacled bears (Tremarctos ornatus) and changes in behavior after relocation and dental treatment. J Vet Behav 8(5):335–341

Mason GJ (2011) Invertebrate welfare: where is the real evidence for conscious affective states? Trends Ecol Evol 26(5):212–213

Massingham N, Schafer WR, Neely GG, Chew YL (2020) Invertebrate models of nociception. In: Wood JN (ed) The Oxford handbook of the neurobiology of pain. Oxford University Press, Oxford, p 61

Mathews K, Kronen PW, Lascelles D, Nolan A, Robertson S, Steagall PV et al (2014) Guidelines for recognition, assessment and treatment of pain. J Small Anim Pract 55(6)

McMillan FD (2017) Behavioral and psychological outcomes for dogs sold as puppies through pet stores and/or born in commercial breeding establishments: Current knowledge and putative causes. J Vet Behav 19:14–26

Meder A (1989) Effects of hand-rearing on the behavioral development of infant and juvenile gorillas (Gorilla g. gorilla). Dev Psychobiol 22(4):357–376

Mogil JS, Pang DSJ, Silva Dutra GG, Chambers CT (2020) The development and use of facial grimace scales for pain measurement in animals. Neurosci Biobehav Rev 116:480–493

Müller BR, Soriano VS, Bellio JCB, Molento CFM (2019) Facial expression of pain in Nellore and crossbred beef cattle. J Vet Behav 34:60–65

Nakatani JY, Greer LL, Wilson GL, Cabrera SY (2019) Surgical and postoperative treatment of intervertebral disc disease in a Mountain Coati (*Nasuella olivacea*) using behavioral management and physical therapy. J Zoo Wildlife Med 50(3):727–730, 724

Page GG (2013) The immune-suppressive effects of pain. Madame Curie Bioscience Database [Internet]. Retrieved from https://www.ncbi.nlm.nih.gov/books/NBK6140/

Paul-Murphy J, Ludders JW, Robertson SA, Gaynor JS, Hellyer PW, Wong PL (2004) The need for a cross-species approach to the study of pain in animals. J Am Vet Med Assoc 224(5):692–697

Pester BD, Crouch TB, Christon L, Rodes J, Wedin S, Kilpatrick R et al (2022) Gender differences in multidisciplinary pain rehabilitation: The mediating role of pain acceptance. J Contextual Behav Sci 23:117–124

Powell DM, Ardaiolo M (2016) Survey of US zoo and aquarium animal care staff attitudes regarding humane euthanasia for population management. Zoo Biol 35(3):187–200

Raja SN, Carr DB, Cohen M, Finnerup NB, Flor H, Gibson S, Keefe FJ, Mogil JS, Ringkamp M, Sluka KA, Song XJ, Stevens B, Sullivan MD, Tutelman PR, Ushida T, Vader K (2020) The revised International Association for the Study of Pain definition of pain: concepts, challenges, and compromises. *Pain 161*(9):1976–1982. https://doi.org/10.1097/j.pain.0000000000001939

Reid J, Nolan A, Hughes J, Lascelles D, Pawson P, Scott E (2007) Development of the short-form Glasgow Composite Measure Pain Scale (CMPS-SF) and derivation of an analgesic intervention score. Anim Welf 16:97

Reid J, Scott E, Calvo G, Nolan A (2017) Definitive Glasgow acute pain scale for cats: validation and intervention level. Vet Rec 108(18)

Rioja-Lang F, Bacon H, Connor M, Dwyer CM (2020) Prioritisation of animal welfare issues in the UK using expert consensus. Vet Rec 187(12):490–490

Ritzman TK (2015) Therapeutic laser treatment for exotic animal patients. J Avian Med Surg 29(1):69–73, 65

Shipley H, Guedes A, Graham L, Goudie-DeAngelis E, Wendt-Hornickle E (2019) Preliminary appraisal of the reliability and validity of the Colorado State University Feline Acute Pain Scale. J Feline Med Surg 21(4):335–339

Shipton K, van Herpt S (2022) A physiotherapy program for bears at animals Asia's Vietnam Bear Rescue Centre. IAABC J 22:1

Sneddon LU, Elwood RW, Adamo SA, Leach MC (2014) Defining and assessing animal pain. Anim Behav 97:201–212

Sparrow S (2020) Clinical application of photobiomodulation therapy in a zoological setting. Vet Nurse 11(10):460–464

Swaisgood RR, Shepherdson DJ (2005) Scientific approaches to enrichment and stereotypies in zoo animals: what's been done and where should we go next? Zoo Biol 24(6):499–518

Tschoner T (2021) Methods for pain assessment in calves and their use for the evaluation of pain during different procedures—a review. Animals 11(5):1235

Vachon P, Millecamps M, Low L, Thompsosn SJ, Pailleux F, Beaudry F et al (2013) Alleviation of chronic neuropathic pain by environmental enrichment in mice well after the establishment of chronic pain. Behav Brain Funct 9(1):22

van de Water E, Oosterlinck M, Dumoulin M, Korthagen NM, van Weeren PR, van den Broek J, van Doorn DA (2017) The preventive effects of two nutraceuticals on experimentally induced acute synovitis. Equine Vet J 49(4):532–538

Walraven E, Andrew P, Boyle N, Duffy S (2018) Euthanasia and small-population management: development of transparent processes, good governance and a practical application. Int Zoo Yearb 52(1):258–266

Witkoś J, Hartman-Petrycka M (2020) Gender Differences in Subjective Pain Perception during and after Tattooing. Int J Environ Res Public Health 17(24):9466

Wolfe TC, Stringer E, Krauss S, Trout T (2015) Physical therapy as an adjunctive treatment for severe osteoarthritis in a Komodo Dragin (*Varanus komodoensis*). J Zoo Wildlife Med 46(1):164–166

Zimmerman M (1986) Physiological mechanisms of pain and its treatment. Klinische Anaesthesiol Intensivether 32:1–19

Zwart P (2001) Pathophysiology: Assessment of the husbandry problems of reptiles on the basis of pathophysiological findings: A review. Vet Q 23(4):140–147

Physiotherapy and Management of the Musculoskeletal Health of Ageing Wild Animals in Human Care

Matthew Shackleton and Louise Lefrere

M. Shackleton (✉) · L. Lefrere
Shackleton Veterinary Physiotherapy, Chester, UK
e-mail: info@shackletonvetphysio.com

S. Brando, S. Chapman (eds.), *Optimal Wellbeing of Ageing Wild Animals in Human Care*, https://doi.org/10.1007/978-3-031-30659-4_10

Mauricette. Mauricette is a 15-year-old red panda (*Ailurus fulgens*) living at Zoo de Bordeaux Pessac in France. She has a curious personality and likes exploring her habitat and interacting with her friend and partner Maurice. After the refurbishment of their habitat, she kept exploring all new areas and heights. She was a very good mother to her babies, easygoing, and showing and teaching them about climbing and other important aspects of being successful in life. Initially, Maurice was not allowed near the youngsters, as she was very protective of her cubs. Her last baby was Makalu who was born during the pandemic. Mauricette prefers all the young and small leaves of bamboo, discarding the larger older ones. Her favourite treats are raisins. Her favourite sleeping spot, curled in a ball with her tail over her head, is at the top of a high tree. Sometimes she can be a bit cheeky, making soft calls calling over Maurice, but when he arrives, she jumps on him and pushes him around. Over the years she has changed her behaviour towards Maurice and her caregivers, being more reserved and spending time on her own. However, when it rains a lot, both will be seen sleeping together in their undercover area. She continues to observe everything and everyone around her, in particular what the caregivers have put into her bowl. If it is not up to par, she stays up in the trees! She currently does not require specific care for her age. She is very active, has a nice healthy coat, and is using all the places in the habitat. Every evening the zoo's peacock 'Panette' is seen to sleep up in the trees with both pandas.

Abstract

Physiotherapy and occupational therapy are an intrinsic part of the treatment and management of numerous types of disease in the human patient. It is still in its infancy in veterinary sciences with very few countries even having qualified veterinary physiotherapists. This chapter discusses the current regulation of this role worldwide, the variation in regulation and training standards, how this field is developing both in domestic animals and at its beginnings in exotic animals. It covers approaches in human medicine and general veterinary medicine and some early examples of physiotherapy being used with wild animal patients. This chapter makes suggestions to facilitate a thorough assessment of ageing wild animals and how best practice zoo protocols can be utilised to successfully apply low-stress rehabilitation. Protocols follow the human model of exercise prescription, lifestyle and environmental adaptation, and behavioural health. The focus should be creating an environment for lasting change rather than just treating symptoms. Case studies with some early examples of physiotherapy being used with wild animal patients are included.

Keywords

Veterinary physiotherapy · Physical therapy · Rehabilitation · Exercise prescription · Manual therapy · Osteoarthritis

1 Introduction

As animals in our care age, they become susceptible to age-related diseases, which may lead to suffering. In certain circumstances, it can be a challenge for animal care professionals to know how best to manage them and continue to provide the highest standard of care. There are many lessons to be learnt from research in ageing human patients about how a holistic approach of exercise, lifestyle management, environment, and attention to mental health can have a lasting impact on the management of musculoskeletal disease. These approaches are beginning to be reflected in recent research in veterinary sciences and are showing lasting results in our domestic animals. Can these approaches be successfully replicated in our population of ageing captive wild animals? Physiotherapy is still a developing field in veterinary sciences, and in this chapter, we review the current scientific approaches used in the human and veterinary fields and consider their use with captive elderly exotic animals in our care.

2 Musculoskeletal Therapy for Animals

The musculoskeletal system of animals includes soft tissues (muscles, tendons, ligaments, fascia, and skin) and hard tissues (bones and joints). As an animal ages, he or she may experience a host of different afflictions and diseases, such as osteoarthritis (degenerative joint disease) and sarcopenia (loss of muscle mass), which may result in discomfort and pain. If an underlying condition causing inflammation is not resolved, the joint inflammation (arthritis) becomes chronic and leads to osteoarthritis; this progressively leads to further damage of the joint structures, which reduces the smooth range of motion of the joint and causes pain (Loeser 2010). Weakening of the ligaments surrounding the joint is also common, which increases joint instability and further exacerbates osteoarthritis (Loeser 2010). Combined, these changes increase pain and cause a decrease in function and range of motion (ROM), which again alters locomotion and behaviour (Aulakh et al. 2020; Ramírez-Flores et al. 2017). Overall, in cases of acute or chronic musculoskeletal pathology, veterinary physiotherapists (VP) must focus on controlling pain and inflammation and restoring normal movement in order to allow healing, prevent the occurrence or progression of osteoarthritis through degeneration due to inflammation, and manage current pathology (Kennedy et al. 2018; Kirkby-Shaw et al. 2020; Looney et al. 2018).

The prevalence of musculoskeletal pathologies in captive wild animals is a complex matter. The development of osteoarthritis is multifactorial and degenerative, and there is a considerable inflammatory component (Tangredi and Lawler 2019). Osteoarthritis has been identified in a variety of species in zoos (Krebs et al. 2018), but there are many deficits in our knowledge and understanding of its occurrence. For example, the occurrence of osteoarthritis in captive animals may be due to the increased longevity of many captive species within our care, alongside age-related degradation of tissue. However, studies have commonly found osteoarthritis in young animals who would not normally be expected to develop these conditions, such as giraffes (Dadone 2019). Other factors such as reduced activity levels, physical fitness, behavioural changes, and environment could all play a major part in the occurrence of these pathologies (Krebs et al. 2018). Managing captive wild animals brings with it the added complexity of not only caring for varying taxa and species, but individuals with drastically different adaptations brought about via adaptive pressures of many varying habitats. In conjunction with limited research in this field, it can be difficult to gain a clear picture of causation across all animals in our care, or the efficacy of interventions to improve welfare. However, when musculoskeletal disease is diagnosed early, we have more of an opportunity to improve animal welfare and delay disease progression.

Due to natural protective bodily responses, pain may reduce or change the normal movement of the musculoskeletal system to avoid possible harm to painful areas (Haussler 2018); this can be termed 'compensatory gait'. If the pathology is not resolved quickly or becomes chronic, the changes to central pattern generators that initially allowed pain relief through compensatory gait become normal for the animal, leading to soft tissue asymmetry and an asymmetric or lame gait. Over time, these soft tissue changes can be followed by hard tissue changes due to the changes in weight-bearing, as the musculoskeletal system increases bone density in areas of increased weight-bearing; this again increases musculoskeletal asymmetry. This can lead to excessive or asymmetrical loading of joints in other areas, which alters the bearing surfaces, and can predispose them to high levels of percussive force (Dykgraaf et al. 2008; MacNicol et al. 2018; Ramírez-Flores et al. 2017). Where pain is present, pain management is vital during rehabilitation to allow healing and compliance with the treatment plan (Kirkby-Shaw et al. 2020) and reduce suffering. From both an ethical and practical point of view, using analgesia to allow pain-free, safe, and controlled movement for rehabilitation is important. As gait can be normalised more quickly with combined analgesia and prescription of a specific exercise programme (see below for more details), rehabilitation should also be more effective.

Achieving success in veterinary physiotherapy is a team effort, with all external and internal specialists working together. Initially, diseases and disorders of the musculoskeletal system are treated by veterinary surgeons, and subsequent complementary treatment for rehabilitation may then be carried out by other musculoskeletal therapists. These can include animal hydrotherapists, animal osteopaths, animal chiropractors, and veterinary or others specialised in animal physiotherapy/physical therapy. Whilst the treatment aims of these professionals may be similar, the techniques used vary. Animal osteopaths and chiropractors focus on manual therapy: using their hands to assess and treat (Haussler et al. 2021), and manual manipulation and mobilisation of hard and soft tissues are used to treat musculoskeletal pathology (Haussler et al. 2021). Animal osteopathy may involve the use of sedation or anaesthesia during treatment (Colles et al. 2014), which may the possibility to impact treatment viability in easily stressed wild animals. Additional exercise-based treatment techniques may also be used, although more emphasis is usually placed on manual therapy. VPs frequently use exercise-based treatment to target pathology (Dybczyńska et al. 2022) as well as manual therapies, but the level of force applied is less than by chiropractors and osteopaths.

Effective team communication regarding assessment findings, medication, treatment recommendations, contraindications, and protocols must take place as a combined approach will improve the successful outcome of the case. Case notes should be documented and made available to the referring veterinarian or specialist, and information on exercise prescription and environmental and management recommendations should be provided for the animal's caregivers. During treatment, caregiver communication with veterinarians and VPs regarding the animal's progress is also vital, particularly if there are further questions regarding treatment or difficulties with treatment implementation. Understanding the progression of osteoarthritis reinforces the need for continual musculoskeletal and behavioural assessment from animal caregivers. Early identification of osteoarthritis can facilitate early treatment and preventative measures to break the cyclic inflammatory processes that progress the condition. Care staff should utilise regular observational and recording techniques to identify signs of lameness and/or behavioural changes that may warrant closer assessment. The approach to assessment can be scaled depending on practicality, heightened risk factors, or historical observations. This could mean that less time-intensive assessment methods, e.g. visual gait assessment, may be carried out on a more regular basis and other more intensive methods, e.g. video recording, functional tasks, or behavioural data collection, may be carried out at an interval which is realistic for the collection and staffing levels. Continual assessment will help to facilitate continuity in communication between keeping and veterinary staff, which is an important feature in the management of any case. If keepers can identify early presentation and provide a greater level of evaluation (as discussed above), this will help to inform veterinarians of current presentation and provide a wealth of detail that may otherwise be difficult to obtain during a consultation or due to time constraints.

3 History of Human and Veterinary Physiotherapy

Physiotherapy is a long-established therapy in human medicine with numerous specialisms and a large evidence base to advise treatment protocols. A multi-faceted treatment approach is used with doctors, physiotherapists, occupational therapists, podiatrists, and cognitive behavioural therapists working in conjunction to support patients with musculoskeletal pathology and any associated pain(s). Veterinary physiotherapy has only been established over the last 30 years with practitioners worldwide (McGonagle et al. 2015). Veterinary physiotherapy training and practice is mainly focused on musculoskeletal and neurological conditions in domestic animals, e.g. horses and dogs, and occasionally small animals, livestock, and wild animals in captivity. In this rapidly evolving field, most equine and small animal referral centres now employ VPs to provide rehabilitation after diagnoses and surgeries. Whilst research is still in its infancy compared to human physiotherapy, there is some evidence in support of the value of physiotherapy to animals (McGonagle et al. 2015) and providing protocols for evidence-based practice. A significant component of physiotherapy practice focuses on geriatric animals, their function, and quality of life in this phase of life (Frye et al. 2022; Marcellin-Little et al. 2015).

4 Veterinary Referral, Regulation, and Specialist Insurance

Legislation related to veterinary physiotherapy treatment varies by country and may also vary by region such as in the USA. VPs may or may not also be veterinarians, and common qualifications undertaken to become a VP vary widely by country. Additionally, some countries may allow a veterinarian only to carry out rehabilitation, rather than a non-veterinarian physiotherapist. Most countries, including the UK and some regions of the USA, legally require referral to the VP by a veterinarian, who is, in effect, 'prescribing' physiotherapy as treatment once a diagnosis is established. However, if the VP is a non-veterinarian, the referring veterinarian maintains the overall responsibility for care and treatment undertaken by the VP (Levine and Millis 2015). This may impact the viability of referral if veterinarians are unable to trust that the VP will be able to treat wild animals effectively and safely. Regulation and membership of professional bodies for VPs also varies by country, as does whether *physiotherapist*, *physical therapist*, etc. are protected terms requiring licensing from an external body (Levine and Millis 2015). Overall, it is imperative for the veterinarian, zoo, or private collection seeking a VP to research relevant legislation and regulation in their area, alongside checking the suitability of the individual by looking at their qualifications, experience with wild animals, and membership of professional bodies. It is also important to note that insurance cover for wild animal treatment is likely to vary by country and by insurer. For example, in the UK, very few musculoskeletal therapists specialised in animals are insured to treat non-domesticated animals, as at the time of writing, one of the main two insurers does not cover these animals at all, and the other only covers the very few therapists with a background in working with wild animal species. This is because wild animal treatment is not generally covered in animal musculoskeletal therapist training courses, so the insurers see it as being beyond the therapist's scope of practice. Until wild animal treatment is covered in this training, this situation is unlikely to change. Unfortunately, most of the prior wild animal treatment in the UK had probably been done unknowingly uninsured, which has placed the musculoskeletal therapist and collection in a poor position.

5 Veterinary Physiotherapy Assessment and Treatment Approaches

5.1 General Aims of Treatment

The overarching aims of veterinary physiotherapy include the reduction of pain, improvement of musculoskeletal function and quality of life, and support of the healing process during rehabilitation of pathology (Dybczyńska et al. 2022). VPs work holistically and view the animal's body as a whole during assessment and treatment; where applicable a therapist will utilise manual therapies to reduce associated symptoms and pain; this reduction will help an animal function in a more comfortable manner and help facilitate recovery. Alongside the primary pathology, *compensatory gait* is likely; this can be defined as alteration to the stride pattern and weight-bearing of the other limbs, which occurs to support the limb(s) with the primary pathology (Goldner et al. 2018). Veterinary physiotherapy treatment also aims to minimise the impact of compensatory gait on the other limbs and body to improve quality of life and reduce the risk of associated pathology occurring elsewhere. The use of exercise prescription will help to strengthen anatomy and return function to help correct gait, maintain tissue health, and reduce compensatory gait patterns. Other strategies and modalities, such as management adaptations/lifestyle changes, electrotherapies, owner education, and behavioural considerations, are also important tools to help realise these goals. There are pathological factors in non-mammals that may not be considered by most human or domestic musculoskeletal therapists, e.g. manual therapies have been applied to reptiles. Metabolic bone disease (MBD) is a common condition in captive reptiles, and it has a variety of symptoms such as weakening of the skeleton, muscle tremors, muscular weakness, and micro-fractures (Carmel and Johnson 2017; Hedley 2012). There is no published research on how exerting significant manual treatment forces on pathological hard tissues, such as those affected by MBD, may affect reptiles, and so it is likely that the possible short-term benefit of manual treatment does not justify the unknown risk.

5.2 Traditional Approaches to Assessment and Treatment

Unfortunately, in the authors' experience physiotherapeutic intervention also for zoo animals is very often employed as a 'last resort' attempt at resolving problems. Treatment with analgesics and anti-inflammatories may no longer be effective, and the VP is called in at a point where the animal's condition has deteriorated to such a degree that a significant improvement is less likely. Cases with severe presentation can still very much benefit from rehabilitation, as it may be possible to increase quality of life, maintain function, and prolong the comfortable lifespan of the animal. However, the earlier rehabilitation is begun, the more profound the change will be and the better the outcome.

Historically, the lack of specialist, experienced wild animal VPs impacted clinical reasoning and a full consideration of behaviour and stress. This led to procedures involving crush cages and sedation for manual therapy (Colles et al. 2014; Nevin et al. 2020), free contact manual therapy for potentially dangerous animals, usually kept in protective contact, such as Komodo dragons (Goldberg 2019; Wolfe et al. 2015), to exercise prescription based on negative reinforcement (Unwin et al. 2008), and with little or no adaptation of the environment or management to support rehabilitation (Goldberg 2019). As a result, modern zoo standards and protocols were not adhered to, and the outcome may have led to increased stress and therefore limited rehabilitative progress.

With wild animal treatment, this is amplified by the vast behavioural diversity between taxa regarding behaviour associated with stress and pain (Goldberg 2017). Eliciting stress in a patient is extremely detrimental for the overall recovery of the animal. Many wild animals are extremely susceptible to the detrimental impacts of stress, which can be damaging to overall health and in extreme cases lead to the development of new pathologies or even to death. For example, capture myopathy, which is a stress response, will reduce the efficacy of therapy on tissues and the overall result will be detrimental to the individual. Behaviour is an important consideration for both assessment and treatment, especially where an animal may be required to remain stationary for a prolonged period (and ideally in a symmetrical posture) and exposed to prolonged periods of physical contact for both assessment and physical treatment. These areas of assessment and treatment have historically been completed via the methods mentioned above. Commonly, in historical cases of rehabilitation, musculoskeletal therapists have been unable to utilise many points of assessment due to the nature of wild animals, alongside limited understanding of viable assessment techniques, so they have relied purely on history taking or visual assessment of gait. Whilst gait assessment is very important, physiotherapists are taught whilst gait assessment is important, VPs should also assess other measures to inform clinical reasoning and treatment to inform clinical reasoning and treatment to inform clinical reasoning and treatment. In a human or domestic animal, this would involve numerous aspects of anatomical pathology and symmetry, gait, posture, function (tasks and movements), and ability to perform required behaviours in their normal environment. Whilst comprehensive assessment has historically been challenging within the zoo environment, we suggest that there are adaptations or other modes of assessment that can be utilised to build a more detailed clinical picture before applying treatment or management adaptations.

Whilst working with numerous taxonomic groups that require specialist care, the zoo industry has consistently demonstrated its dedication to welfare and scientific advancement. In the treatment of any patient, building trust is important. In certain animals, it can allow the animal to accept physical contact (where appropriate and safe), and to undertake exercise prescription. Zoos often make use of animal learning and training theories and practices, where positive reinforcement training is used to shape behaviours allowing the animal the choice to participate in protocols for their care and treatment, such as health assessment, routine care, and diagnostics (Dadone et al. 2016; Joyce-Zuniga et al. 2016). This approach has multiple benefits, as it reduces potential stress/fear responses to situations and equipment, and avoids the need for sedation or anaesthesia (Whittaker and Laule 2012). The focus on predominately positive reinforcement training should be the cornerstone of treatment and assessment techniques for modern veterinary physiotherapy treatment for wild animals in captivity.

5.3 Assessment Techniques and Technologies

Assessment of the animal is undertaken at the start of a veterinary physiotherapy session and includes different types of assessment. Assessment should begin with a detailed conversation with the animal's caregiver(s) regarding the animal's history, pathology, behaviour and management, record keeping, and overall quality of life perceptions and assessment(s). This gives the VP a useful starting point to plan other assessments which look in more detail at the animal's current musculoskeletal function and pain levels. Together, the subjective and objective data obtained through assessment can give insights into quality of life. Techniques vary depending upon the pathology and any handling limitations.

Behavioural assessment(s) can help to determine how the pathology may be behaviourally and psychologically impacting the animal and what other assessment may be possible. Again, we emphasise that a continued collaboration with the animal care staff is fundamental. Behavioural assessments are routinely carried out in zoos and research is increasingly demonstrating the inherent link between discomfort, pain, and behaviour. Recent research in domestic dogs and cats showed a strong association between 'problem' behaviours and the presence of pain, with at least a third of cases referred for behavioural consultations also having a painful health condition (Mills et al. 2020). The presence of pain has wide-ranging effects on behaviour and varies by species. Alongside gait changes and other factors discussed previously, these may include reductions in/changes to normal behaviour (activity levels, sleeping behaviour, temperament, posture, grooming, feeding or drinking, interactions with the environment, conspecifics, or keepers); increases in pain-associated behaviours (facial expressions and ear positions, flinching and muscular contractions, restlessness, photosensitivity, over-grooming, vocalisations, respiratory rate); and an increase in abnormal or stereotypical behaviour (Goldberg 2017). Behavioural assessment regarding pain is complex as behavioural changes may be subtle, particularly in certain taxa such as reptiles (Malik 2018), and the human observer's presence may also inhibit or reduce the animal's response to pain-eliciting stimuli (Carbone 2021; Fleming and Robertson 2012). Assessing species-specific pain-related behaviours can give the team information about current pathology and changes over time, which provides important information on the success of treatment and the impact of environmental changes such as seasonal variation or enclosure parameters. However, due to the subtleties of pain-related behaviours and difficulties assessing pain through behavioural change, findings should be critically reviewed and used alongside other assessment techniques. See also chapter 'Pain Physiology, Recognition, and Management in Zoo Animals' for more information on the physiology of pain.

Static assessment assesses the animal's conformation and stance, and observes any increased/decreased limb weight-bearing, musculoskeletal asymmetry, or functional changes. Following this, gait assessment quantifies any lameness or other changes in movement during gait, such as compensation in other limbs or the torso. The animal may also be observed carrying out other functional movements, such as eating, toileting, enrichment use, sitting, or lying down. Active ROM of joints may also be observed by watching the natural joint function as the animal interacts with objects in the environment such as poles on the ground, inclines/declines, and steps.

With a domestic animal, it is possible to directly influence them to move at a variety of speeds, in various gaits, over varied terrain or obstacles, turning or backing up, and changing posture (Goff 2016). However, with wild animal assessment, observation may be limited to naturally occurring movement within the enclosure, or perhaps through training. Assessment is also adversely impacted by the relative infancy of the specialism in the treatment of wild animals. There is less understanding generally of what constitutes a 'normal' ROM, how lameness may present, how anatomy may impact movement, and less understanding of locomotion generally in such animals due to a lack of research on gait in most wild animal species. Making comparisons between the gait and behaviour of other animals in the same exhibit or in other collections may not be accurate, as the captive environment may impact this more generally and an abnormality may seem 'normal' when it is not truly representative of movement in the wild.

If a VP can touch the animal, further assessment can occur. Passive ROM can be assessed by the VP manually moving the animal's joints to look at limitations in movement, pain, end feel, and crepitus, irregularities in sensation felt or heard during manipulation of the joint, which may be due to pathological changes in the joint cartilage. Palpation involves assessment of orthopaedic symmetry, muscle mass, and tone and any other abnormalities, e.g. heat and swelling, using the VP's hands and fingers. Finally, neurological testing of reflexes, proprioception, and pain responses may occur to assess neurological function. It can be important to note other behavioural changes that may be observed within a geriatric patient, and other important aspects of assessment can give insights into the psychological impacts of ageing which may include reduced confidence in certain activities or functional tasks and/or a reduced engagement with social behaviours, e.g. play. These may indicate pain or reduced functionality within a specific area of anatomy, which is impacting the animal's ability to complete or participate in specific behaviours. Sudden behavioural changes or deferred behaviours such as signs of frustration or aggression may be directly associated with pain, or frustration may occur due to the animal being unable to complete behaviours; these are important to monitor throughout assessment or history taking.

Where possible, the VP will employ further measures of function, such as measuring muscle mass, lameness scoring, stance analysers, validated questionnaires regarding pain or quality of life, and using goniometers, a tool which measures joint angles to assess joint range of motion. These objective measures allow assessment of the current impact of pathology and provide a baseline measure to quantify changes over time. Once assessment has occurred, the VP considers the findings, creates a problem list, and devises a treatment plan whilst staying in close communication with the animal care staff. There are a variety of tools and utilities that can be employed to help facilitate more reliable and reflective assessment which are described below.

Static and moving assessment by means of photos or videos is a valuable assessment tool for wild animals in captivity. By observing presentations, videos, or images of animals, an assessor can observe the animal's general symmetry and posture, which can provide a wealth of information regarding weight-bearing, conformational changes, and musculature (Goff 2016). Care must be taken to check that the stance or photos taken truly represent the animal's general stance rather than just how he happened to stop during gait. Environmental conditions, light, location, and other factors can affect the scoring and must be considered to obtain valid results. Using slow-motion footage allows an in-depth scrutiny of the animal's gait to assess subtle gait abnormalities that may not be seen in real time (Millis and Ciuperca 2015), and is particularly helpful for smaller animals or those with fast movement. Numerous variables such as head and neck movement, pelvic excursion, stance/flight time, weight-bearing, flight arc, joint ROM, posture, and symmetry can be closely assessed from footage, allowing the VP to thoroughly assess the degree of lameness and compensation. Video footage can also be scrutinised by numerous professionals to see if additional abnormalities are identified, and it can serve as a record of the animal's condition that can be revisited in the future and compared to future points of assessment. A process as with photos needs to be put in place so different types of videos can be reviewed over time.

It is important to note that although gait scoring frameworks are well established for domestic animals, they are not available for most exotic animals. A gait scoring framework presents a number of descriptors with a numerical value attached (e.g. 1–5) where higher values usually represent increasing lameness. The observer chooses the descriptor and numerical value which best describes the animal's gait quality. There are significant limitations to gait scoring. The reliability and validity of gait scoring of domestic animals can be poor even in experienced veterinarians. It is poorer still for hindlimb lameness (Starke and Oosterlinck 2019) or when lameness is subtle (Quinn et al. 2007). Many zoo species do not have validated lameness scoring systems and normal gait parameters do vary across taxa. If gait scoring frameworks are developed by professionals who lack a biomechanical or veterinary background, details of more subtle lameness may be overlooked. Therefore, more work is needed in this area.

5.4 Common Treatment Techniques

The end goal of most treatment modalities is to support correct movement to allow the musculoskeletal system to restore joint ROM, soft-tissue strength, and coordination (Haussler 2018). Treatment can be broken down into manual therapies, exercise- and management-based treatment, and other therapeutic modalities. Manual therapies are used to treat hard and soft tissues and are carried out on domestic animals and wild animals managed in free contact (without barriers). Soft tissue manual therapies may include massage, static stretches, and release, which are used to treat abnormal muscles, tendons, and fascia (the tissue around muscle). Other manual therapies such as passive ROM and joint mobilisations are used to improve joint range of motion and health. In passive ROM treatment, the VP applies simple joint motion that replicates the animal's natural movement. During joint mobilisations, joints are manipulated using small accessory movements such as glides and oscillations, which cannot be performed actively by the animal. Joint mobilisations are graded I–IV in terms of treatment technique, and grades III and IV may cause discomfort (Saunders et al. 2015); grades III and IV may therefore be contraindicated in wild animal treatment due to the stress and risk involved. Whilst stress is an important consideration for all animals, the increased proximity, continual physical contact, and application of pressure could be particularly aversive for wild animals. Preferably, specific exercise prescription should assist in the treatment of musculoskeletal physiotherapy instead of the higher grades of joint mobilisation.

5.4.1 Exercise Prescription

Exercise prescription, also known as therapeutic exercise or a home exercise programme, is the most effective and common modality used by VPs (Dybczyńska et al. 2022). It involves the use of tailored exercises that are amended over time to improve the animal's musculoskeletal function, and can have goals such as reducing pain, improving joint ROM, increasing muscle strength and mass, improving balance, and supporting healing (Kirkby-Shaw et al. 2020). Techniques used to train and guide these behaviours vary, but often involve the use of positive or negative reinforcement training with dogs and horses, respectively. We believe and encourage the focus on positive reinforcement for all animals; see chapter 'The Role of Learning and Training for Ageing Animal Care and Wellbeing' for more on animal training. Specialist equipment such as poles and wobble cushions may be used alongside features of the environment, such as inclines/declines and platforms. Exercise prescription must be carefully planned and amended considering the animal's current state of health and fitness, and to avoid exacerbating pathology or causing fatigue (Kirkby-Shaw et al. 2020).

5.4.2 Environmental and Management Modifications

Environmental and management modifications should be used to reduce the negative impact of the pathology on the animal, minimise stress, and allow the correct conditions for musculoskeletal rehabilitation. These may include changes to exercise parameters, activity levels, and equipment; interactions with conspecifics; enclosure fixtures and fittings, bedding, substrates, and flooring; diet, feeding techniques, and food enrichment and supplementation; and environmental parameters such as temperature, lighting/UV, humidity, and the weather. Unfortunately, this area is often underutilised due to time constraints, circumstances, and limitations in veterinary physiotherapy training. For example, when patients are seen in a veterinary clinic rather than in the home environment/enclosure, this reduces the VP's ability to observe and rectify any possible issues. Similarly, if a thorough history is not taken regarding management and a thorough check of the environment is not made, opportunities for supporting rehabilitation may be missed. See chapters 'Facility Design for Disabled and Ageing Wild Animals' on design and 'Environmental Enrichment for Ageing Zoo Animals' on enrichment for more in-depth information. Other veterinary physiotherapy treatments include thermal modalities, electrotherapies, and laser therapies which are described below.

5.4.3 Thermal Modalities

Thermal modalities involve the use of hot and cold to manage pathology and support healing, in order to improve the animal's musculoskeletal function and allow rehabilitative exercise. Thermotherapy involves the superficial application of heat to the animal's skin to increase circulation and soft tissue extensibility, relax spasmed muscles, and provide pain relief. Cryotherapy involves the superficial application of cold to the animal's skin to decrease circulation, reduce the impact of inflammation, and again relax spasmed muscles and provide pain relief. Thermotherapies must be applied with care to ensure treatment efficacy and avoid potential harm to the tissues. Cryotherapy should not be used for poikilothermic animals such as reptiles and amphibians, as it reduces immune function, which may impact healing (Rychel et al. 2011). Similarly, it should be used with care where tropical endothermic species are kept in cooler habitats.

5.4.4 Electrotherapies

Electrotherapies are used to influence a variety of responses in hard and soft tissues to support healing, reduce pain, and improve musculoskeletal function. All physiological processes in animals involve the maintenance of ion/electrical gradients across cell membranes, and electrical cues from these natural currents impact cellular processes such as healing (McCaig et al. 2009). Electrotherapies transfer various types of energy, at a high or low level, to tissues to influence this bioelectricity and cause various biological effects (Watson 2010). Common electrotherapies include phototherapy such as therapeutic lasers, therapeutic ultrasound, pulsed-magnetic field therapy, neuromuscular electrical stimulation (NMES), and transcutaneous electrical nerve stimulation (TENS). Each has different biological effects, although there is some crossover, and they vary in terms of ease of use, safety, contraindications, whether contact is required with the body, and sensation experienced by the animal during treatment.

5.4.5 Laser Therapy

Laser therapy has grown in popularity in zoos in recent years, but treatment protocols for non-domestic species are often not researched and care should be taken with animals with photosensitive skin. Secondary modalities should be used in conjunction with the core treatment approach, and not as standalone treatments, as they treat the symptoms rather than causal factors and have limited impact on long-term recovery.

5.4.6 Use of Sedation and Anaesthesia

Sedation, or general anaesthesia, has been employed to reduce stress during manual treatment and ensure therapist safety; this is more common for osteopathic adjustments as it allows for close proximity and the application of physical force to a potentially dangerous animal. However, this approach carries an associated risk to the animal. Behavioural feedback from the patient is very important during the application of force onto tissues to ensure a therapist does not cause damage. Effective sedation can also be difficult to apply and monitor in captive wild animals due to the variation in species' physiology (Almeida et al. 2022; Bennett and Lewis 2022), lack of research in certain species (Ozeki and Caulkett 2014), and variations in the individual's response to sedation due to factors such as health status (Lamont and Grimm 2014). Additionally, elderly animals undergoing rehabilitation may have a variety of age-related pathologies; as many sedatives impact respiratory and cardiovascular function, this increases the risk of sedation for elderly animals (Larsen and Kreeger 2014). These risks need to be weighed up against the short-term benefit of manual therapy. In addition, if the causal factors are not dealt with, and if exercise prescription to strengthen tissues is not employed, manual therapy may have to be continually reapplied for a long period.

5.4.7 Treatment Response

The animal's responses to treatment modalities should be monitored during and after treatment to provide further information about treatment efficacy and the impact of pathology. Further assessment of function and pain then occurs at each subsequent session, and modifications are made to treatment over time, allowing further functional improvements for the animal. Treatment sessions can then be reduced in number or ceased, leaving management and exercise prescription in place if required, when musculoskeletal function has returned or improved to a reasonable level.

6 Environmental Design and Animal Training

It is well known that maintaining activity is beneficial for elderly or osteoarthritic animals, and should therefore be consistently encouraged and not limited due to limits in keeper time or environment (Brando and Buchanan-Smith 2018). Whilst physiotherapy is of significant benefit to more severe presentations of progressive musculoskeletal pathologies (such as osteoarthritis), which are prevalent in ageing wild animals in captivity, it can be even more beneficial when used at the initial onset of lameness and musculoskeletal disease. Further, reducing the risk factors leading to the development of these conditions in captivity, such as poor habitat design, may prevent or significantly delay the progression of pathology. Environmental design, nutrition, behaviour (including human animal interactions, and health have a direct influence on animal wellbeing) (Mellor et al. 2020). Together with pharmacological and physical interventions, it is extremely important to consider all aspects of a patient's life in order to manage chronic pain and support recovery. It is also important to consider that whilst husbandry guidelines are well established for many species, what is considered 'correct' for a healthy or young individual may no longer be applicable for an older animal, or one experiencing chronic disease (Brando and Buchanan-Smith 2018).

The overuse or underuse of the musculoskeletal system due to feeding behaviours is an important consideration. A study comparing reindeer bones demonstrated obvious musculoskeletal changes between captive and wild populations; captive individuals showed changes linked to increased use of the subscapularis which is the shoulder-stabilising muscle used during standing, whereas wild individuals showed increased use of the deep digital flexor and biceps brachii muscles, most likely due to the foraging behaviours, such as pawing, exhibited by wild animals (Niinimäki and Salmi 2016). Increased activity should be introduced gradually, should be of the right kind, and can become detrimental if overdone. Certain types of activity (such as intense or repetitive training techniques or movement on unsuitable substrates) should not be encouraged because of the impact they may have on tissues, potentially exacerbating inflammatory processes, increasing pain, or accelerating the progression of pathology. We can encourage and shape movement through enrichment that can be beneficial for the therapeutic management of disease. It is of the utmost importance that enrichment and the resulting physical activity are clinically reasoned and focus on functional choices in the context of the animal's unique anatomy, behaviours, and pathology. Non-functional enrichment, creating patterns of movement that the animal would not perform in the wild, such as hanging up carcasses or hay nets for species who do not feed in this way, may have detrimental effects on the musculoskeletal system over time (Hodgson et al. 2022). Similarly, giraffes have a very specific anatomy (Endo et al. 1997) and ecological niche utilising browse-feeding, yet they do not feed solely at this height, and should therefore be provided with feeding opportunities at different heights, locations, and activities, with particular attention to elderly giraffes. The use of enclosure features can encourage specific activities. For example, shift areas or channels within an enclosure can be created, and objects such as poles can be added. Enrichment, such as novel objects or suitable food enrichment items, may also be employed to encourage movement around an enclosure for rehabilitation (Flower et al. 2013).

Positive reinforcement training in free, semi-protected, and protected contact is now beginning to be employed to facilitate exercises for rehabilitation (Nakatani et al. 2019; Neal Webb et al. 2020). Training an animal to actively participate in its own exercises offers a variety of benefits over traditional approaches. Targeting is a commonly trained behaviour that allows a lot of flexibility for exercise prescription, such as to encourage the animal to engage with exercise prescription equipment, or to encourage specific movements. The quality of movement is an important feature and shaping a behaviour can offer greater fluidity of movement as the animal moves freely and independently. This enables the VP to shape a wide variety of movements to create the desired impact on tissue; for example, this may be asking an animal to extend a requested movement to create a stretch through targeted tissues, encourage increased ROM of joints, or increase weight-bearing. Similarly, shaping behaviours offers many possibilities for a VP to select clinically reasoned desired movements and to impact anatomical areas (Neal Webb et al. 2020).

An approach using external force, for example using negative reinforcement or physically blocking and controlling movement, may increase stress, risk of injury, and potential confusion for wild animals, leading to less effective rehabilitation. Using training for exercise prescription may also increase cognitive awareness of the environment and movement. This is particularly important as many exercises hope to achieve gait or postural re-education and increased cognitive awareness of limb placement, and movement can help to achieve this more rapidly. Ultimately, shaped behaviours may become part of an animal's behavioural repertoire and may encourage use of areas of the enclosure or equipment that is beneficial for musculoskeletal health, e.g. ramps or ledges.

7 Case Studies

Below there are four case studies which display some of the approaches highlighted within this chapter. These cases utilised positive reinforcement training and environmental and management changes to varying degrees with little or no manual therapies being applied. All cases have recorded maintained improvements within the patients and utilised a team-based approach to embed the necessary training and management changes. Whilst these approaches can sometimes be difficult within the increased demands placed on animal care teams, they often result in sustained change.

7.1 Case Study 1: Multimodal Physiotherapy Approach for Toby, a 21-Year-Old Male South American Tapir (*Tapirus terrestris*) at Noah's Ark Zoo Farm, Bristol in the UK

Toby was diagnosed with bilateral osteoarthritis in interphalangeal joints, i.e. digit joints, via radiographs, with additional osteoarthritis suspected in other hindlimb joints, and was on analgesics during the initial consultation. As he aged, his activity levels had significantly reduced, he was taking rest breaks whilst feeding, and he was showing intermittent bilateral hindlimb lameness which worsened during the colder months. Toby presented with a narrow hindlimb posture in static and gait assessment, and had a reduced hindlimb stance phase, protraction, and retraction in walk (see Figs. 1 and 2). General disuse atrophy of hindlimb musculature was present, and limited passive range of motion was found in hindlimb joints giving reduced hindlimb retraction. Compensatory changes due to hindlimb osteoarthritis were seen in Toby's gait, forelimbs, and back musculature: the forelimb stance phase was increased, and hypertonicity/hypertrophy (high muscle tone and mass) were present in a number of forelimb and back muscles.

Toby was treated using a combination of manual therapy, exercise prescription, and environmental adaptation, with data recorded on activity levels, sleep quality, and general behaviour before and during treatment. Due to his history and habituation to physical contact, manual treatment included stretches, massage, ischaemic pressure, and joint mobilisations. Massage and the application of thermotherapy/cryotherapy were taught to and subsequently applied by the keepers, as needed. Exercise prescription made use of target training that was already in place, and included rhythmic stabilisations, baited stretches (see Fig. 3), and ground pole-work. Environmental adaptations included utilisation of enclosure features, e.g. inclines, feeding enrichment variation to encourage gentle weight-bearing without muscular fatigue, and modifications to bedding and indoor substrates to support

Fig. 1 Gait assessment of Toby (right) the male South American tapir (*Tapirus terrestris*) from the caudal view. Image credit: Louise Lefrere

joints. During Winter months, indoor activity was increased due to the benefit of the warm temperatures, with more encouragement used to increase outdoor activity.

Improvements have been consistent since initial treatment in the spring of 2021. Toby has showed significantly increased activity levels and increased engagement with enrichment devices. His behavioural repertoire broadened, with behaviours such as climbing and browsing at more varied heights occurring. Analgesics were withdrawn after several months of treatment and have not been required since.

7.2 Case Study 2: Multimodal Physiotherapy Approach for a 7-Year-Old Male Argentine Black and White Tegu (*Salvator merianae*)

This case was a 7-year-old male tegu who had been diagnosed with lumbar spine arthritis and spondylosis. The lizard had a varied presentation with intermittent non-weight-bearing of the pelvic limbs, and when able to stand and locomote, would present with a very 'weak' and broadened (lateral) stance. Keepers also noted

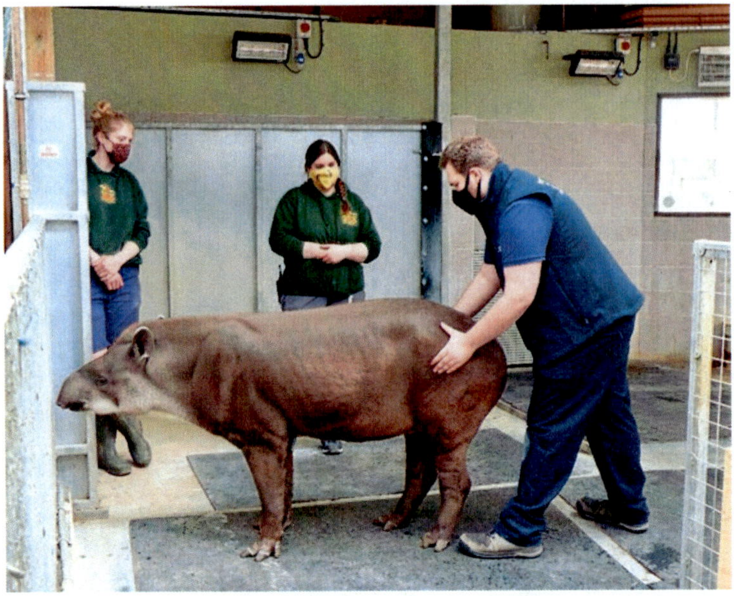

Fig. 2 Assessment of musculoskeletal symmetry, mass, and tone of Toby the male South American tapir (*Tapirus terrestris*). Image credit: Louise Lefrere

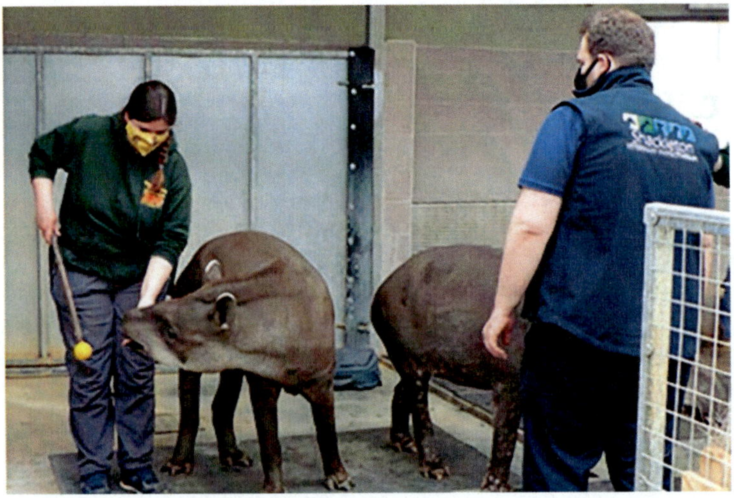

Fig. 3 Target training used on Toby the male South American tapir (*Tapirus terrestris*) to apply lateral baited stretches. Image credits: Louise Lefrere

greatly reduced activity levels. The approach involved laser therapy that was applied on the lumbar spine (whilst feeding) to reduce pain and to help cell function—only introduced as environmental parameters were correct, as there is no research on how laser may impact a hypo-calcaemic reptile. Some minimal physical intervention was used in the form of rhythmic stabilisations (applying a gentle rocking motion), to encourage gentle variation in weight-bearing and neuromuscular re-education. Keepers were encouraged to create an exercise area and to use it daily, as the original enclosure was smaller and restricted opportunities for activity. The importance of using the correct surface was communicated to reduce risk, and usual feed times were used to encourage exercise by scattering food in the exercise area and utilising live food at later stages, rather than feeding in a single bowl. To introduce more specific exercises, target training was used to encourage movement on different undulating and varying surfaces for strength and proprioceptive input. It was also employed to encourage specific movement within the spine: a small amount of lateral flexion (see Fig. 4), which increased in degree and duration as improvements were shown. This was particularly important due to spinal pathology and the reduced lateral movement in the axial skeleton, which is fundamental to lizard locomotion.

Important modifications were made to the animal's enclosure which included an increase in size, increased undulation, and substrate changes. The general aim was to increase movement and improve the complexity of the environment which was achieved not just through physical fixtures and a water feature within the enclosure, but also by using more feeding stations and avoiding tweezer feeding. Additionally, increasing the number of basking lamp locations encouraged movement between them, whilst still maintaining the thermal gradient. Over the treatment period, a more normal, upright posture was seen and a reduction in non-weight-bearing episodes occurred, until these episodes eventually stopped. Keepers also noticed increased activity levels during exercise periods, which became more consistent upon introduction to the new enclosure, and he seemed to show a greater interest in utilising enclosure features. On closer examination, it was observed that that his proprioception continually improved through the recovery period.

Fig. 4 The use of target training to elicit lateral movement in the axial skeleton of a male Argentine black and white tegu (*Salvator merianae*). Image credit: Louise Lefrere

7.3 Case Study 3: Training-Based Exercise Prescription and Enclosure Modification, for an 8-Year-Old Female Coati (*Nasuella olivacea*) at Moorpark College in the USA

This female coati was diagnosed with intervertebral disc disease (Nakatani et al. 2019) which, in this case, was characterised by mineralised disc material which had made its way into the thoracolumbar spinal canal, causing neurological deficits. The animal initially received surgical intervention (hemilaminectomy) but was continuing to present with intermittent paresis and intermittently delayed proprioceptive positioning of the pelvic limbs, bilaterally. Whilst considering physiotherapy, it was suggested by the keepers that manual therapies and hydrotherapy would be too stressful for the animal due to its nature and management, and so was deemed impractical. However, the animal had a strong reinforcement history, and this was a key factor in the safe management of the animal.

Utilising the animal's historical training sessions, the team were able to shape numerous behaviours to build an exercise regime. Firstly, the animal was trained to locomote up shallow inclines and over/through obstacle courses, e.g. ball pits, boxes, logs, tyres, and varying surfaces to encourage motor function and proprioceptive activity.

The animal was also trained to complete much more challenging tasks to target different muscle groups and engage them in challenging ways that would also intensify the need for enhanced balance and coordination. These tasks included pushing a ball up a gentle incline and being trained to open a jar; the team believed these tasks would help target muscles of the spine. This programme saw continual improvements in ambulation and proprioception, with the animal being weaned off analgesics between 19 and 22 months, remaining stable, and showing gradual improvement. However, 30 months post-surgery, the animal was unfortunately euthanised due to severe pneumonia.

7.4 Case Study 4: An Exercise Prescription Programme for Male and Female Geriatric Chimpanzees (*Pan troglodytes*) at the National Centre for Chimpanzee Care in the USA

A group of 18 chimpanzees were studied at the National Centre for Chimpanzee Care (Neal Webb et al. 2020). Eight geriatric chimpanzees (aged 39–53) and one non-geriatric individual (aged 30), all with mobility impairments, took part in personalised exercise prescription regimes and were compared to nine control chimpanzees in an unblinded, matched pair study design. Mobility impairments in the intervention group included diagnosed arthritis in varying joints, previous strokes causing neurological damage, and previous injuries. Prior to the study, these chimps showed significantly slowed or limited movement and mobility, showed joint dysfunction, and had been prescribed analgesics. The control group had not been classified by a veterinarian as being mobility impaired, although they had some mild mobility impairments due to their age.

For treatment, positive reinforcement techniques were adapted for exercise prescription and tailored to each individual's pathology, fitness, prior training, and treatment aims and practicalities. Exercises took place twice a week and included squats, weight shifts, and resistance exercises, which were increased in intensity as the chimpanzees' fitness and comfort advanced over the 25 weeks of the study. The control group received comparable body examination training using positive reinforcement during the latter 15 weeks of the study, to attempt to isolate any effects of human interaction or training from the physiotherapy treatment.

A team approach involving caregivers, trainers, behavioural scientists, and veterinary professionals was used throughout in order to apply and monitor the effect of the exercise prescription regime and compare it to the controls. Caregiver wellness ratings, mobility scoring, and physical examinations by veterinarians were compared to baseline measures and against control animals. Prior to the beginning of the physiotherapy programme, baseline measures for the chimpanzees receiving the exercise prescription intervention compared to the control group were significantly lower on ratings of physical health, activity levels, ease of movement, and wellbeing, and were higher for depressive behaviour. After 25 weeks of treatment, there were no significant differences in these ratings when comparing the intervention and control groups, suggesting that the chimpanzees taking part in the exercise prescription showed a rated improvement in physical health parameters, wellbeing, and activity levels, whereas the control chimpanzees did not show any significant improvements.

8 Potential Implications of Non-specialist Treatment Approaches

In other documented cases of rehabilitation, veterinary professionals or caretakers have used a variety of exercise prescription techniques, but with limited consideration of behaviour leading to possible stress. Harnesses or slings have been used to encourage walking in patients with significant limb pathologies, such as a flamingo with capture myopathy; unfortunately, this animal became tangled due to a stress response, requiring the use of anxiolytic medication to enable its use (McEntire and Sanchez 2017). Animals have been placed in/encouraged into water features to encourage movements in the water, including those who primarily or only use terrestrial gaits (Unwin et al. 2008). Resistance bands have been used around limbs to apply restrictive force to increase the effort required for limb protraction in order to strengthen muscles (Goldberg 2019). Negative reinforcement has also been used, such as the application of aversive and 'unpleasant stimuli' to encourage an animal to move across an enclosure (Unwin et al. 2008); unfortunately, this is inherently stressful for any animal, including wild animals. There has also been a case of hydrotherapy used with a Gila monster (*Heloderma suspectum*) and Komodo dragon (*Varanus komodoensis*) who were experiencing neurological deficits and musculoskeletal pathologies. These species are naturally terrestrial and may swim in their natural habitat. This animal was placed in a water-filled container, or within the enclosure's water feature to elicit swimming or moving in the water. Hydrotherapy can be beneficial for recovery, and it is important to provide support to an animal that may be physically disabled to reduce physical risk. It is also important to consider movement on land to aid full recovery for normal locomotion, as there is a fundamental difference in locomotion in water compared to land. Whilst all these animals did demonstrate an improvement, clinical reasoning suggests that the same physiological conditions could have been treated with lower stress methods to encourage the same healing processes. Physiotherapy should never follow a 'cut and paste' approach. It is discouraged to apply the same approaches to numerous cases that may share a pathology, as patients' precise presentation, needs and circumstances vary widely.

9 General Recommendations

Physiotherapy is an exciting new field within veterinary science and holds much potential for its use within zoological facilities. However, due to the lack of standardisation and regulation currently within the field there are great deficits in knowledge and understanding of exotic species and zoo management. A greater standardisation for this brand-new field, with additional continuous professional development, training, and qualifications, is key to creating and maintaining high levels of professionalism, in line with other specialisms. Recommendations for the continual advancement of this field and how to approach physiotherapy in wild animals are listed below.

1. The use of multiple modes of assessment. Using numerous points of assessment will both increase the reliability of the findings and may give greater depth of evaluation in both clinical presentation and areas in which the animal may be struggling. Manual therapies are not the primary form of treatment for most wild animals and should only be applied if the animal is fully habituated to physical contact or can be trained to accept manual treatment through a well-designed, ethical training routine, to avoid negative association with VP treatment. A management protocol such as protected contact may preclude manual treatment, as may an animal's size. Muscle mass and high muscle tone in megafauna may make it difficult to correct or treat specific issues for example.
2. A patient experiencing a chronic condition is unlikely to obtain functional recovery using manual therapy alone. VPs therefore utilise exercise prescription to help improve strength and re-educate function, such as symmetry of weight bearing and gait, to combat this, creating the optimum environment for tissue recovery.
3. Some key factors to consider when encouraging or shaping movement for pathological anatomy include controlling limb loading, repetition, and level of exertion.

4. A thorough understanding of special anatomy, behaviour, common pathologies, and management is vital to reason any form of treatment correctly clinically.
5. With chronic cases of joint disease, an animal may reduce the use of a limb causing disuse atrophy of muscle, asymmetry, and instability of the joint. The primary goal of a physiotherapist is to return symmetrical function.
6. Assessment of all environmental and management aspects to support health and reduce risk factors in a patient is a vital part of the treatment process. If not tackled, this may reduce the quality of treatment outcomes or miss causal factors of pathology. These considerations must be taken in the context of the pathology, species anatomy, and ecological niche.
7. A VP could be an important part of long-term management of geriatric animals within zoological collections, helping to inform decision on training regimes, environmental adaptation, and enrichment devices for the long-term help of animals in our care.
8. Implementing a team-based approach, which includes specialist experience and knowledge, allows clinical reasoning of the most effective approach for the animal's pathology, presentation, management, anatomy, and behaviour. Any deficits in any of these areas will be detrimental to the patient and their recovery.

If the above criteria can be met, this field will continue to advance, develop, and become an intrinsic part of managing and treating geriatric animals in captivity.

10 Conclusion

Bringing in a VP to aid rehabilitation or give musculoskeletal support would greatly enhance animals' wellbeing, if in collaboration with veterinary staff and caregivers. It is important to find a wild animal specialist, as the likelihood of treatment efficacy is higher, and unlike a domestic practitioner, a specialist will definitely be insured to treat these animals. A specialist wild animal VP will be able to carry out clinical reasoning based on a full understanding of rehabilitation, collection and training protocols, wild animal behaviour, and taxon-specific pathology and anatomy. Physiotherapy is beneficial at most stages of life and disease, although more positive changes can be made in earlier phases of disease before the animal is severely affected. In the future, an ideal scenario would involve a specialist VP's input during habitat refurbishment and/or design, or where some animals in a group have developed musculoskeletal disease. This strategy should reduce the likelihood or severity of musculoskeletal disease development and promote good welfare across the lifespan of captive wild animals.

References

Almeida D, Kennedy M, Wend-Hornickle E (2022) Snake sedation and anesthesia. Vet Clin North Am Exot Anim Pract 25(1):97–112. https://doi.org/10.1016/j.cvex.2021.08.003

Aulakh KS, Dongaonkar KR, Barnes K, Gines AJ, Bordelon JT, Hulse D et al (2020) Influence of orthopedic examination on lameness scores and interobserver and intraobserver agreement in dogs with naturally occurring elbow osteoarthritis. Vet Surg 49(3):455–462. https://doi.org/10.1111/vsu.13390

Bennett K, Lewis K (2022) Sedation and anesthesia in rodents. Vet Clin North Am Exot Anim Pract 25(1):211–255. https://doi.org/10.1016/j.cvex.2021.08.013

Brando S, Buchanan-Smith HM (2018) The 24/7 approach to promoting optimal welfare for captive wild animals. Behav Processes 156(SI):83–95. https://doi.org/10.1016/j.beproc.2017.09.010

Carbone L (2021) Do "prey species" hide their pain? Implications for ethical care and use of laboratory animals. J Appl Anim Ethics Res 2(2):216–236. https://doi.org/10.1163/25889567-bja10001

Carmel B, Johnson R (2017) Nutritional and metabolic diseases. In: Doneley B, Monks D, Johnson R, Carmel B, Wiley J (eds) Reptile medicine and surgery in clinical practice. Wiley Blackwell, Oxford, UK, pp 185–195. https://doi.org/10.1002/9781118977705.ch15

Colles CM, Nevin A, Brooks J (2014) The osteopathic treatment of somatic dysfunction causing gait abnormality in 51 horses. Equine Vet Educ 26(3):148–155. https://doi.org/10.1111/eve.12122

Dadone L (2019) Lameness diagnosis and management in zoo giraffe. In: Miller RE, Lamberski N, Calle PP (eds) Fowler's zoo and wild animal medicine current therapy, vol 9. Elsevier, St. Louis, MO, pp 623–629. https://doi.org/10.1016/B978-0-323-55228-8.00088-6

Dadone LI, Schilz A, Friedman SG, Bredahl J, Foxworth S, Chastain B (2016) Training giraffe (Giraffa camelopardalis reticulata) for front foot radiographs and hoof care. Zoo Biol 35(3):228–236. https://doi.org/10.1002/zoo.21279

Dybczyńska M, Goleman M, Garbiec A, Karpiński M (2022) Selected techniques for physiotherapy in dogs—a systematic review. Animals 12(14). https://doi.org/10.3390/ANI12141760

Dykgraaf S, Firth EC, Rogers CW, Kawcak CE (2008) Effects of exercise on chondrocyte viability and subchondral bone sclerosis in the distal third metacarpal and metatarsal bones of young horses. Vet J 178(1):53–61. https://doi.org/10.1016/j.tvjl.2007.08.016

Endo H, Yamagiwa D, Fujisawa M, Kimura J, Kurohmaru M, Hayashi Y (1997) Modified neck muscular system of the giraffe (Giraffa camelopardalis). Ann Anat Anatomischer Anzeiger 179(5):481–485. https://doi.org/10.1016/S0940-9602(97)80055-6

Fleming GJ, Robertson SA (2012) Assessments of thermal antinociceptive effects of butorphanol and human observer effect on quantitative evaluation of analgesia in green iguanas (Iguana iguana). Am J Vet Res 73(10):1507–1511. https://doi.org/10.2460/ajvr.73.10.1507

Flower JE, Lynch K, Clark-Price SC, Welle KR, O'Brien R, Whittington JK (2013) Antemortem diagnosis and successful management of noncompressive segmental myelopathy in a Siberian-Bengal mixed breed tiger. J Zoo Wildl Med 44(4):1115–1119. https://doi.org/10.1638/2013-0044r2.1

Frye C, Carr BJ, Lenfest M, Miller A (2022) Canine geriatric rehabilitation: considerations and strategies for assessment, functional scoring, and follow up. Front Vet Sci 9:1–11. https://doi.org/10.3389/fvets.2022.842458

Goff L (2016) Physiotherapy assessment for the equine athlete. Vet Clin North Am Equine Pract 32(1):31–47. https://doi.org/10.1016/j.cveq.2015.12.002

Goldberg ME (2017) How to be a pain management advocate for exotic and zoo animals. The Veterinary Nurse 8(7):389–397. https://doi.org/10.12968/vetn.2017.8.7.389

Goldberg ME (2019) A walk on the wild side: a review of physiotherapy for exotics and zoo animals. Vet Nurs J 34(2):33–47. https://doi.org/10.1080/17415349.2018.1529547

Goldner B, Fischer S, Nolte I, Schilling N (2018) Kinematic adaptions to induced short-term pelvic limb lameness in trotting dogs. BMC Vet Res 14(1):1–11. https://doi.org/10.1186/S12917-018-1484-2/TABLES/2

Haussler KK (2018) Equine manual therapies in sport horse practice. Vet Clin North Am Equine Pract 34:375–389. https://doi.org/10.1016/j.cveq.2018.04.005

Haussler KK, King MR, Peck K, Adair HS (2021) The development of safe and effective rehabilitation protocols for horses. Equine Vet Educ 33(3):143–151. https://doi.org/10.1111/eve.13253

Hedley J (2012) Metabolic bone disease in reptiles: Part 1. Companion Anim 17(6):52–54. https://doi.org/10.1111/j.2044-3862.2012.00210.x

Hodgson S, Bennett-Skinner P, Lancaster B, Upton S, Harris P, Ellis AD (2022) Posture and pull pressure by horses when eating hay or haylage from a hay net hung at various positions. Animals 12(21). https://doi.org/10.3390/ani12212999

Joyce-Zuniga N, Newberry R, Robbins C, Ware J, Jansen H, Nelson OL (2016) Positive reinforcement training for blood collection in grizzly bears (Ursus arctos horribilis) results in undetectable elevations in serum cortisol levels: a preliminary investigation. J Appl Anim Welf Sci 19:1–6. https://doi.org/10.1080/10888705.2015.1126523

Kennedy KC, Martinez SA, Martinez SE, Tucker RL, Davies NM (2018) Effects of low-level laser therapy on bone healing and signs of pain in dogs following tibial plateau leveling osteotomy. Am J Vet Res 79(8):893–904

Kirkby-Shaw K, Alvarez L, Foster SA, Tomlinson JE, Shaw AJ, Pozzi A (2020) Fundamental principles of rehabilitation and musculoskeletal tissue healing. Vet Surg 49(1):22–32. https://doi.org/10.1111/vsu.13270

Krebs BL, Marrin D, Phelps A, Krol L, Watters JV (2018) Managing aged animals in zoos to promote positive welfare: A review and future directions. Animals 8. https://doi.org/10.3390/ani8070116

Lamont LA, Grimm KA (2014) Clinical pharmacology. In: West G, Heard D, Caulkett N (eds) Zoo animal and wildlife immobilization and anesthesia. Blackwell, Ames, IA, pp 1–41. https://doi.org/10.1002/9781118792919.ch1

Larsen RS, Kreeger TJ (2014) Canids. In: West G, Heard D, Caulkett N (eds) Zoo animal and wildlife immobilization and anesthesia, 2nd edn. Blackwell, Ames, IA, pp 585–598. https://doi.org/10.1002/9781118792919.ch40

Levine D, Millis D (2015) Regulatory and practice issues for the veterinary and physical therapy professions. In: Millis DL, Levine D (eds) Canine rehabilitation and physical therapy, 2nd edn. Saunders Elsevier, Philadelphia, PA, pp 8–15

Loeser RF (2010) Age-related changes in the musculoskeletal system and the development of osteoarthritis. Clin Geriatr Med 26(3):371–386. https://doi.org/10.1016/j.cger.2010.03.002

Looney AL, Huntingford JL, Blaeser LL, Mann S (2018) A randomized blind placebo-controlled trial investigating the effects of photobiomodulation therapy (PBMT) on canine elbow osteoarthritis. Can Vet J 59(9):959–966

MacNicol JL, Lindinger MI, Pearson W (2018) A time-course evaluation of inflammatory and oxidative markers following high-intensity exercise in horses: A pilot study. J Appl Physiol 124(4):860–865. https://doi.org/10.1152/japplphysiol.00461.2017

Malik A (2018) Pain in reptiles: a review for veterinary nurses. Vet Nurs J 33(7):201–211. https://doi.org/10.1080/17415349.2018.1468291

Marcellin-Little DJ, Levine D, Millis DL (2015) Physical rehabilitation for geriatric and arthritic patients. In: Millis DL, Levine D (eds) Canine rehabilitation and physical therapy, 2nd edn. Saunders Elsevier, Philadelphia, PA, pp 628–641

McCaig CD, Song B, Rajnicek AM (2009) Electrical dimensions in cell science. J Cell Sci 122(23):4267–4276

McEntire MS, Sanchez CR (2017) Multimodal drug therapy and physical rehabilitation in the successful treatment of capture myopathy in a lesser flamingo (Phoeniconaias minor). J Avian Med Surg 31(3):232–238. https://doi.org/10.1647/2015-128

McGonagle L, Blythe L, Levine D (2015) History of canine physical rehabilitation. In: Millis D, Levine D (eds) Canine rehabilitation and physical therapy, 2nd edn. Saunders Elsevier, Philadelphia, PA, pp 1–7

Mellor DJ, Beausoleil NJ, Littlewood KE, McLean AN, McGreevy PD, Jones B, Wilkins C (2020) The 2020 Five Domains Model: Including Human–Animal Interactions in Assessments of Animal Welfare. Animals 10(10):1870. https://doi.org/10.3390/ani10101870

Millis DL, Ciuperca IA (2015) Evidence for canine rehabilitation and physical therapy. Vet Clin North Am Small Anim Pract 45:1–27. https://doi.org/10.1016/j.cvsm.2014.09.001

Mills DS, Demontigny-Bédard I, Gruen M, Klinck MP, McPeake KJ, Barcelos AM, Levine E (2020) Pain and problem behavior in cats and dogs. Animals 10(2). https://doi.org/10.3390/ani10020318

Nakatani JY, Greer LL, Wilson GL, Cabrera SY (2019) Surgical and postoperative treatment of intervertebral disc disease in a mountain coati (Nasuella olivacea) using behavioral management and physical therapy. J Zoo Wildl Med 50(3):727. https://doi.org/10.1638/2018-0063

Neal Webb SJ, Bridges JP, Thiele E, Lambeth SP, Schapiro SJ (2020) The implementation and initial evaluation of a physical therapy program for captive chimpanzees (Pan troglodytes). Am J Primatol 82(3):e23109. https://doi.org/10.1002/ajp.23109

Nevin T, Brooks J, Elbrond V, Schultz R, Atkin B, Colles K (2020) An introduction to the osteopathic treatment of animals and birds. In: Nevin T, Colles C, Tozzi P (eds) Animal osteopathy: A comprehensive guide to the osteopathic treatment of animals and birds, 1st edn. Handspring, Pencaitland

Niinimäki S, Salmi AK (2016) Entheseal changes in free-ranging versus zoo reindeer-observing activity status of reindeer. Int J Osteoarchaeol 26(2):314–323. https://doi.org/10.1002/oa.2423

Ozeki L, Caulkett N (2014) Monitoring. In: West G, Heard D, Caulkett N (eds) Zoo animal and wildlife immobilization and anesthesia, 2nd edn. Blackwell, Ames, IA, pp 43–51. https://doi.org/10.1002/9781118792919.ch2

Quinn MM, Keuler NS, Lu Y, Faria MLE, Muir P, Markel MD (2007) Evaluation of agreement between numerical rating scales, visual analogue scoring scales, and force plate gait analysis in dogs. Vet Surg 36(4):360–367. https://doi.org/10.1111/j.1532-950X.2007.00276.x

Ramírez-Flores GI, Del Angel-Caraza J, Quijano-Hernández IA, Hulse DA, Beale BS, Victoria-Mora JM (2017) Correlation between osteoarthritic changes in the stifle joint in dogs and the results of orthopedic, radiographic, ultrasonographic and arthroscopic examinations. Vet Res Commun 41(2):129–137. https://doi.org/10.1007/s11259-017-9680-2

Rychel JK, Johnston MS, Robinson NG (2011) Zoologic companion animal rehabilitation and physical medicine. Vet Clin North Am Exot Anim Pract 14(1):131–140. https://doi.org/10.1016/j.cvex.2010.09.008

Saunders DB, Walker JR, Levine D (2015) Joint mobilization. In: Millis DL, Levine D (eds) Canine rehabilitation and physical therapy, 2nd edn. Saunders Elsevier, Philadelphia, PA, pp 447–463

Starke SD, Oosterlinck M (2019) Reliability of equine visual lameness classification as a function of expertise, lameness severity and rater confidence. Vet Rec 184(2):63. https://doi.org/10.1136/vr.105058

Tangredi BP, Lawler DF (2019) Osteoarthritis from evolutionary and mechanistic perspectives. Anat Rec 1–10. https://doi.org/10.1002/ar.24339

Unwin S, Sanderson S, Chantrey J (2008) Physiotherapy treatment and outcome of a neurologic dysfunction in a Komodo Dragon (Varanus komodoensis). In: AAZV ARAV joint conference, pp 237–239

Watson T (2010) Key concepts with electrophysical agents. Phys Ther Rev 15(4):351–359

Whittaker M, Laule G (2012) Training techniques to enhance the care and welfare of nonhuman primates. Vet Clin North Am Exot Anim Pract 15:445–454. https://doi.org/10.1016/j.cvex.2012.06.004

Wolfe TC, Stringer E, Krauss S, Trout T (2015) Physical therapy as an adjunctive treatment for severe osteoarthritis in a Komodo dragon (Varanus komodoensis). J Zoo Wildl Med 46(1):164–166. https://doi.org/10.1638/2014-0103r.1

Euthanasia of Geriatric Zoo Animals: Decision-Making and Procedure

Sarah Chapman, Julian Chapman, and James Chatterton

S. Chapman (✉) · J. Chapman
Chapman Zoo Consultancy, Birmingham, UK

J. Chatterton
Aukland Zoo, Auckland, New Zealand
e-mail: James.Chatterton@aucklandzoo.co.nz

S. Brando, S. Chapman (eds.), *Optimal Wellbeing of Ageing Wild Animals in Human Care*, https://doi.org/10.1007/978-3-031-30659-4_11

HaHa. HaHa, a female North American river otter (*Lontra canadensis*), lived at the Wildfowl and Wetlands Trust, Slimbridge Centre in the UK, for 13 years. She outlived her mother, Flo, and her sister, Minnie. Minnie and HaHa's names were a composite of the Native American woman "Minnehaha" featured in an epic poem. In Dakota, Minnehaha translates to waterfall, which very much suited HaHa. She had a waterfall at the summit of her habitat that she loved to block with all manners of leaves, twigs, branches, and weeds! She was always mischievous. Although senior, she was completely oblivious to her age. Whenever she was presented with bowls, either as enrichment or accidentally, she would find a way to get a hold of them. She would hoard them underneath a wooden boardwalk where the centre's visitors could stand and see into her pond. Her most recent caregiver, Lauren, would routinely put on a pair of waders and service the area underneath the boardwalk, HaHa's lair, and pull heaps of them back out whilst shaking her head. HaHa had a very generous rotation of enrichment items handmade by Lauren, including a firehouse platform that allowed her to lounge and noisily eat her fish heads without ever having to leave her pond. Whenever she did finally exit the water, HaHa loved to roll and luxuriate in leaf litter and wood chippings, drying herself in funny contortions drawing a crowd around the habitat's viewing window to photograph her. One of her favourite fish was smelt, and on Christmas Day she would often get an entire trout or salmon head to enjoy.

Abstract

The majority of animals cared for by modern zoos now live longer than their wild counterparts. This increase in both mean and maximum lifespan has resulted in a greater number of individuals which can be termed aged or geriatric (henceforth geriatric) resulting in an increased number of individuals with age-related pathologies. A wide range of these have been identified in zoo animals which can present subtly and are typically debilitating and painful and negatively impact the overall welfare of the animal. End-of-life decision-making in geriatric animals takes place in most modern zoos and can be an emotive subject involving multiple stakeholders. Effective decision-making requires clear communications between stakeholders and is influenced by the culture of the people making the decisions, as well as the relevant applicable legislation. Modern zoos make end-of-life decisions in geriatric animals based on a detailed welfare assessment, including as much objective data as possible, with an overall focus on preventing unnecessary suffering of the animals involved. This chapter goes through the stages of decision-making and planning with examples of cases and scenarios we have experienced.

Keywords

Euthanasia · Welfare · Geriatric · Aged · Zoo · Communication

1 Introduction

Species cared for by modern zoos now live longer than their wild counterparts (Kitchener and MacDonald 2004; Krebs et al. 2018; Vogelnest and Talbot 2019). This is due to factors such as reduced exposure to mortality risk factors (e.g.

infectious disease, trauma, predation, human conflict) and advances in the care provided for the animals (Kitchener and MacDonald 2004; Vogelnest and Talbot 2019; Collins and Kays 2011). This care includes utilising animal welfare models to assess welfare and facilitate improvements, as well as improvements in both preventative/proactive and reactive veterinary medicine. A result of this longevity is increased exposure to senescence which is characterised by a range of physiological changes that occur in advancing age. These act cumulatively over time to increase the risk of morbidity and/or mortality and may reduce certain other factors (e.g. fertility). Senescence is not a disease process in itself, though the changes involved may increase the risk factors for the development of pathology (Krebs et al. 2018; Raiti 2019).

Previous publications have described the wide range of common pathologies identified in geriatric non-domestic animals, and they are also discussed elsewhere in this book (Bays 2020; Boylan 2020; Chitty 2020; Dutton 2020; Johnson 2020; Kitchener and MacDonald 2004; Kumar et al. 2017; Pellett et al. 2020; Raiti 2019; Reavill and Imai 2020; Scheelings 2020). Many of these pathologies that develop in advanced age are debilitating, painful, and incurable and negatively impact the welfare of affected animals (Kitchener and MacDonald 2004; Krebs et al. 2018; Vogelnest and Talbot 2019). Early identification of these pathologies can be challenging due to subtle presenting signs and/or the deterioration occurring over a protracted period (Willems et al. 2017; Vogelnest and Talbot 2019). Abnormal behaviour may be especially hard to detect in non-domestic species, which typically mask signs of illness and may also change how they respond to human interaction (Raiti 2019; Vogelnest and Talbot 2019). Early detection of this reduction in welfare requires highly skilled staff operating within a cross-departmental approach including animal caregivers and veterinary teams (Bacon 2019; Krebs et al. 2018). Typically, a range of templates and/or electronic records may be used to objectively assess certain physical and/or behavioural aspects of the animal which aids overall welfare assessment (for more on record keeping, see chapter on record keeping). Repeating these standardised assessments can then help detect any deterioration in animal welfare over time in a more objective manner. Action points can then be set which may include implementation of palliative treatment and ultimately euthanasia (Follmi et al. 2007; Vogelnest and Talbot 2019; Wolfensohn et al. 2018).

2 Definition of Euthanasia

The term euthanasia is derived from the Greek "eu" (good) and "Thanatos" death (Nevarez 2019). The ability to provide a "good death" for animals (i.e. one that minimises the negative effects experienced by the animal), especially those facing ongoing compromised welfare, is a core part of zoo veterinary work. Guidelines for animal euthanasia are provided by a range of professional and national organisations (AVMA 2020; BIAZA 2014; ZAA 2015). Making the decision to euthanise an animal is an important and often emotive subject as the people involved are often attached to the animals and taking a decision on another life is no small matter. The involvement of multiple stakeholders and cultural differences between countries and communities means that there is no single approach that suits every situation in every society. Nevertheless, the authors believe that veterinarians, and the wider animal care teams, are in a privileged position in which they can utilise euthanasia to prevent unnecessary suffering.

2.1 Decision-Making

End-of-life decision-making in a modern zoo is a function of the positive welfare state of the individual animal(s); the current welfare assessment of the animal(s) (including the presence of pathological processes); the welfare of conspecifics (members of the same species) and the wider population and/or sympatric animals (individuals within the same living space); the relevant legislation; and the culture of the humans providing animal care (Krebs et al. 2018; Lennox 2020; Vogelnest and Talbot 2019; WAZA 2003; ZAA 2015).

Defining an individual animal as geriatric first requires a definition of normal or expected longevity for the same species. Data collected by modern zoos can be used to calculate the expected longevity of species when under human care, using institutional, regional, or global records (e.g. Species 360™—Zoo Information Management System: ZIMS™) and/or evidence from published literature. This longevity may also be compared with published data from the same species in the wild (Follmi et al. 2007; Vogelnest and Talbot 2019). Vogelnest and Talbot (2019) suggest "expected lifespan" be derived from the 90th percentile of age at death of the population in regional studbooks.

Whilst accurately defining the term geriatric for each species can still be problematic, there is no single accepted definition, even in human medicine (Singh and Bajorek 2014). Vogelnest and Talbot (Vogelnest and Talbot 2019) suggest an individual animal be classified as geriatric (or aged) once it has exceeded 80% of the "expected lifespan". Krol et al (Krol et al. 2020) adopts a different approach, highlighting that there is no agreed definition for geriatric age for Sphenisciformes and that Magellanic penguins can live up to 30 years of age; in their study they classified individuals >15years old as geriatric. Ultimately, in the absence of agreed terminology in peer-reviewed literature and/or by regional studbooks, it is useful for individual institutions to record the age at which they consider each species to be geriatric, to at least provide consistency within the institution.

To enable effective end-of-life decision-making, it is important to first define the positive welfare state for the animal(s) and the wider population (institutional, national, and/or regional) to which he or she belongs. In addition to the individual animal's overall welfare, there are a range of other factors to consider, including the welfare of other conspecifics and/or sympatric animals in the enclosure, institution, and/or region, as well as logistical and resource constraints.

Geriatric zoo animals suffer pathological processes which are often insidious in nature, typically debilitating and incurable, and may be expressed through changes in a wide range of physical, neurological, and behavioural parameters (Bays 2020; Boylan 2020; Chitty 2020; Dutton 2020; Johnson 2020; Kitchener and MacDonald 2004; Pellett et al. 2020; Raiti 2019; Reavill and Imai 2020; Scheelings 2020). There is a range of welfare assessment tools available, and these are discussed elsewhere in this book (see chapter "Holistic Approaches for Promoting Good Wellbeing for Ageing Wild Animals"). The use of appropriate welfare assessment tools is critical to providing objective welfare assessments and to enable tracking of changes in welfare over time. These methodologies, together with meticulous data collection and records, maximise the objectivity of animal welfare assessments which aids subsequent decision-making (Follmi et al. 2007; Krebs et al. 2018; Mellor et al. 2020; Vogelnest and Talbot 2019; Wolfensohn et al. 2018).

Identification of a welfare issue should trigger a process which includes defining the presenting issue, outlining areas that require further information (e.g. further medical assessment under general anaesthesia), and summarising the detailed results (including the working diagnosis, likely disease trajectory and prognosis) in a comprehensive and accessible report. Shearer (2011) suggests there are four groups of disease trajectories that apply to geriatric animal patients, terms which originate in human medicine:

1. Acute disease: deterioration in welfare occurs over a relatively short period of time (weeks) prior to death.
2. Acute-on-chronic disease: decline in welfare occurs over an extended period (weeks/months), with varying severity of clinical signs and/or periods of relatively static welfare, prior to acute decompensation (e.g. chronic renal failure, congestive heart failure).
3. Chronic disease: welfare state declines over a protracted length of time, typically months/years (e.g. chronic mobility issues due to osteoarthritis), often with secondary complications occurring (e.g. skin lesions and/or urinary tract infections).
4. Peracute disease: sudden onset (hours) of severe disease that is extremely debilitating (e.g. intervertebral disc herniation).

If a chronic and/or incurable disease is diagnosed, then subsequent assessment needs to be made about current overall welfare and whether palliative care options should be considered. End-of-life care in human medicine focuses predominantly on relief of symptoms, and quality of life assessment tools have been developed to help measure health outcomes in such patients (Bergner et al. 1976; Devlin and Brooks 2017). Animal welfare assessment tools have been developed for use in companion animal species, including those based on the five domains of animal welfare as well as other methods. These models provide a framework with which to assess animal welfare, to estimate the efficacy of subsequent medical interventions, and to aid in euthanasia decisions (Mellor et al. 2020; Sapkota et al. 2020; Villalobos 2011). These models have been further adapted for use with zoo/wildlife patients, with similar aims as those in domestic species (Follmi et al. 2007; Krebs et al. 2018; Mellor et al. 2020; Vogelnest and Talbot 2019; Wolfensohn et al. 2018). Palliative treatment plans in zoo/wildlife patients should aim to mitigate the effects of the primary disease, alleviate clinical symptoms, address any secondary complications, and discuss end-of-life decision-making (Jessup and Scott 2011; Lennox 2020; Shearer 2011). Discussing end-of-life decision-making matters in advance, including clearly outlining trigger points for euthanasia, and creating an electronic/written summary of the plan greatly aids decision-making later in this process.

The overall decision-making process which results in the euthanasia of any animal within an institution should be available to staff and be part of a transparent process. This provides a framework within which end-of-life discussions about an animal(s) can occur, builds trust in the decision-making process, and may be a legal obligation depending on the country involved. There are several euthanasia decision trees and guidelines available in publications to aid decision-making (BIAZA 2014; WAZA 2003; ZAA 2015). Figure 1 shows an example flowchart which is used at Auckland Zoo.

It is important to consider not only the welfare of the individual animal, but also the potential effects on any sympatric animals, and to consider what actions will be required if an acute, severe exacerbation is seen in the chronic illness. For example, consider a geriatric carnivore experiencing hindlimb lameness due to arthritis for several months, which is receiving medication, but subsequently collapses on exhibit and is intractable. There is no doubt that this individual will be suffering at this point, no matter what the species. However, intervening in this situation is far more challenging with a large felid, e.g. a lion (*Panthera leo*), than it would be with a small terrestrial carnivore, e.g. bush dog (*Speothos venaticus*): the former situation likely resulting in a more protracted suffering for the lion whilst the procedure is planned and undertaken, with more inherent risks to staff carrying out the procedure.

Effective communication is important with animal euthanasia, especially with cases of insidious illness, slow deterioration in overall welfare, and/or when high-profile animals are affected (see more on communication in chapter "The View from Beyond the Fence: Ageing Zoo Animals and Communicating with the Outside World"). Additional challenges may be present when discussing euthanasia of an animal suffering deteriorating mental health, e.g. negative welfare due to behavioural problems and/or decreasing mental faculties. All stakeholders need to be identified and include animal caregivers (current, previous), veterinary team, zoo management, zoo association staff (e.g. programme managers, Taxon Advisory Group (TAG), chairs, and association management/board members), internal media department, zoo visitors, zoo animal adopters, local communities, landowners, wider community, and external media. Communication early in the process is helpful and provides time for effective two-way communication with stakeholders and for directly affected staff to feel agency in the decision-making process. Electronic/written records of discussions, meetings, and action points, as well as diarised times for reassessments, are important parts of this overall process. One aim of the communications with directly affected staff is to build a consensus in support of a decision. There may be animals for which veterinarians are recommending euthanasia,

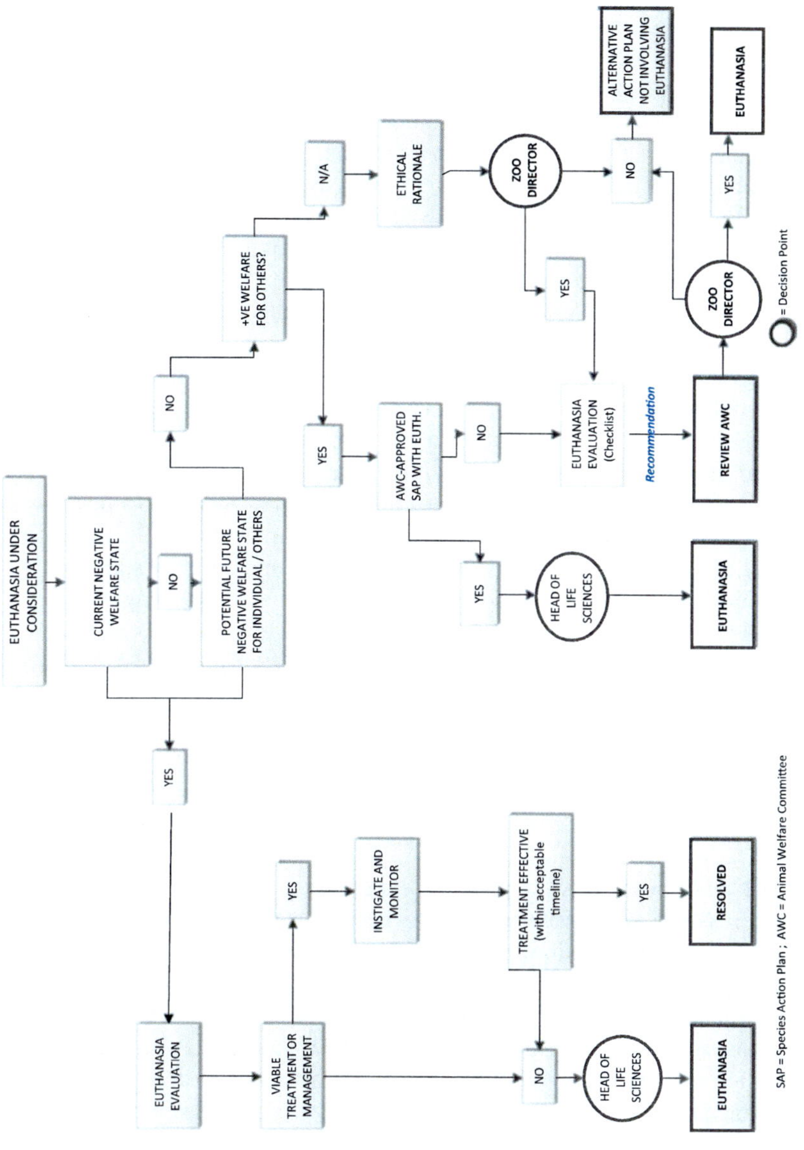

Fig. 1 Example euthanasia decision-making flowchart from Auckland Zoo in New Zealand

but some zookeepers and/or zoo management do not agree with the decision. Alternatively, there may be occasions where veterinary staff do not agree with the euthanasia decision which has been requested by zoo management. Unanimous decisions are not always possible, but with a transparent decision-making process it is hoped that all staff understand that a decision is required and respect the process used, even if some staff might not agree with the specific decision itself.

Many modern zoos create committees (e.g. Animal Ethics and/or Animal Welfare Committee) to provide input into euthanasia decisions, which may also fulfil legislative requirements, depending on the location and the situation involved. These committees can provide external validation to, and demonstrate transparency of, the overall decision-making process. These committees usually consist of a wide range of people representing diverse backgrounds and eclectic points of view, for example: both senior and junior staff from all zoo teams (i.e. not just animal care teams); non-zoo members with experience in the subject matter (e.g. animal welfare organisations, veterinarians); and/or local community representatives (e.g. local head of school, community leaders).

In addition to animal welfare assessments, there are also ethical components to the euthanasia decision-making process. The culture of a specific institution can be influenced by the staff but is often also bound by the cultural beliefs and religion(s) of the people in the country within which the institution is located. It is beneficial for an institution to develop an ethical decision-making framework to provide staff and other stakeholders with clarity over how ethical decisions are made (Knesl et al. 2017). In addition to providing a framework for staff making these decisions, this process also provides transparency and helps engender trust in the overall process. It is also important for institutions to recognise the potential impact of euthanasia decisions on the mental health of staff involved and to provide access to support where required.

2.2 Euthanasia Procedure

The euthanasia procedure should be performed in such a way that minimises the stress and/or other negative welfare states experienced by the animal(s), always maintains staff safety, and satisfies relevant legislation (AVMA 2020; BIAZA 2014; Woodbury 2014). With these aims in mind, the overall euthanasia procedure requires careful planning and effective preparation and can be divided into three main phases: pre-euthanasia, euthanasia, and post-euthanasia. Zoo veterinarians should have an effective euthanasia plan outlined for all species under their care, which allows for implementation at short notice in a reactive situation when required; e.g. see disease trajectory 4 mentioned previously in Shearer (Shearer 2011). However, many geriatric zoo animals will follow disease trajectories 1–3 (Shearer 2011), and there are significant benefits to performing euthanasia in a well-planned and proactive manner in these cases. This is especially true when dealing with potentially dangerous and/or high-profile individual animals.

2.2.1 Pre-euthanasia

This phase includes effective communication with all stakeholders, together with detailed planning of the euthanasia procedure itself. Effective and timely cross-departmental communication is vital for the overall success of a euthanasia procedure. All relevant stakeholders should have been identified in the decision-making process and, where appropriate/possible, will have received advanced communications about the planned procedure. The timing and location of the euthanasia procedure will, as outlined earlier, have implications for staff, including other animal care sections, maintenance, visitor services, and marketing/media teams. Euthanasia of geriatric zoo animals, especially those that are long-lived and/or iconic, commonly leads to a response from zoo visitors and wider community. In geriatric animals suffering from diseases with expected trajectories 1–3 (see earlier in this chapter), it may be beneficial to consider external communications at an early stage and throughout the decision-making process. This can give time for people to understand and trust the decisions being made and the process being followed, and ultimately support any subsequent euthanasia decision. Responsibility for these communications can fall to a range of zoo staff, depending on the organisational structure. The communication typically acknowledges the details of the animal and the rationale for the euthanasia decision. Communications at the time, or subsequently, may also seek to celebrate the life of the animal and the role it played in the conservation of the species (Vogelnest and Talbot 2019; WAZA 2003).

Veterinarians should communicate clearly with zoo staff involved in the euthanasia procedure, avoiding ambiguity, and providing a forum for non-veterinary staff to ask questions. Information provided should include roles and responsibilities; identifying hazards and/or health and safety aspects; method(s) of restraint and euthanasia; expected reaction of the animal (including potential reactions after euthanasia has been performed); reactions of conspecific and/or sympatric animals; necessary contingencies; and the post-euthanasia plan (see later). There are a number of public/client-facing documents available through various veterinary organisations, which outline how euthanasia is typically carried out with domestic animals, and these can provide a useful framework for discussions with zoo staff (Animal Welfare Foundation 2013; British Veterinary Association 2016; World Association of Zoos and Aquariums 2015).

The procedural plan for euthanasia should identify the necessary people, equipment, location, timing, and restraint required for euthanasia to be performed. It should also include sufficient contingency planning so that the staff involved have a clear understanding of their roles and responsibility as the procedure progresses. The relevant skilled staff should include experienced veterinary staff and animal caregivers but may also include other zoo staff including those from maintenance, visitor services, management, and/or communication teams. In addition to the required veterinary equipment, a range of supportive equipment may be required, e.g. manual lifting equipment, gurneys/trolleys, and vehicles such as a flatbed truck, a forklift, or a crane (McAloose et al. 2018). Noisy, large, or heavy equipment should arrive on site prior to the procedure commencing, so as not to disturb the animals or otherwise interfere with the smooth running of the procedure.

A suitably controlled environment should be identified which is preferably indoors, or at least protected from inclement weather. The planned location and timing of the procedure should allow the animal to remain as calm as possible whilst permitting suitable access for veterinary staff. Veterinary staff should liaise with other zoo staff to identify risks of increased auditory or visual stimulation of the animal, e.g. maintenance work, zoo-hosted parties, and/or music in the vicinity. Additional consideration should be given to the chosen location and timing if this will unavoidably result in the procedure being visible to the zoo visitors. This may require certain areas/walkways within the zoo to be temporarily closed, which creates challenges for other parts of zoo operations, and so forward planning and effective communication are required to allow the rest of zoo operations to action any necessary mitigations.

Immediately prior to the euthanasia phase of the procedure, there will usually be behavioural, physical, or chemical restraint used to facilitate euthanasia itself. Behavioural restraint may involve changes to the social setting of the animal and the technique and timing of this can be crucial to minimise the stress to the animal(s). However, the specific methodology will vary depending on the species, individual animal, and setting; e.g. a geriatric goat living in a herd should be separated prior to euthanasia, to avoid herd-mates reacting negatively during euthanasia. However, this separation can itself lead to increased stress for the individual; thus, mitigations should be in place to minimise this. An operant conditioning plan is often beneficial to the pre-euthanasia and the euthanasia phases, and this can require weeks or more of work in advance of the procedure itself. This is especially useful with larger species, e.g. helping a great ape remain calm when separated from the rest of the group and/or permitting hand injection of an anaesthetic and an elephant in a protective contact situation being trained to lay down and permit intravenous catheter placement. With geriatric long-lived species, these behaviours may already be well established. In geriatric animals, decreasing physical health may require enclosure adaptations to facilitate the same behaviour; e.g. the lowering of platforms, installing steps where the animal previously would have been able to leap, changing the position of access hatches, and decreasing mental faculties may similarly negatively impact the ability of the animal to participate fully in these training events.

Physical restraint may be appropriate and necessary in some circumstances, e.g. restraining a bird or small mammal whilst anaesthesia is induced, and staff should have access to relevant protective equipment to minimise injury during this process. Geriatric animals may have physical health conditions that will cause pain on manual restraint, e.g. osteoarthritis, necessitating altered techniques and/or inclusion of effective analgesia prior to the restraint occurring. Ultimately, the result of geriatric health and welfare assessments should be considered when planning the restraint technique.

With zoo and wildlife species, some form of chemical restraint is commonly used as a precursor to euthanasia. Oral anxiolytics, neuroleptics, and/or sedatives can be useful, though there is a wide range of inter- and intra-species variation in efficacy. Furthermore, some individual animals may exhibit an idiosyncratic paradoxical (rare, unpredictable) reaction to some of these drugs (Lennox 2020;

Woodbury 2014). Where possible, the authors advocate trialling a chosen drug in the individual animal(s) in advance of the euthanasia procedure, often more than once, to allow an appropriate dose rate to be chosen and to minimise the risk of unexpected paradoxical reactions during the euthanasia procedure itself.

General anaesthesia may also be necessary to suitably chemically restrain the animal prior to euthanasia. Experienced zoo veterinarians should choose anaesthetic drugs which provide quick and reliable induction of anaesthesia. If animals are unable to be hand-injected, then restraint equipment, e.g. nets, chutes, squeeze-cages, or darting equipment, may be used in the administration of anaesthetic drugs. The use of restraint equipment carries some inherent risks, e.g. darting, can temporarily increase stress for the animal during the pre-euthanasia phase, and may require a higher dose, therefore increasing the risks. However, experienced zoo veterinary teams, working effectively with animal caregivers, will be able to minimise the severity and duration of this stress. Effective drug combinations and techniques for zoo animal anaesthesia, together with the relevant health and safety and legislative requirements, are well described elsewhere and are beyond the scope of this chapter (Heard 2016; Mans et al. 2019; Sladakovic and Divers 2019; West et al. 2014). Higher doses of anaesthetics may be considered in euthanasia procedures to ensure a deeper plane of anaesthesia and because successful recovery is not required. However, higher doses of some drugs can negatively affect blood pressure, complicating intravenous access and euthanasia drug administration. Additionally, geriatric animals may be more likely to be suffering from diseases causing poor circulatory function, further complicating intravenous access, especially during general anaesthesia.

2.2.2 Euthanasia

To meet the definition of euthanasia the procedure must meet certain criteria, which are often also stipulated by regional/professional governing bodies. These criteria include that the procedure must be humane, minimise pain and/or distress experienced by the animal, and ultimately lead to permanent loss of neurological/brain function (AVMA 2020; Greenacre 2016; Nevarez 2019; Woodbury 2014). In most mammal and avian species there will also be a loss of cardiac and respiratory function immediately preceding the loss of brain function; however, this order and timeline may not hold true in other taxa. In ectotherms, it is necessary to physically damage the brain to ensure permanent loss of function, which may involve pithing and/or decapitation, but must only be performed once the animal is unconscious, most commonly as a result of general anaesthesia (AVMA 2020; Nevarez 2019; Woodbury 2014). In modern zoos, euthanasia of zoo species is generally performed by a veterinary surgeon with the correct expertise, support staff, equipment, and/or drugs relevant for the species. However, depending on the legislation of the country, non-veterinarians may be allowed to perform euthanasia under certain specified circumstances.

Euthanasia can be performed with a range of physical and chemical techniques and the chosen method will depend on the species of animal, availability of experienced zoo veterinarians, relevant skill and experience of non-veterinary staff, available equipment, location, health and safety aspects, and relevant legislation (AVMA 2020). The choice of methodology will also be dependent on the post-euthanasia plan: for example, if full necropsy is planned, including histopathology of the brain, then pithing is not appropriate (Nevarez 2019), and in some specific circumstances, some body parts may be used as food for other animals, e.g. feeding deceased rodents to carnivores.

Physical euthanasia methods commonly used in zoos include the use of manual methods, captive bolt, or firearms. Manual methods include blunt force trauma to the head, which can cause death outright, or might render the animal unconscious, requiring an additional procedure to complete the euthanasia such as cervical dislocation, intra-cranial brain destruction (e.g. pithing), or decapitation. Manual methods should only be used in small species with thin craniums, e.g. small birds, rodents, and neonates. The use of captive bolt is an effective and appropriate technique when used by trained staff but requires close contact with the animal, which may present additional health and safety concerns with non-domestic species, though these may be mitigated using operant conditioning; this procedure may also require an additional procedure (such as pithing) to complete euthanasia. The use of firearms may also be appropriate in certain situations, e.g., for a megavertebrate species collapsed in a mud wallow with no viable option for darting, but requires a high level of skill, training, situational awareness, and health and safety practices for the staff involved, the details of which are beyond the scope of this chapter (AVMA 2020; Woodbury 2014). With any method chosen, there needs to be a contingency plan in place, so that staff can still effectively deal with the procedure if it is not progressing as initially intended.

Chemical euthanasia involves initial restraint followed by administration of the euthanasia drug, typically intravenously, and suitable agents are well described elsewhere (AVMA 2020; Woodbury 2014). Barbiturates, e.g. sodium pentobarbital and secobarbital, are commonly used veterinary drugs, though large volumes may be required in large animals, depending on the concentrations available locally, and there is a significant risk of secondary toxicity with the carcass. Agents that induce cardiac arrest, such as potassium chloride and magnesium chloride, can also be administered intravenously, carry reduced risk of secondary toxicities, and can have a more practical volume in large animals, but the animal must be adequately anaesthetised prior to administration of these agents to avoid negative welfare and/or reactions that are hazardous to staff (AVMA 2020; Woodbury 2014). Contingency plans should again be in place and well understood by staff; with large/potentially dangerous species, the presence of a skilled emergency-response/firearms-trained member of staff is advisable.

Whilst the euthanasia procedure typically involves veterinary staff, animal caregivers may also be required for some/all of the procedure. Additionally, some non-veterinary staff may wish to be present immediately prior and/or during the euthanasia procedure for training purposes, to aid their own grief process, and/or to facilitate internal and external communications. Perhaps somewhat counterintuitively, it is the authors' experience that euthanasia procedures carried out in a professional and empathetic manner can improve the working relationships between veterinary and non-veterinary teams. The psychological wellbeing of all staff should be taken into consideration, including the diverse way in which grief is experienced, and where possible, staff should feel agency in their degree of involvement in the euthanasia procedure. The effects on veterinary staff of carrying out animal euthanasia should not be underestimated and can be more pronounced when veterinary staff have worked for many years with the zookeepers and/or the animals, as well as the accumulative effects of procedures. A constructive, supportive working relationship between the veterinary team, zookeepers, and zoo management is crucial to ensuring all staff feel adequately supported through what can be an extremely challenging time (Brown 2011; Cohen 2007; Vogelnest and Talbot 2019).

2.2.3 Post-euthanasia

The post-euthanasia phase contains four key elements: post-mortem examination; carcass disposition; relevant grieving process for staff; and the plan for conspecific/sympatric animals. The carcass should be handled with respect throughout this phase. Discussion and planning for this phase should occur in the pre-euthanasia phase; however, for ease of reference, some of the planning aspects are included in this section.

Post-mortem examinations (necropsies) are an important part of disease surveillance in zoos and provide data of direct relevance to the healthcare provision for the remaining animals. Necropsies require staff expertise, dedicated facilities, and a range of specialised equipment which are well described elsewhere (McAloose et al. 2018). Necessary staff includes experienced zoo veterinarians and/or veterinary anatomical pathologists, veterinary support staff, and other staff with necessary expertise, e.g., to operate lifting equipment and/or a chainsaw (McAloose et al. 2018). In the authors' experience, some animal caregivers wish to be involved with the necropsy and these occasions can provide a useful forum for increasing knowledge in areas such as anatomy, animal welfare assessments, and refining decision-making processes. In geriatric animals especially, some staff can find it useful to see any pathological lesions present as this helps them understand more about the disease processes that were present. In turn, this can positively influence subsequent decision-making in similar cases. Other non-veterinary staff may not want to be present for the necropsy and, as with the euthanasia procedure itself, wherever possible staff should feel agency in their personal degree of involvement.

Creation of a clear list of sample requests is useful, which may be in addition to those necessary for pathological investigations, e.g. tissue banking, teaching materials, "frozen ark", museum requests, cultural requests from indigenous groups. The necropsy itself may take some considerable time depending on the species and the degree of sample collection. Lifting equipment may be required for transferring the carcass and/or during the necropsy itself, and other specialised equipment (e.g. chainsaws) may also be required, especially with megavertebrate species (McAloose et al. 2018).

The grieving process of affected staff should be discussed, well understood, and empathetically addressed by their peers, veterinary staff, and zoo management (Brown 2011; Cohen 2007). Animal caregivers, veterinary teams, and other stakeholders from within the organisation may request time with the animal after death which can be an important part of their grieving process. Grief may be more pronounced with the death of geriatric animals, especially with long-lived and/or charismatic species. Mementos such as ink prints of hands or noses may be taken by staff and some cultures may require return of the carcass for ceremonial purposes (New Zealand Conservation Authority 1997).

Grief may also be expressed by members of the public, and as mentioned in the pre-euthanasia phase, effective communication with zoo visitors and the wider community about the death of the animal is an important part of the work of modern zoos (Vogelnest and Talbot 2019). The communications may reference some of the relevant necropsy results and ultimately should strike a balance between professionalism and empathy whilst keeping conservation messaging at their core.

The type of post-euthanasia plan required for conspecific/sympatric animals will depend on the individual and species involved. For instance, the necropsy of an elephant may need to occur within an enclosure, necessitating moving the conspecifics away from a certain area which may affect their access to shelter, food, and water. Studies of comparative thanatology (the scientific study of death) describe the behavioural response of animals following the death of an individual animal, typically a conspecific within their group. Some behaviours appear to be grief-like and other behaviours include increased affiliative behaviours towards the bereaved female (Anderson 2020; Goldsborough et al. 2020; Jakucińska et al. 2020). To allow similar behaviour to occur, some zoos elect to provide conspecifics with the choice to see and/or touch the dead body post-euthanasia. Allowing conspecifics access to the carcass post-euthanasia is not without risks, including the loss of post-mortem examination data through autolysis and/or missing body parts, as well as secondary toxicity if body parts are ingested. It may be practical for staff to agree a location in which the contact can occur that facilitates easy observation of the animals and subsequent separation of conspecifics if unwanted behaviour occurs. In some cases, allowing this behaviour may be prioritised over the risk of not being able to subsequently perform a necropsy. It is helpful to agree in advance a time limit, by which time the body is returned to the veterinary team for necropsy. The time required for veterinary staff to complete the necropsy should be considered when setting this time limit. The checklist, shown in Table 1, can be used to ensure all aspects of the euthanasia process have been considered and documented.

3 Conclusion

In conclusion, the reality of working with animals in a modern zoo is that there will be a significant number of geriatric individuals being cared for in a population. Utilising a welfare-centric approach will inevitably lead to better end-of-life decision-making for many of these individuals. These decisions include multiple stakeholders and can be emotive. Nevertheless, modern zoos should follow a transparent process, with a focus on detailed welfare assessment which utilises as much objective data as possible. Ultimately, proactive informed decision-making leads to euthanasia of geriatric animals at a time which prevents unnecessary suffering, giving the individual animals a respectful and dignified end to their lives.

Table 1 Checklist to be used prior to the euthanasia of a geriatric zoo animal

	Consideration	Y/N
Pre-euthanasia	Identify affected animal(s): individual(s) for euthanasia; conspecifics; sympatrics	
	Obtain approval for decision by relevant staff and/or confirmed through decision tree	
	Identify location and timing: animal welfare, vet access, staff health and safety, visitor view?	
	Communicate the decision: – Internal (keepers, vets, wider zoo, volunteers) – External (studbook, prior collections, visitors, sponsors, wider community)	
	Communicate the procedure plan (e.g. location/timing): – Vets, keepers, wider zoo (e.g. maintenance, marketing/PR)	
	Outline the plan for required animal movements and/or relevant operant conditioning	
Euthanasia	Identify the method of restraint and euthanasia to be used: staff, equipment, veterinary drugs/anaesthetics; staff briefing; and, where needed, debrief afterwards	
Post-euthanasia	Discuss plan for conspecific/sympatric animals: sight, sound, contact following euthanasia?	
	Identify and discuss plan for staff access (e.g. keepers, volunteers)	
	Identify and discuss the plan for *post-mortem* examination: – Location, timing, staff, equipment, sample requests, cultural requests, staff health and safety	
	Implement support plan for staff mental health awareness; access to support services	

References

Anderson JR (2020) Responses to death and dying: primates and other mammals. Primates 61(1):1–7. https://doi.org/10.1007/s10329-019-00786-1

Animal Welfare Foundation (2013) Saying goodbye—the ultimate kindness. 2021 (01/07/2021):1–11

AVMA (2020) AVMA guidelines for the euthanasia of animals. American Veterinary Medical Association

Bacon H (2019) Welfare of geriatric dogs and cats. In: BSAVA congress proceedings 2019. British Small Animal Veterinary Association. https://doi.org/10.22233/9781910443699.32.6

Bays TB (2020) Geriatric care of rabbits, guinea pigs, and chinchillas. Vet Clin North Am Exot Anim Pract 23(3):567–593. https://doi.org/10.1016/j.cvex.2020.05.006

Bergner M, Bobbitt RA, Pollard WE, Martin DP, Gilson BS (1976) The sickness impact profile: validation of a health status measure. Med Care 14(1):57–67. https://doi.org/10.1097/00005650-197601000-00006

BIAZA (2014) BIAZA euthanasia policy. British and Irish Association of Zoos and Aquaria

Boylan S (2020) Geriatric freshwater and marine fish. Vet Clin North Am Exot Anim Pract 23(3):471–484. https://doi.org/10.1016/j.cvex.2020.05.001

British Veterinary Association (2016) Guide to euthanasia, pp 1–6

Brown MV (2011) How they cope: a qualitative study of the coping skills of hospice volunteers. Am J Hosp Palliat Care 28(6):398–402. https://doi.org/10.1177/1049909110393946

Chitty J (2020) Care of the geriatric raptor. Vet Clin North Am Exot Anim Pract 23(3):503–523. https://doi.org/10.1016/j.cvex.2020.05.003

Cohen SP (2007) Compassion fatigue and the veterinary health team. Vet Clin North Am Small Anim Pract 37(1):123–134. Abstract ix. https://doi.org/10.1016/j.cvsm.2006.09.006

Collins C, Kays R (2011) Causes of mortality in North American populations of large and medium-sized mammals. Anim Conserv 14(5):474–483. https://doi.org/10.1111/j.1469-1795.2011.00458.x

Devlin NJ, Brooks R (2017) EQ-5D and the EuroQol group: past, present and future. Appl Health Econ Health Policy 15(2):127–137. https://doi.org/10.1007/s40258-017-0310-5

Dutton M (2020) Selected veterinary concerns of geriatric rats, mice, hamsters, and gerbils. Vet Clin North Am Exot Anim Pract 23(3):525–548. https://doi.org/10.1016/j.cvex.2020.04.001

Follmi J, Steiger A, Walzer C, Robert N, Geissbuhler U, Doherr MG, Wenker C (2007) A scoring system to evaluate physical condition and quality of life in geriatric zoo mammals. Anim Welf 16:309–318

Goldsborough Z, Van Leeuwen EJC, Kolff KWT, De Waal FBM, Webb CE (2020) Do chimpanzees (Pan troglodytes) console a bereaved mother? Primates 61(1):93–102. https://doi.org/10.1007/s10329-019-00752-x

Greenacre CB (2016) Euthanasia. In: Speer BL (ed) Current therapy in avian medicine and surgery, 1st edn. Elsevier, St. Louis, MO

Heard D (2016) Chapter 19—Anesthesia. In: Speer BL (ed) Current therapy in avian medicine and surgery. W.B. Saunders, St Louis, MO, pp 601–615. https://doi.org/10.1016/B978-1-4557-4671-2.00028-8

Jakucińska A, Trojan M, Sikorska J, Farley D (2020) Reaction to the death of the oldest female in a group of chimpanzees at the Municipal Zoological Garden, Warsaw. Primates 61(1):103–109. https://doi.org/10.1007/s10329-019-00772-7

Jessup DA, Scott CA (2011) Hospice in a zoologic medicine setting. J Zoo Wildlife Med 42(2):197–204

Johnson DH (2020) Geriatric hedgehogs. Vet Clin North Am Exot Anim Pract 23(3):615–637. https://doi.org/10.1016/j.cvex.2020.05.005

Kitchener A, MacDonald A (2004) The longevity of legacy. European Association of Zoos and Aquaria, Kolmarden, Sweden, 21/09/2004, pp 132–137

Knesl O, Hart BL, Fine AH, Cooper L, Patterson-Kane E, Houlihan KE, Anthony R (2017) Veterinarians and humane endings: when is it the right time to euthanize a companion animal? Front Vet Sci 4. https://doi.org/10.3389/fvets.2017.00045

Krebs BL, Marrin D, Phelps A, Krol L, Watters JV (2018) Managing aged animals in zoos to promote positive welfare: a review and future directions. Animals (Basel) 8(7):116. https://doi.org/10.3390/ani8070116

Krol L, Moore RP, Mutlow AG, Brady SM, Dorsa D (2020) A retrospective analysis of mortality in captive Magellanic penguins (Spheniscus magellanicus) in the United States, 2008–2018. Zoo Biol 39(6):405–410. https://doi.org/10.1002/zoo.21561

Kumar S, Laurence H, Owston MA, Sharp RM, Williams P, Lanford RE, Hubbard GB, Dick EJ Jr (2017) Natural pathology of the captive chimpanzee (Pan troglodytes): A 35-year review. J Med Primatol 46(5):271–290. https://doi.org/10.1111/jmp.12277

Lennox AM (2020) End-of-life decisions: palliative care, hospice, and euthanasia for exotic animals. Vet Clin North Am Exot Anim Pract 23(3):639–649. https://doi.org/10.1016/j.cvex.2020.06.003

Mans C, Sladky KK, Schumacher J (2019) 49—General anesthesia. In: Divers SJ, Stahl SJ (eds) Mader's reptile and amphibian medicine and surgery, 3rd edn. W.B. Saunders, St. Louis, MO, pp 447–464.e442. https://doi.org/10.1016/B978-0-323-48253-0.00049-0

McAloose D, Colegrove KM, Newton AL (2018) Wildlife necropsy. In: Terio KA, McAloose D, Leger JS (eds) Pathology of wildlife and zoo animals. Elsevier, San Diego, CA, pp 1–19

Mellor DJ, Beausoleil NJ, Littlewood KE, McLean AN, McGreevy PD, Jones B, Wilkins C (2020) The 2020 five domains model: Including human–animal interactions in assessments of animal welfare. Animals 10(10):1870

Nevarez JG (2019) Euthanasia. In: Divers SJ, Stahl SJ (eds) Mader's reptile and amphibian medicine and surgery, 3rd edn. Elsevier, St. Louis, MO, pp 437–440

New Zealand Conservation Authority (1997) Maori customary use of native birds, plants and other traditional materials. New Zealand Conservation Authority, Wellington, New Zealand. https://www.legislation.govt.nz/act/public/1953/0031/latest/whole.html#DLM277890

Pellett S, O'Brien M, Kennedy B (2020) Geriatric invertebrates. Vet Clin North Am Exot Anim Pract 23(3):595–613. https://doi.org/10.1016/j.cvex.2020.05.002

Raiti P (2019) Geriatric medicine. In: Divers SJ, Stahl SJ (eds) Mader's reptile and amphibian medicine and surgery, 3rd edn. Elsevier, St. Louis, MO, pp 960–966

Reavill DR, Imai DM (2020) Pathology of diseases of geriatric exotic mammals. Vet Clin North Am Exot Anim Pract 23(3):651–684. https://doi.org/10.1016/j.cvex.2020.06.002

Sapkota S, Laven R, Müller K, Kells N (2020) Animal welfare assessment: can we develop a practical, time-limited assessment protocol for pasture-based dairy cows in New Zealand? Animals 10(10):1918

Scheelings TF (2020) Geriatric reptiles and amphibians. Vet Clin North Am Exot Anim Pract 23(3):485–502. https://doi.org/10.1016/j.cvex.2020.05.004

Shearer TS (2011) Pet hospice and palliative care protocols. Vet Clin N Am Small Anim Pract 41(3):507–518. https://doi.org/10.1016/j.cvsm.2011.03.002

Singh S, Bajorek B (2014) Defining 'elderly' in clinical practice guidelines for pharmacotherapy. Pharm Pract (Granada) 12(4):489–489. https://doi.org/10.4321/s1886-36552014000400007

Sladakovic I, Divers SJ (2019) 52—Amphibian anesthesia. In: Divers SJ, Stahl SJ (eds) Mader's reptile and amphibian medicine and surgery, 3rd edn. W.B. Saunders, St. Louis, MO, pp 480–485.e481. https://doi.org/10.1016/B978-0-323-48253-0.00052-0

Villalobos AE (2011) Quality-of-life assessment techniques for veterinarians. Vet Clin North Am Small Anim Pract 41(3):519–529. https://doi.org/10.1016/j.cvsm.2011.03.013

Vogelnest L, Talbot JJ (2019) Quality of life assessment and end of life planning for geriatric zoo animals. In: Miller RE, Lamberski N, Calle PP (eds) Fowler's zoo and wild animal medicine current therapy, vol 9. Elsevier, St. Louis, MO, pp 83–91. https://doi.org/10.1016/C2016-0-01845-4

WAZA (2003) WAZA code of ethics and animal welfare. Wold Association of Zoos and Aquaria

West G, Heard D, Caulkett N (2014) Zoo animal and wildlife immobilization and anesthesia, 2nd edn. Wiley, Hoboken, NJ

Willems A, Paepe D, Marynissen S, Smets P, Van de Maele I, Picavet P, Duchateau L, Daminet S (2017) Results of screening of apparently healthy senior and geriatric dogs. J Vet Intern Med 31(1):81–92. https://doi.org/10.1111/jvim.14587

Wolfensohn S, Shotton J, Bowley H, Davies S, Thompson S, Justice WSM (2018) Assessment of welfare in zoo animals: towards optimum quality of life. Animals 8(7):110

Woodbury M (2014) Euthanasia. In: West G, Heard D, Caulkett N (eds) Zoo animal and wildlife immobilization and anesthesia. Wiley, Hoboken, NJ, pp 149–153. https://doi.org/10.1002/9781118792919.ch10

World Association of Zoos and Aquariums (2015) Caring for wildlife: The world zoo and aquarium animal welfare strategy, pp 1–45

ZAA (2015) ZAA euthanasia guidelines. Zoos and aquariums Australasia

The Longevity Legacy: The Challenges of Old Animals in Zoos

Andrew C. Kitchener

A. C. Kitchener (✉)
National Museums Scotland, Edinburgh, Scotland, UK
e-mail: a.kitchener@nms.ac.uk

S. Brando, S. Chapman (eds.), *Optimal Wellbeing of Ageing Wild Animals in Human Care*, https://doi.org/10.1007/978-3-031-30659-4_12

Coco. Coco, a female chimpanzee (*Pan troglodytes*), was probably born in 1990 in the wild. When she was a baby, she was sold in Las Ramblas, Barcelona, Spain, to a private owner, who had two other chimpanzees, Tom and Bea. All of them became friends and shared their lives together until 2012, when they came to Fundació Mona, north of Barcelona, where they now live and met other chimpanzees. From the very beginning of her arrival in Mona, Coco showed how curious she was; she needed to check, touch, and investigate every place, surface, or plant within her new home. Even now, she spends a lot of hours playing with the sand, the water, and everything she finds interesting in her environment. She enjoys solving puzzles and being focused in the enrichments that the keepers provide for them. She is incredibly smart and gives the keepers a hard time trying to find new and more difficult enrichments that can be challenging for her. Coco is also a really sensitive, empathic chimpanzee and a key individual in her group considering her strong social attitude. She is always ready to start or participate in grooming sessions with other group members, to offer consolation and give a big hug after a stressful moment. Even when Bea became older, Coco took care of her, helping her to climb, find food, or just with company and love, staying always by her side. Roger Belis Manyós

Abstract

As knowledge of husbandry has improved with the keeping of wild animals in zoos over the last 200 years, so longevity has also improved, bringing with it challenges owing to the development of pathologies associated with ageing. In this chapter, the principal skeletal and dental pathologies of aged zoo mammals are described, and the results of previous studies are summarised. Data are presented on pathologies from zoo specimens in National Museums Scotland that were collected over the last 30 years in a range of large mammals, including bears, big cats, great apes, babirusas, and pygmy hippopotamuses. The causes of these pathologies and the possible role of enrichment in reducing their prevalence are discussed. Awareness of the presence of these pathologies and their possible impact on behaviour and welfare is highlighted, as well as the need to monitor and assess individuals effectively, in order to minimise negative welfare.

Keywords

Ageing · Behaviour · Enrichment · Euthanasia · Pathology · Skeleton · Teeth

1 Introduction

When many zoos were established in the nineteenth century, the main aim was to maximise the number of species held as new parts of the world were explored by the European colonial powers. Lists were kept and updated of the many new species being held over the decades (e.g., Sclater 1879, 1896). Many new species and subspecies were first described from animals that lived often only a short time in zoological gardens (Blunt 1976). Long journeys over land and sea for many weeks and months left animals in a weak condition on arrival and there was often little or no knowledge of wild diets or possible alternatives that could be offered in captivity, so that many animals died after only a few days or weeks before being transferred to museums to be studied by museum curators and researchers. The longevity of animals was extended as knowledge of captive husbandry developed mostly by trial and error. For example, the first captive gorilla (*Gorilla gorilla*) at London Zoo was fed sausages, cheese sandwiches, boiled potatoes, mutton, and beer, and not surprisingly, it survived for only a few weeks in the autumn of 1887 (Blunt 1976). The maximum longevity of gorillas has increased from 33 years 5 months (Jones 1962) to 54 years only 30 years later (Nowak 1991), and the record today is now 56 years (Erwin et al. 2002; Weigl 2005). For some species improvements have been dramatic, such as for Goeldi's monkey (*Callimico goeldii*), where maximum longevity increased from only 2 years 4 months to 17.9 years in the same period (Jones 1962; Nowak 1991). It is still common for zoos to demonstrate their skills in captive husbandry by announcing longevity records (see numerous records in Manville 1957; Wolstenholme and O'Connor 1959; Rabb 1960; Snyder and Moore 1968; Jones 1968, 1977, 1993; Mallinson 1971; Nowak 1999; Geissmann et al. 2009). For birds see Stott (1948), and Brouwer et al. (1992, 1994, 2000). Tidière et al. (2016) analysed zoo record data and amongst 50 mammal species, longevity was increased in 84%, and particularly those with a fast pace of life (short life span, high reproductive capacity, and high mortality in the wild), whereas species with the opposite characteristics benefited less in captivity from extended longevity (see Fig. 1). Hatt et al. (2011) found that whilst many captive mammal species live longer than in the wild, some have shorter longevities than in the wild, such as elephants, some deer,

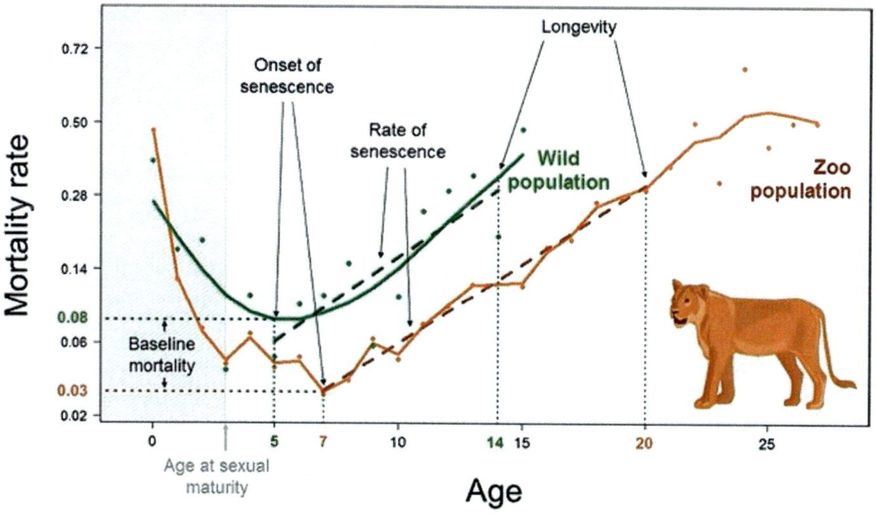

Fig. 1 Survival and senescence in wild and captive female lions (Tidière et al. 2016). Female zoo lions live longer (age in years), have a lower baseline annual mortality (log%), and show later onset of senescence (years) and a lower rate of actuarial senescence (measured as the exponential rate of mortality increase per year). Reproduced under Creative Commons CC BY licence and with permission of the author

and cetaceans. Roller et al. (2021) also showed that adult longevity amongst captive carnivorans increased in the period 1950–2019.

Much has changed over recent decades as zoos have become much more focused on animal welfare (Cole and Fraser 2018; Ward et al. 2018; Wolfensohn et al. 2018; Fernandez and Martin 2021), conservation, research, and education (Packer and Ballantyne 2010; Roe et al. 2014; Barongi et al. 2015). The perception of an endless source of wild animals has largely disappeared as populations in the wild have declined and become increasingly threatened. The introduction of CITES in the 1970s to control trade in endangered species has effectively banned trade in the most threatened species. In addition, the success of managed captive breeding programmes has resulted in increasing independence from the wild for some species and essential conservation support for others (Ballou et al. 1995; Conde et al. 2011; Fa et al. 2011). However, as captive husbandry has continued to improve, longevity has also continued to increase and the need to monitor and assess ageing in old animals and ways to manage them have also developed (Föllmi et al. 2007; Krebs et al. 2018; see chapter "Holistic Approaches for Promoting Good Wellbeing for Ageing Wild Animals" for an overview). Although longer lives may benefit some species in the development of social groups, successful reproduction, and opportunities for social learning, challenges and problems may also arise in the long-term management of long-lived species, which compromise the effectiveness of captive conservation programmes, hence the importance of a preventive and proactive approach. Age-related diseases are becoming common, some of which may be difficult to diagnose and treat, thereby compromising welfare. The use of humane euthanasia to alleviate suffering in old animals has also been increasingly questioned, particularly in great apes, by the public and zoo professionals (Erwin et al. 2002; Browning 2018; see chapter "The View from Beyond the Fence: Ageing Zoo Animals and Communicating with the Outside World" for an overview).

Ageing may have a variety of consequences that affect welfare, including dental and skeletal deterioration, muscle wastage, reproductive senescence, memory loss, and behavioural and cognitive decline. Although ageing and its consequences are well studied in humans and some domestic and laboratory mammals (Colman and Binkley 2002; Colman et al. 1999; Coni et al. 1992; Herndon et al. 1997, 1999; Lacreuse et al. 1999; Lane 2000; Mohr et al. 2001; Obregon and Ramirez 1997; Packer et al. 1998; Phelan 2001; Smucny et al. 2001), it is unusual for other species. Perhaps, this is because wild animals rarely live long enough to suffer from age-related problems and zoo animals have not been well studied at this stage of their lifespan. An example of an early study of wild mammals was the investigation of hyaline arteriosclerosis (a common vascular lesion resulting in a thickening of the arteriolar wall, particularly in the kidneys, caused by the accumulation of various serum proteins in the subendothelial space) in Africa by McKinney (1970), which determined that although this condition occurred in wild mammals and was age-related, its prevalence was far less than in humans. Prevalence of skeletal pathologies has been recorded in other wild mammals in a few other studies, e.g., Fox (1939), Schultz (1935, 1939, 1940, 1941, 1942, 1944, 1956, 1960), and Greer et al. (1977). However, these kinds of studies are still comparatively rare and do not go beyond reporting prevalence of mostly skeletal deterioration, often with very small sample sizes (Canfield and Spencer 1993; Morbeck et al. 2002). Rothschild and his collaborators have investigated the prevalence of skeletal pathologies, such as osteoarthritis and spondyloarthrosis, in a wide array of mostly wild taxa based on museum collections (Rothschild 1993, 2005, 2009, 2010, 2021; Rothschild and Panza 2005, 2006; Rothschild and Rühli 2005a, b, 2007; Rothschild and Woods 1989, 1991, 1992a, b, c, 1993, 1996; Rothschild et al. 1993, 1998, 2001a, b, 2005). In general, the prevalence of skeletal pathologies was low, probably because of the poor survival of affected animals, which would be susceptible to predation or unable to hunt or forage. Some of these studies (e.g., Rothschild et al. 1998) also investigated prevalence of skeletal pathologies in captive animals.

The prevalence of age-related conditions in domestic mammals is recorded and methods for their alleviation are being developed as they have for humans (e.g., Giel et al. 2000). There is a growing realisation that captive great apes may have to be looked after in old age as longevities increase and because culling is seen as unacceptable (Erwin et al. 2002; Lowenstine et al. 2016). For example, the first hip replacement was carried out on a female gorilla (*Gorilla g. gorilla*) called Beta at Brookfield Zoo, Chicago, in 1986. Studies of ageing in zoo animals are rare. For example, Nichols and Zihlman (2002) have recorded the prevalence of dental and skeletal deterioration in captive gorillas ($n = 5$) and have shown that females are potentially more likely to suffer these problems, although it must be noted that the sample size was very small. Lowenstine et al. (2016) reviewed age-related problems in wild and captive great apes, including dental and skeletal pathologies, and found that whilst only humans suffer from gout, spondyloarthropathy is more prevalent in great apes than in humans. Kitchener and Macdonald (2002, 2005) and Kitchener (2004) recorded the prevalence of dental and skeletal problems in captive bears; more than 90% of individuals from seven species older than 16 years old had severe osteoarthrosis (degeneration of cartilage in synovial joints with associated changes in adjacent bone) or spondyloarthrosis (inflammatory diseases that affect the vertebral column and peripheral joints). This is of great concern given that bears often live more than 20 years in captivity. Kitchener et al. (2000) carried out a preliminary study of the effect of captivity on the flight musculoskeletal system of Rodrigues fruit bats (*Pteropus rodricensis*). They observed that captive animals continue to accumulate subcutaneous fat deposits throughout life, so that individuals more than 20 years old had 30% subcutaneous fat and were probably incapable of flight. Not only did this have an important effect on a key behaviour of this species, but also it is likely to be having other effects on health and welfare that are associated with obesity, such as heart disease and diabetes.

As wild habitats are increasingly destroyed and fragmented, and resources for *ex situ* conservation are severely limited, the International Union for Conservation of Nature (IUCN) has developed the One Plan approach, which aims to manage both wild and captive populations together (Byers et al. 2013). This is especially important for large species, which live at low population densities. Thus, as the importance of captive populations increases, it is equally important that the welfare of captive individuals is assessed critically to minimise suffering, maintain good health status into old age, and maximise the efficient management of captive populations in synchrony with wild populations. The problem of what to do with old zoo mammals, which may occupy limited enclosure space intended for supporting sustainable captive populations, is pressing (Browning 2018).

This chapter is an update of Kitchener and Macdonald (2002) and Kitchener (2004). It briefly reviews the range of skeletal pathologies that have been observed in zoo vertebrates and previous studies on wild and zoo mammals, with a particular emphasis on primates and large felids. It also presents data on dental and skeletal pathologies based on specimens acquired from zoos over more than 30 years by National Museums Scotland, emphasising the need for long-term collection of skeletons for comparative studies. Ways in which the longevity legacy can be resolved for the benefit of species and individual animals alike are discussed. However, this will require a greater understanding by all zoo staff and the wider public of vital issues concerning natural life cycles, behaviour, and the welfare of captive and wild animals.

2 Bears in Zoos

Bears are long-lived and may produce many cubs during their lives (Table 1), and they are often housed in expensive enclosures designed to be strong enough to hold such powerful animals. In addition, because other aspects of husbandry, including veterinary care, have continued to improve, longevities of bears have also increased. As a result, for most species, the proportion of the captive population made up of bears of more than 20 years of age is often very high. The captive polar bear population is an extreme case with 32% of individuals more than 20 years old compared with only 3% of this age in the well-studied population in Hudson Bay, Canada (Ramsay and Stirling 1988). Bears begin breeding at a relatively young age (c. 3–6 years; Table 1) and may produce several litters of often more than one cub, so that they may have very long breeding lives with the potential to produce many young even

Table 1 Mean life expectancy, ages to which 25% of the captive population survives, maximum-recorded captive age, first and maximum ages of female breeding in zoos in years

	Mean life expectancy at birth (years)		Age to which 25% survive (years)		Maximum-recorded age (years)		Female first and maximum ages of zoo breeding (years)
Species	Males	Females	Males	Females	Males	Females	
Polar bear	12.1	16.4	19.5	25.9	35.9	38.5	4–28
Andean bear	16.3	19.1	26.5	27.7	32.1	33	3–23
Sloth bear	14.6	13.9	19	20.7	27	32.1	2–20
Sun bear	12.9	21.3	21.9	29.6	31.7	35.8	3–23

Modified from Kolter et al. (2021)

with a long inter-birth interval. For example, a pair of European brown bears (*Ursus arctos*) at the Highland Wildlife Park produced 18 young in 8 litters before the male died at the relatively young age of 28 (J. Usher-Smith personal communication). With no prospect of being able to house cubs, many zoos have given up breeding bears and have decided to no longer keep bears when their old animals die. The consequence of bears surviving for many years beyond breeding age is that breeding is severely limited by the lack of opportunity to place cubs in the collections that otherwise continue to house these old non-reproducing bears. The Ex-Situ Programme (EEP) Ursid Taxon Advisory Group (TAG) has long recognised that the lack of suitable cage space is compromising breeding programmes and the long-term survival of captive populations in Europe (Kolter 1995).

For more than 30 years, National Museums Scotland has been receiving dead bear specimens from several European zoos, in order to preserve their skins and skeletons for taxonomic and other research as part of the European Association of Zoos and Aquaria (EAZA) Ursid TAG post-mortem protocol. Preliminary results were presented by Kitchener et al. (2001), and this chapter is intended as an update now that more data are available. After preparing the skeletons of these bears, it soon became evident that most were elderly (i.e. 20+ years old) and that they were suffering from severe skeletal and dental pathologies, which in humans would be regarded as painful. Given the potential for bears to live well into their thirties, zoos may be unwittingly condemning their bears to long painful lives as their teeth and skeletons degenerate, if left unmanaged.

2.1 Skeletal Pathologies

Vertebrate skeletons are subject to various skeletal pathologies that are caused by trauma, congenital disorders (Fig. 2), nutritional deficiencies, infection, and ageing (wear and tear). Bone fractures, tumours, infections, and nutritional deficiencies

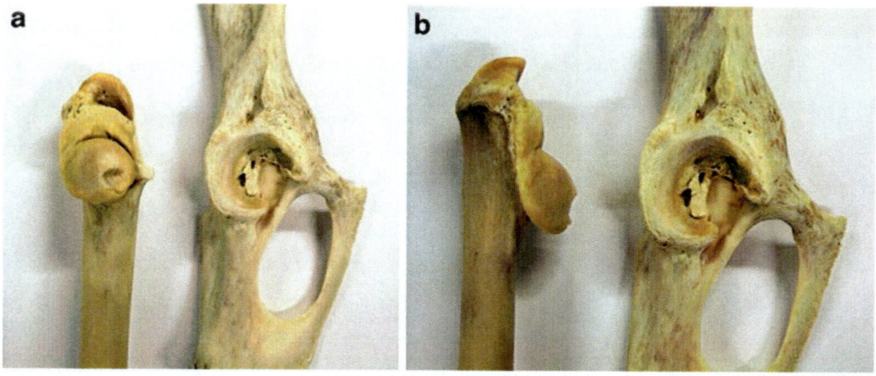

Fig. 2 Hip dysplasia in an Amur leopard (*Panthera pardus orientalis*) (NMS.Z.1999.201); a congenital condition that may occur at high frequencies in captive breeding programmes owing to inbreeding: (**a**) medial and (**b**) anterior view of femoral head © National Museums Scotland

may affect any part of any bone, whereas joints become affected by arthrosis. This chapter reviews the range of principal skeletal and dental pathologies that affect captive and wild vertebrates, including their known and possible causes, as well as the prevalence of these pathologies in captive populations of some larger mammal species. There is a glossary of the terms used to describe skeletal pathologies in this chapter in Table 2. Finally, the importance of these findings is discussed in relation to welfare, husbandry, and captive conservation.

Various diseases are caused by bacterial infections of bones, including tuberculosis and leprosy. The resulting osteitis (infection of bones), including periostitis (infection of the periosteum surrounding bones) and osteomyelitis (infection of the bones themselves), may cause a range of effects from degradation of bone to irregular (rough) thickening of the bone surface (Brothwell 1981). Neoplastic bony tumours may develop on the surface of the bone, resulting in irregular bony growths.

Nutritional deficiencies, such as calcium, phosphorus, and vitamin D, result in rickets in juveniles with their characteristic bent long bones and osteomalacia or soft bones in adults (Fig. 3) (Brothwell 1981). There have been numerous studies of metabolic bone disease in zoo vertebrates, e.g., Du Boulay and Crawford (1968), Freedman et al. (1976), McWilliams and Leeson (2001), Adkesson and Langan (2007), Chesney and Hedberg (2010), and Farrell et al. (2015).

Fractures may reveal rough broken surfaces, if peri-mortem, or smooth surfaces and the presence of callus where broken bones have healed (Fig. 4) (Duckworth 1911; Halstead 1974). In a recent study of the skeletons of 107 wild Japanese macaques, *Macaca fuscata*, Nakai (2003) found the commonest pathology was 80 healed fractures in 31 of 52 males and 71 healed fractures in 26 of 55 females.

Table 2 Glossary of skeletal pathologies

Pathology	Definition
Spondyloarthropathy	Includes inflammatory and non-inflammatory spinal changes
Spondyloarthritis	Any inflammatory arthritis affecting the spine, including ankylosing spondylitis and arthritis associated with other autoimmune disorders, such as inflammatory bowel diseases, psoriasis, and Reiter's Syndrome. It could also include spinal infection, such as TB or *Brucella*, but those are always called by their own names
Ankylosing spondylitis	An inflammatory arthritis of the spine as a result of an abnormal immune reaction, which progresses to fusion of the vertebrae. It may additionally affect other joints in the appendicular skeleton
Spondylosis	A degenerative spinal disorder due to wear and tear in the intervertebral discs. It includes osteoarthrosis of the posterior zygapophyseal joints. Severe cases are often called spondylosis deformans
Osteoarthrosis	Degenerative wear-and-tear disorder of synovial joints often erroneously called osteoarthritis
Rheumatoid arthritis	An autoimmune inflammation, but mostly only affects the peripheral joints
Calcium pyrophosphate deposition disease	An arthritis caused by deposition of calcium pyrophosphate crystals in joint cartilage

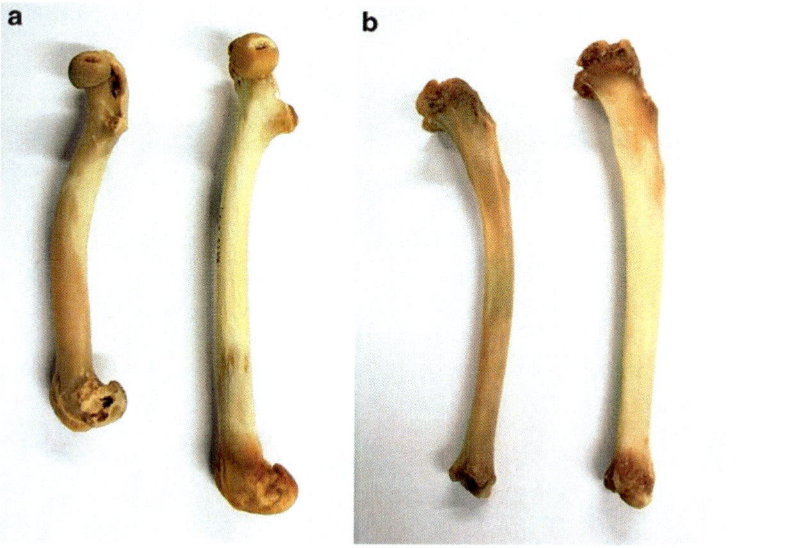

Fig. 3 Rickets in the femur (**a**) and tibia (**b**) of a male stump-tailed macaque (*Macaca arctoides*) (NMS.Z.2018.41.6) on the left compared with normal bones from a normal male on the right (NMS.Z.2017.84.1). © National Museums Scotland

Fractures in males mostly occurred in adulthood and those that affected the trunk were probably caused by rolling down steep slopes or falling out of trees, possibly during intertroop transfers, whilst fractures in females occurred during childhood and senescence. However, secondary osteoarthrosis due to fractures was rare, but degenerative changes were common in old animals. Schultz (1944) recorded old mostly healed fractures in 28% of female and 37% of adult male lar gibbons, *Hylobates lar*, from Thailand, and similar percentages amongst male (36%) and female (30%) orangutans, *Pongo* spp. (Schultz 1941).

Arthropathy affects the vertebral joints (spondyloarthritis and spondylosis) as well as the limb joints and may be due to a number of causes, including ageing, infections, immune response, and unknown causes (Brothwell 1981; Rothschild 2005; Lowenstine et al. 2016). However, it is often difficult to compare studies because of a confusing and changeable nomenclature as well as probable differences between species (Rothschild 2015; Lowenstine et al. 2016).

Osteoarthritis, or more correctly osteoarthrosis, is characterised by destruction of synovial joints with loss of cartilage initiated by a change in the nature of the synovial fluid (Halstead 1974; Gardner 1983; Hutton 1989). This is also commonly known as degenerative joint disease (DJD) as in Fig. 5 (Lowenstine et al. 2016). Osteoarthrosis is often related to age, but is also a consequence of injuries, damage from infections, ageing, nutritional or metabolic disorders, endocrine dysfunctions, etc. (Gardner 1983). Although osteoarthrosis is degenerative, it is also anabolic, resulting in bone sclerosis, capsular fibrosis, and osteophyte formation. When the cartilage is completely lost, the bone is exposed, resulting in eburnation (polishing

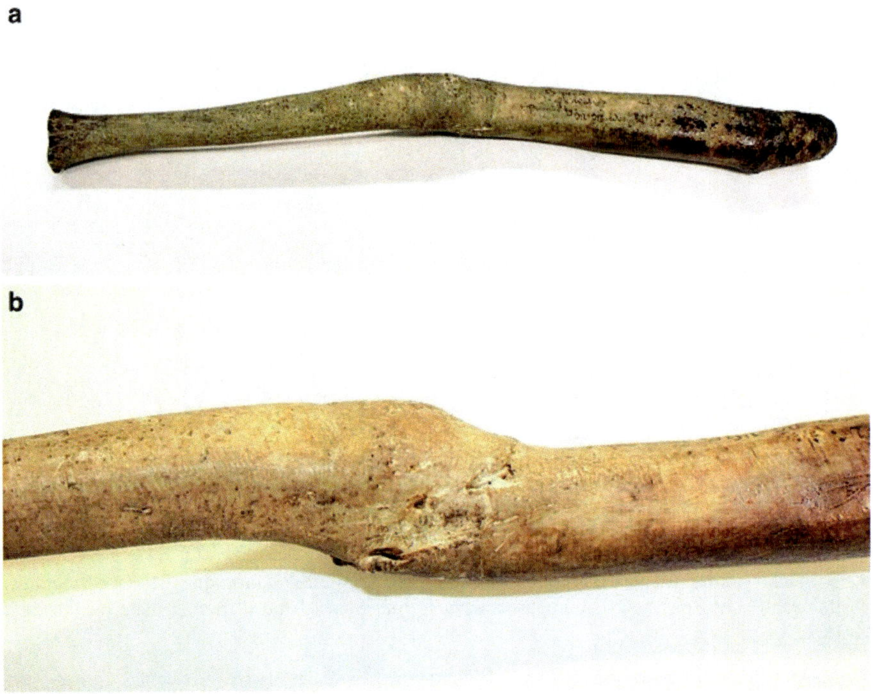

Fig. 4 Healed fracture in the baculum of a walrus, *Odobenus rosmarus* (NMS.Z.1996.83.30): (**a**) baculum, (**b**) close-up of healed fracture. © National Museums Scotland

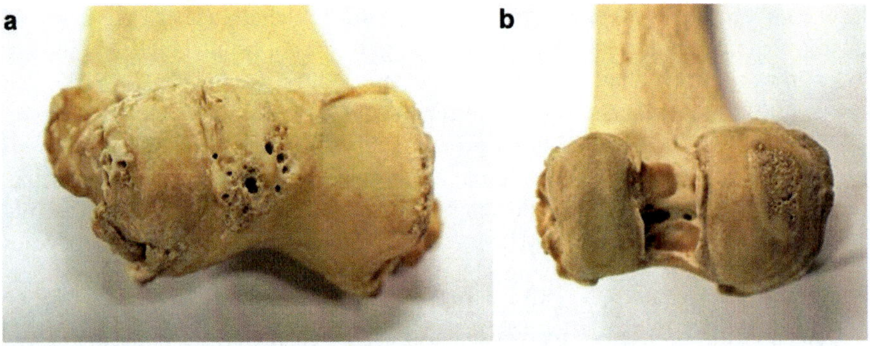

Fig. 5 Degenerative joint disease in the distal right humerus (**a**) and distal left femur (**b**) of a 48-year-old female western chimpanzee (*Pan troglodytes verus*) (N.B. eburnation) (NMS.Z.2018.129.1). © National Museums Scotland

of bones by rubbing against each other), remodelling, and development of osteophytes (bony growths) and cysts in the subarticular bone (beneath the joint surface).

There are several types of autoimmune inflammatory arthritis (IA). Rheumatoid arthritis (RA) occurs in humans and causes stiff swollen joints. This condition is serum-positive for rheumatoid factor and primarily affects the appendicular skeleton. However, in both humans and other animals there are forms of arthritis, which are secondary to inflammatory processes elsewhere in the body, such as inflammatory bowel disease, bacterial enteritis, urinary tract infections, and psoriasis (Lowenstine et al. 2016). These kinds of IA often affect the vertebral column as well as the appendicular skeleton and are known as seronegative, non-erosive, or reactive arthritis. If the spine is involved, the disease is known as spondyloarthritis (SA) (Lowenstine et al. 2016). The disease called "ankylosing spondylitis" in humans is part of this group. As noted above, it primarily affects the spine, but the appendicular skeleton may be affected, and inflammation of other tissues can be one of its associated manifestations.

Therefore, the arthroses of vertebrae that are commonly seen in zoo mammals can be categorised as follows (Kompanje 1999; Lowenstine et al. 2016):

1. Spondylosis is a combination of discarthrosis and zygarthrosis. Also called degenerative disc disease (DDD), it is characterised by degeneration of the intervertebral disc, leading to structural and functional failure of the intervertebral joint (Figs. 6 and 7) (see also Lowenstine et al. 2016). However, it can be difficult to distinguish from normal ageing of the vertebral column, of which it is a progression, and which also involves some intervertebral disc degeneration. Degeneration is often followed by bony changes (discarthrosis), e.g., vertical marginal osteophytes, disc-space narrowing, sclerosis, and erosion of the vertebral endplate.
2. Infectious spondylitis involves large irregular new bone formations associated with destruction of the vertebrae, such as adjacent vertebral endplates, which is characteristic of advanced spondylodiscitis (Kompanje 1999).
3. Not an arthrosis or an arthritis, but needing to be distinguished from both, diffuse idiopathic skeletal hyperostosis (DISH) is characterised by the ossification of the anterior longitudinal or ventral ligament and thus the anterolateral vertebral column. The posterior spine is often affected, particularly in some species (Fig. 8) (Kompanje 1999). The new bone formation in the spine resembles "dripping candle wax", which fuses several vertebrae, but there is no degeneration of the vertebrae or intervertebral discs. DISH and spondylosis may occur in the same vertebral column. DISH may also affect the entheses of peripheral joints as well as muscle attachments to limb and pelvic bones (Lowenstine et al. 2016; Livingstone et al. 2020). It is unclear whether DISH is a disease or a physiological ageing process. In humans, it is frequently asymptomatic and it can be difficult to distinguish from spondyloarthritis and spondylosis (Table 3), although distinguishing criteria have been established by Livingstone et al. (2020). SA may also fuse vertebrae owing to the formation of syndesmophytes (Lowenstine et al. 2016).

4. Spondyloarthritis (SA) (including ankylosing spondylitis). As noted above, this is also known as seronegative spondyloarthritis and spondarthritis. It is a non-rheumatoid inflammatory disease (Rothschild et al. 1993; Kompanje 1999; Kompanje et al. 2000). Spondyloarthritis may be part of a reactive arthritis, such as psoriatic arthritis and spondylitis/arthritis, associated with inflammatory bowel disease (e.g., Crohn's disease and ulcerative colitis in humans) or may be sexually transmitted (non-specific urethritis) (see also Rothschild and Woods 1993). Slender, horizontal bony outgrowths on the vertebrae are called syndesmophytes, resulting in an ossified intervertebral bridge. It almost always affects the sacroiliac joint (Rothschild and Woods 1993). Cases with bony fusion must be distinguished from DISH (Table 2).
5. Calcium pyrophosphate deposition disease (CPPD) results from the deposition of calcium pyrophosphate crystals, which cause calcification of cartilage in and subsequent inflammation and damage to joints. CPPD appears as sheets of calcified cartilage projecting from the articular surface or calcified concretions at the joint margins (Rothschild and Woods 1989, 1992b, 1993). CPPD seems to have only been recorded in captive primates and carnivorans (Rothschild and Woods 1989, 1992b, 1993).

2.2 Previous Studies of Skeletal Pathologies in Zoo Mammals

Rothschild and collaborators have carried out extensive surveys of skeletal pathologies in vertebrates, mainly mammals. Most of these studies are based on wild-caught individuals from museum collections, but some also include captive animals. For example, Rothschild and Woods (1992a, b) examined skeletal pathologies in

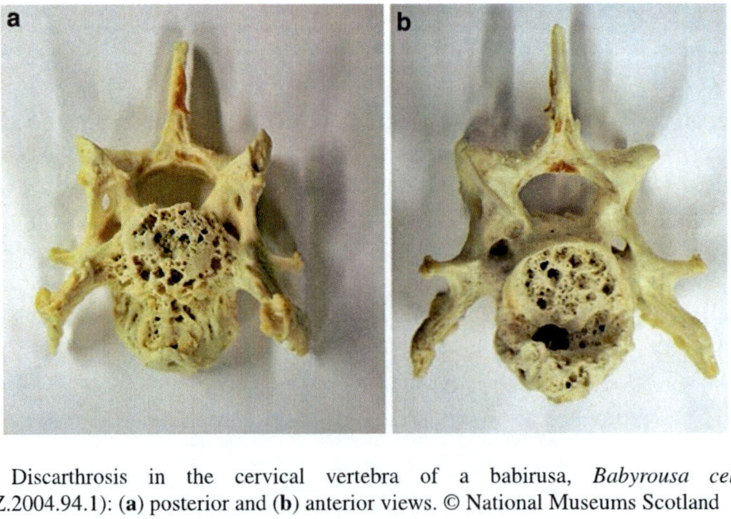

Fig. 6 Discarthrosis in the cervical vertebra of a babirusa, *Babyrousa celebensis* (NMS.Z.2004.94.1): (**a**) posterior and (**b**) anterior views. © National Museums Scotland

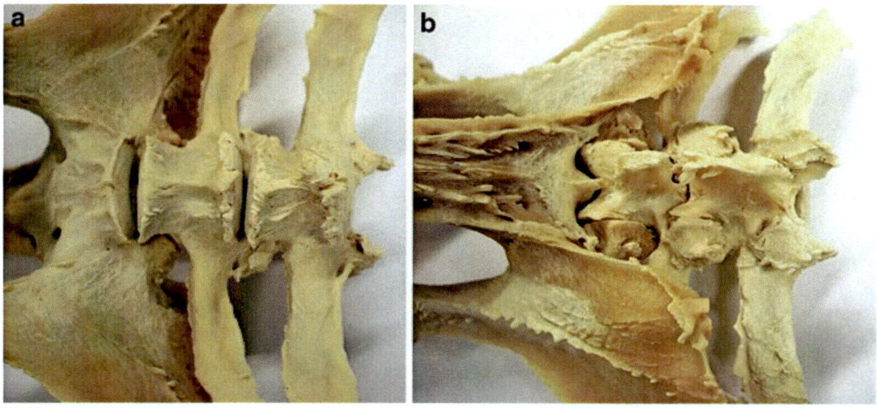

Fig. 7 Spondylosis in the lumbar vertebrae of a male Sulawesi babirusa, *Babyrousa celebensis* (NMS.Z.2004.94.1): (**a**) ventral view and (**b**) dorsal view. © National Museums Scotland

Old World primates, which were categorised into "free-ranging" and "artificially restrained". They examined 267 strepsirrhines and 1250 anthropoid primate skeletons and found that a variety of skeletal pathologies (osteoarthrosis, CPPD, and infectious arthritis/osteomyelitis) affected only 1.7% of free-ranging strepsirrhines and 2.5% of free-ranging Old World anthropoid primates. There was higher prevalence of osteoarthrosis in animals from zoos and experimental stations (4.8% strepsirrhines, 3.7% anthropoids) compared to those that were free-ranging (0.8% strepsirrhines, 0.9% anthropoids). Rothschild et al. (2005) recorded spondyloarthropathies in c. 20% of great apes (western lowland gorillas 20%; chimpanzees (*Pan troglodytes*) 28%; orangutans *Pongo* spp., 17%), baboons (*Papio* spp., 4–20%), and geladas (*Theropithecus gelada*, 18%). Rothschild et al. (2005) suggested that the cause was a habitat-dependent infectious agent, which led to a diarrhoea-induced reactive arthritis on a background of genetic predisposition, although they could not rule out reactive arthritis caused by psoriasis (Group A *Streptococcus* infection). Injecting collagen in Freund's adjuvant, which contains mycobacteria, into rhesus macaques (*Macaca mulatta*) caused 70% to develop arthritis compared with 20% spondyloarthritis in the wild-living Cayo Santiago colony (Rothschild et al. 1999). Rothschild and Rühli (2005a, b) found similar levels of erosive arthritis and joint fusion in lowland (21%) and mountain (20%) gorillas, *Gorilla beringei*.

Rothschild et al. (1998) recorded spondyloarthritis in 3.7% of 386 large cats, including eight species in three genera (*Panthera* spp., *Neofelis* spp., *Puma concolor*, *Acinonyx jubatus*, *Leopardus pardalis*), and it occurred with similar frequencies in wild and captive animals. However, osteoarthrosis and CPPD were only found in zoo animals. Joint erosions of the vertebral column and sacroiliac joint were of the reactive type, so that inflammatory arthritis is almost twice as common as osteoarthrosis amongst all felids. They suggested that antibiotics for treating infectious diarrhoea-related spondyloarthritis could well resolve the disease and prevent joint damage.

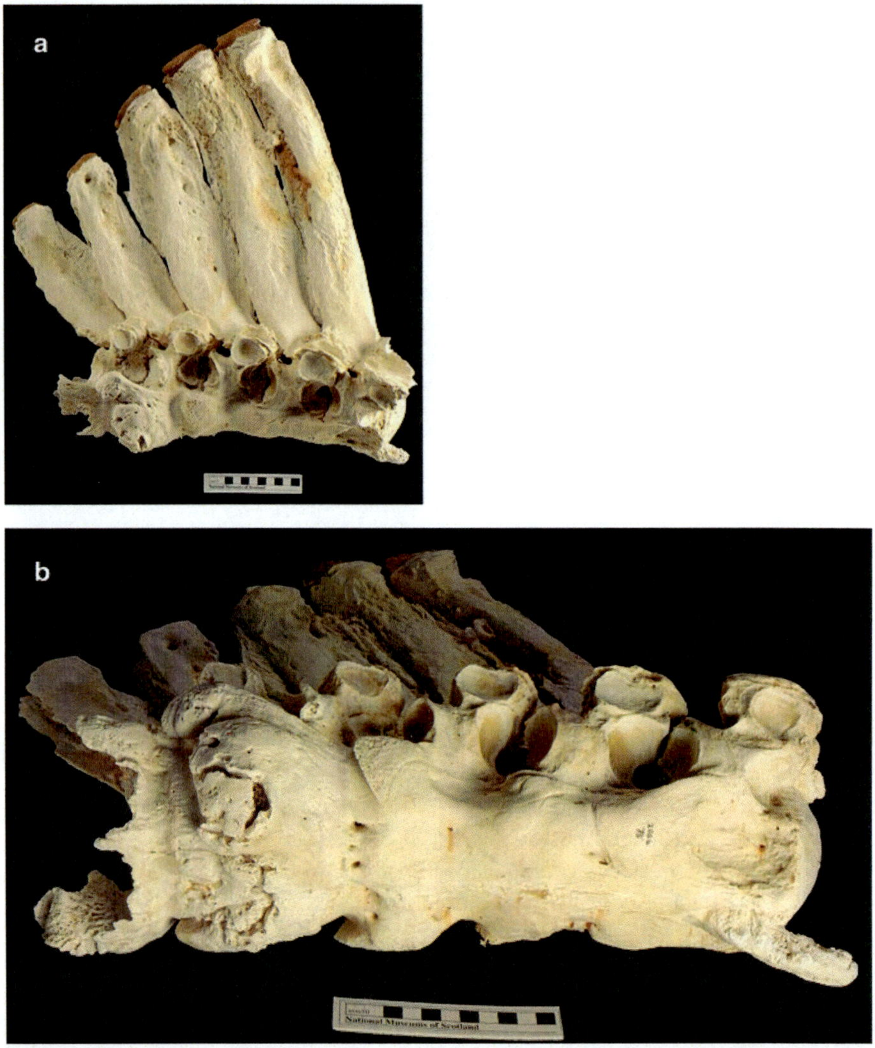

Fig. 8 DISH in the thoracic vertebrae of a white rhinoceros, *Ceratotherium simum* (NMS.Z.2004.75). (**a**) Lateral and (**b**) ventral views. © National Museums Scotland

Other groups surveyed for skeletal pathologies included birds (Rothschild and Panza 2005, 2006), ratites (Rothschild and Rühli 2007), perissodactyls (Rothschild et al. 2001a), and reptiles (Rothschild 2010).

2.3 Dental Pathologies

Dental caries results from the acids produced by plaque bacteria on teeth, which cause demineralisation (Hillson 2005). Spread of infection can lead to periodontal disease that affects the alveolar bone and soft tissues of the mouth (Fig. 9) (Brothwell

Table 3 Criteria for distinguishing between spinal arthroses and DISH (after Resnick and Niwayama 1976, modified for gorillas by Livingstone et al. 2020)

		Differential diagnoses	
Skeletal site	Diffuse idiopathic skeletal hyperostosis (DISH)	Spondyloarthritis	Spondylosis
Synonyms		Spondyloarthrosis	Spondylosis deformans
Vertebral bodies	Flowing ossification of anterior common ligament; hyperostosis bulging in front of disc spaces; bony ankylosis frequent	Thin syndesmophytes; osteitis with squaring (not bulging) of vertebral bodies; extensive bony ankylosis ("bamboo spine")	Vertebral body endplate sclerosis and osteophytes on superior and inferior surfaces of adjacent vertebrae
Intervertebral discs	Normal or mild decrease in height	Normal or convex in shape	Moderate to severe decrease in height; "vacuum" phenomena on plain radiographs
Zygapophyseal joints between intervertebral arches	Normal or mild sclerosis; occasional osteophytes	Erosions, sclerosis, and bony ankylosis	Marginal osteophytes and narrowed irregular joint spaces
Sacroiliac joints	Para-articular hyperostosis; with variable degree of fusion of syndesmosis posteriorly, but synovial joints intact	Erosions, sclerosis, and bony ankylosis of synovial joints and syndesmoses. Para-articular ossification can occur, but uncommon	Normal
Peripheral skeleton	Hyperostosis of ligaments, which may envelop joints with extra-articular fusion. Florid enthesopathy	Destructive arthritis may be with intra-articular bony fusion. Enthesopathy not florid	No specific features

1981). The regression of alveolar bone can lead to loosening and loss of teeth. Chronic abscesses typically occur at the tips of the roots of teeth, resulting from infections that have spread from the pulp cavity or along the periodontal membrane that holds the tooth in the alveolus (Fig. 10). Broken teeth, or carious teeth, or extreme tooth wear exposing the pulp cavity can result in abscesses around the tooth root. Schultz (1944) recorded low levels of caries (7.9%) in old ("senile") wild lar gibbons from Thailand and none in younger animals, but abscesses were more frequent, ranging from 4.3% in juveniles to 57.9% in old adults. Closed alveoli, following tooth loss, ranged from 4.3% in juveniles to 21.1% of old adults. Caries occurred in 4.4% of adult and 28.5% of old adult orangutans (Schultz 1935).

Linear enamel hypoplasia causes ridges across the enamel on teeth where the tooth growth is interrupted during development (Fig. 11). It can be caused by

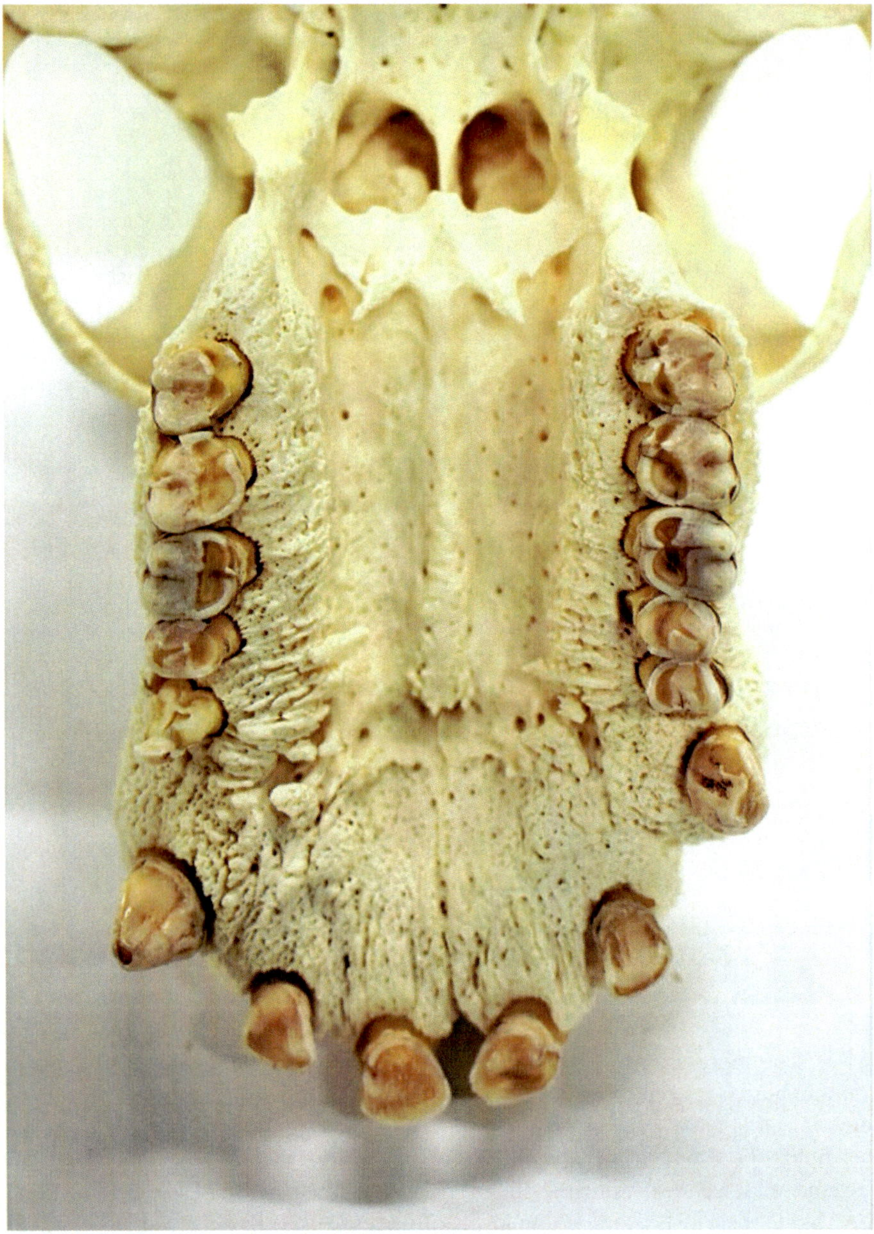

Fig. 9 Periodontal disease in the upper jaws of a 48-year-old female chimpanzee, *Pan troglodytes verus* (NMS.Z.2018.129.1). Note the worn teeth. © National Museums Scotland

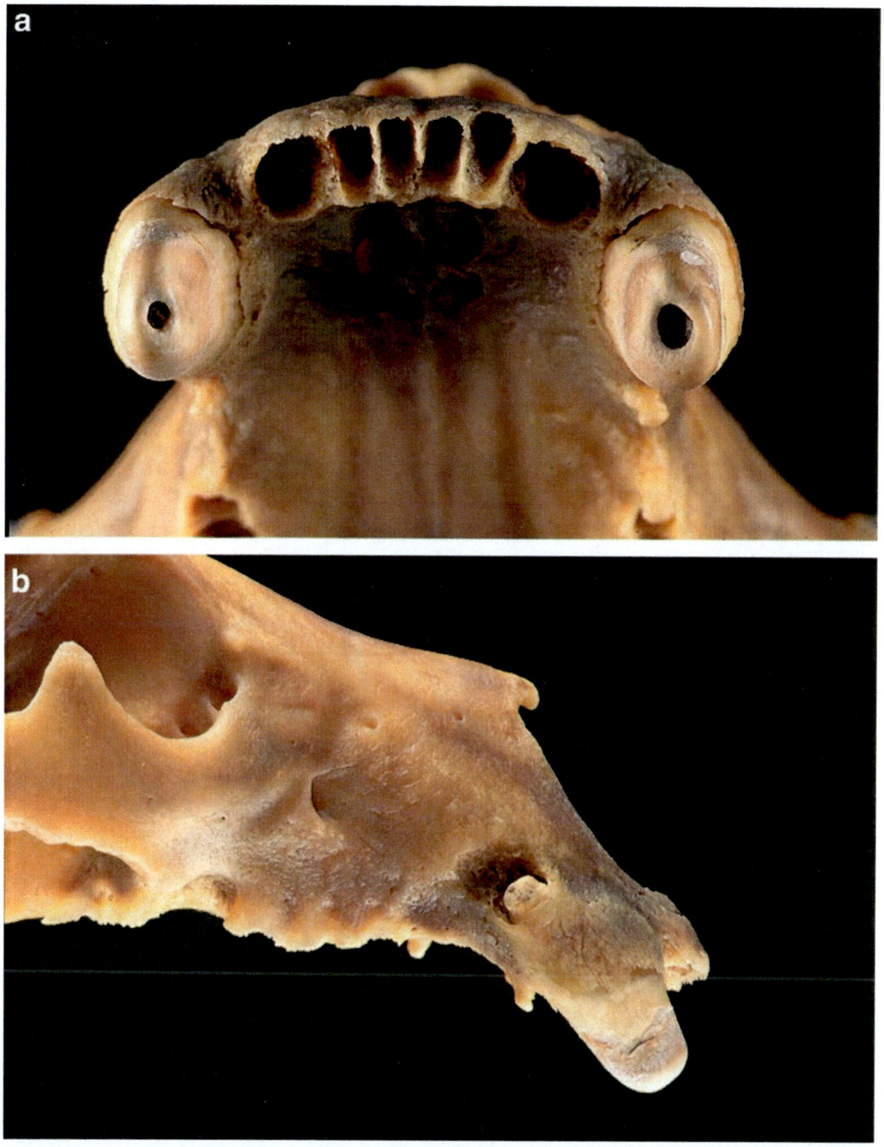

Fig. 10 Ventral (**a**) and lateral (**b**) views of the skull of a 28-year-old female brown bear, *Ursus arctos*, showing worn tips of canines and abscess at the root tip (NMS.Z.2000.236). © National Museums Scotland

disease, vitamin D deficiency, or stress (Guatelli-Steinberg and Skinner 2000; Guatelli-Steinberg et al. 2012).

Dental calculus or tartar is the deposition of calcium on teeth, which can irritate gums and cause periodontal disease (Haberstroh et al. 1984; Willis et al. 1999) (Fig. 12).

Fig. 11 Linear enamel hypoplasia in the canine tooth of a female western lowland gorilla, *Gorilla g. gorilla* (NMS.Z.1881.23.1). © National Museums Scotland

3 Skeletal Pathologies in Zoo Mammals at National Museums Scotland

3.1 Methods

During the routine collection of skeletons of bears (Family Ursidae) from zoos, it was noticed that many have pathologies. Pathologies were recorded on each skeleton, on the basis of whether they might be painful for the bears (in comparison with

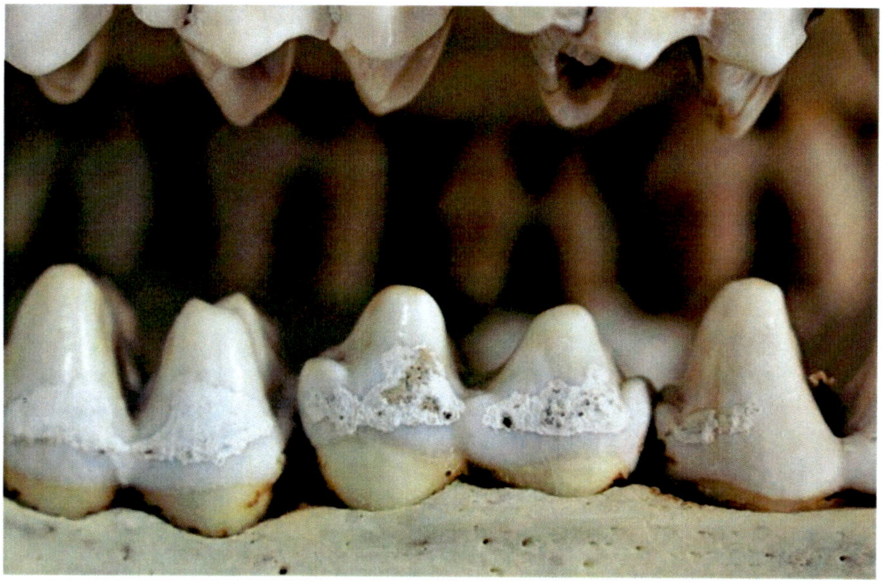

Fig. 12 Dental calculus on the cheek teeth of a Malayan tapir, *Acrocodia indica* (NMS.Z.1822.19). © National Museums Scotland

similar conditions in humans) and which might cause serious impairment of the natural movements and feeding of the bears, and hence compromise their survival in the wild. Therefore, pathologies of the teeth, vertebral column, and limb bones were selected.

For teeth, we recorded the following:

1. Broken or worn canines in the upper and lower jaws (Fig. 10)
2. Abscesses in the upper and lower jaws, especially associated with broken canines (Fig. 10)
3. Dental caries or tooth decay

For the skeleton, we recorded the following skeletal pathologies:

1. Spondyloarthropathies of the vertebral column (where bony growths known as osteophytes or exostoses develop between the centra and the articulations (zygapophyses) of the vertebrae). This includes spondylosis, SA, and DISH. No attempt was made to distinguish between these in the prevalence data (Fig. 13).
2. Complete fusion of any vertebrae as caused by extensive spondyloarthropathies (Fig. 14).

3. Erosion or reabsorption of bone in the centra of the vertebrae (discarthroses) (Fig. 6).
4. Osteoarthroses of the long bones and the pelvis (where osteophytes develop at the joints of the main limb bones) (Fig. 15).
5. Erosion through wear of the joints of the long bones (DJD), often leading to a polishing of the subchondral bone (eburnation) as it rubs against an adjacent bone after the cartilage has been worn away (Fig. 16).

The sample was expanded to include a range of larger mammal species (Kitchener and Macdonald 2002), including lions, tigers, gorillas, orangutans, babirusas, and pygmy hippos. These samples were further increased as more skeletons have been acquired by National Museums Scotland (Table 4).

A total of 170 specimens (including 10 bears from pathology reports) from 15 species of mammal (Table 4) were examined for evidence of skeletal and dental pathologies as previously reported in Kitchener (2004). Therefore, comparisons can be made between species that have different morphological adaptations, phylogenies, diets, behaviours, habitats, etc., but all of which occur commonly in zoos. Pathologies were scored on a presence or absence basis (i.e. prevalence) rather than trying to assess how severe the condition was. For each pathology for each taxon, the percentage of individuals showing a specific pathology was calculated based on the total number of individuals of the minimum age and greater that first showed

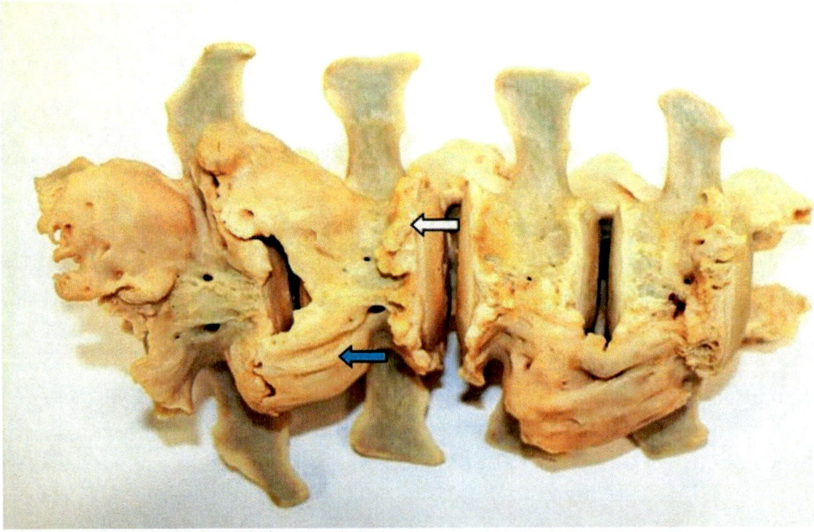

Fig. 13 Ventral view of the lumbar vertebrae of a female Asian black bear, *Melursus thibetanus*, showing osteophytes due to spondylosis (white arrow) and DISH (blue arrow) (NMS.Z.2021.5) © National Museums Scotland

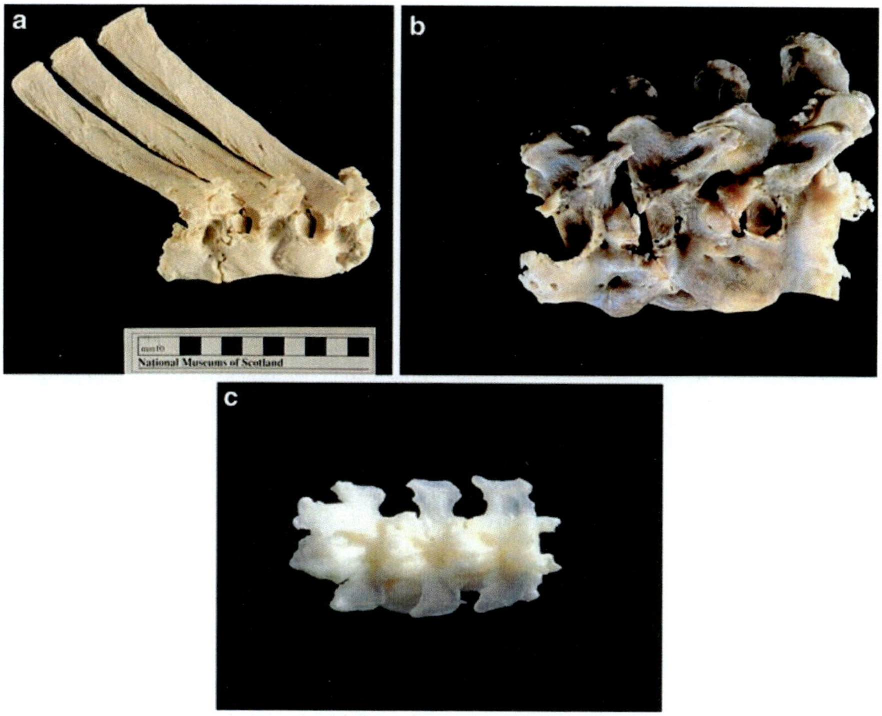

Fig. 14 Lateral views of (**a**) thoracic vertebrae of a babirusa, *Babyrousa celebensis* (NMS.Z.2004.94.1), (**b**) lumbar vertebrae of a female brown bear, *Ursus arctos* (NMS.Z.2000.236), and (**c**) ventral view of the lumbar vertebrae of a 12-year-old grey mouse lemur, *Microcebus murinus* (NMS.Z.2004.98.1), showing fusion of vertebrae by osteophytes probably due to DISH © National Museums Scotland

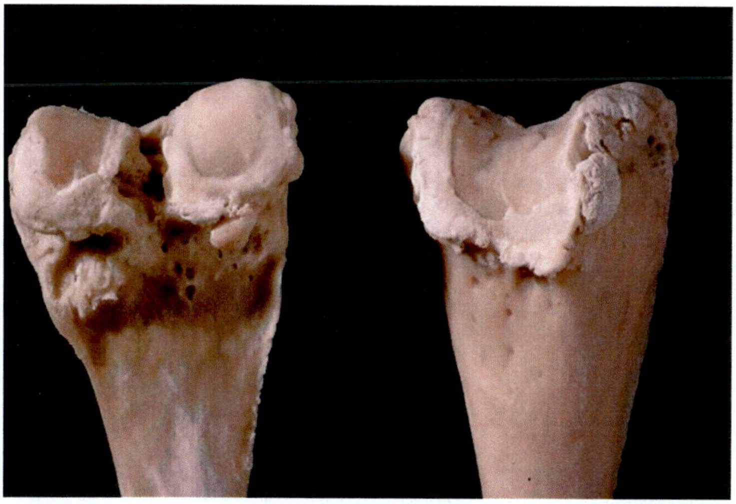

Fig. 15 Distal femora of a male Siberian brown bear (*Ursus arctos*), showing osteophytes, typical osteoarthrosis, as well as eburnation (NMS.Z.2001.165). © National Museums Scotland

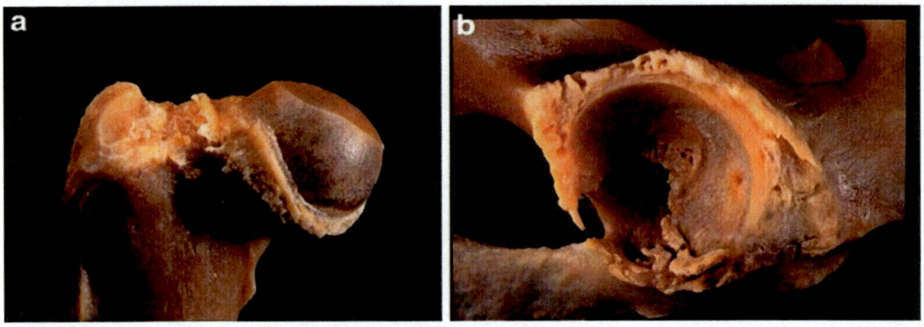

Fig. 16 (**a**) Femoral head of a female brown bear (*Ursus arctos*), showing severe eburnation. (**b**) Acetabulum of the same bear showing proliferation of osteophytes around and erosion of the joint (NMS.Z.2000.236). © National Museums Scotland

Table 4 Sample sizes of zoo mammals that were examined for skeletal and dental pathologies. Figures in parentheses refer to additional data from pathology reports from zoos

Species	*n*
Gorilla, *Gorilla* spp.	16
Orangutans, *Pongo* spp.	12
Bears, Family Ursidae	34 (+10)
Lion, *Panthera leo*	28
Tiger, *Panthera tigris*	32
Babirusas, *Babyrousa celebensis*	27
Pygmy hippopotamus, *Choeropsis l. liberiensis*	11

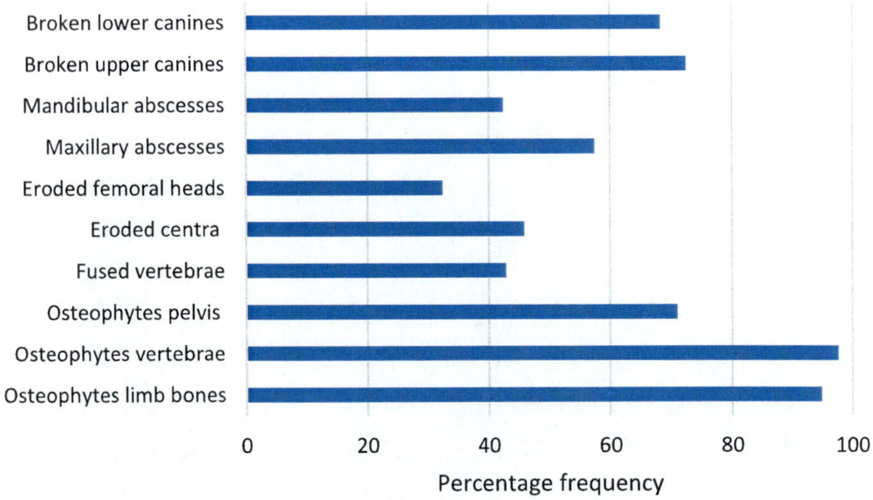

Fig. 17 Prevalence of dental and skeletal pathologies in bears, family Ursidae ($n = 44$), more than 14 years old from zoos

any sign of a skeletal or dental pathology. In other words, the youngest tiger that had skeletal pathologies was four years old, so that our sample for tigers included all animals four years old or greater regardless of whether they displayed any skeletal or dental pathologies.

3.2 Results

Amongst bears the prevalence of skeletal and dental pathologies recorded in this study is summarised in Fig. 17. The commonest pathology was the presence of osteophytes on the vertebral column (c. 98%), which often resulted in fusion of usually lumbar vertebrae (c. 43%), followed by osteophytes on the long bones (c. 95%). Broken or open-tipped upper and lower canines (c. 70% and c. 68%, respectively), which may well have been caused by biting inappropriate and very hard materials or structures such as bars (Fig. 10a), were the commonest dental pathology. This condition may well be very painful because it exposes the pulp cavity within the tooth to infection, which often results in abscesses that cause erosion of the alveolar bone surrounding the roots of the teeth in both maxillae and mandibles (Fig. 10a). More than 57% of maxillae and more than 48% of mandibles were affected by abscesses as a result of either broken canines and other teeth or tooth decay. Tooth decay was found in 41.2% of all bears, but is likely to be under-recorded, because few older bears had complete dentitions and they may well have lost teeth due to decay or dental treatment.

The most common pathological change in the skeletons of the bears was spondyloarthropathy (Figs. 13 and 14b). This change began commonly in the lumbar region and advanced along the vertebral column. Around 98% of bears had this condition, and in c.43% of bears, several lumbar vertebrae were ankylosed (fused together) as a result of bridging by new bone (Kompanje et al. 2000), although this was generally found only in the more elderly bears (>23 years old) (Fig. 14b). Although fusion may well result in a cessation of pain in that part of the vertebral column, other areas either side could still be very painful and the degree of development of the osteophytes may compress the spinal cord resulting in impairment of locomotion. However, spondyloarthropathy is common in elderly mammals, such as pet domestic dogs (*Canis familiaris*), and may not be painful if it is only mild (S. Redrobe, pers. comm.). Erosion or resorption of bone in the centra of the vertebrae (discarthrosis) was also seen in more than 81% of bears and tended to occur in the most severe cases where osteophytes were well developed (Fig. 6).

Osteophytes were also observed at the articular facets of the zygapophyses of the vertebrae (Fig. 13) and in the joints of the limb bones (Fig. 15), most commonly in the hip (femoro-pelvic) joint (Fig. 16b). Osteoarthroses of limb bone joints occurred in around 95% of bears and osteophytes were also common either on the sacral area of the pelvis or at the hip joint (71% of bears). It is unclear whether this condition was a consequence of degenerative or inflammatory joint disease. In severe cases, the articular cartilage of the head of the femur was eroded away (c.32% of bears), resulting in mild to severe eburnation in some cases (Fig. 16a). This condition would cause loss of movement, stiffness, difficulty in walking, sitting and rising, and severe pain (S. Redrobe, pers. comm.).

In comparison with bears, tigers of four years old and above showed lower prevalence of all pathologies (Fig. 18). However, 71.9% showed both skeletal and dental problems. Two individuals had fused vertebrae (7.1%) and five (15.6%) had abscesses in the jaws, although most had broken canines. Lions of similar age showed mostly similar proportions of these pathologies, with 87% of individuals having osteophytes on the vertebrae (but none fused) and two individuals (9.1%) had abscesses in both maxillae and mandibles (Fig. 19). Lions had more osteophytes on the pelvis (c. 67%) and vertebrae (c. 87%) and more broken canines (c. 74%) than tigers.

Gorillas more than 14 years old and orangutans more than 17 years old had very high prevalence of skeletal pathologies. Both gorillas and orangutans had very high prevalence of osteophytes on their vertebrae (93.8% and 100%, respectively) and pelvis (75% and 72.5%, respectively), but gorillas were more likely to have osteophytes on their limb bones (87.5%) than orangutans (66.7%). Gorillas were also more likely to have fused vertebrae (all DISH; Livingstone et al. 2020) and eroded centra (discarthrosis) than orangutans (Fig. 20). Dental problems were roughly equally common amongst gorillas and orangutans, but with double the prevalence of maxillary abscesses in gorillas and more than double the prevalence of broken canines in orangutans (Fig. 20).

Babirusas more than eight years old showed similar prevalence of skeletal pathologies to those of bears (Fig. 21), with all individuals exhibiting osteophytes on their vertebrae, except bears were more likely to have osteophytes on their pelvis (71% bears, 38.5% babirusas) and babirusas were more likely to have fused vertebrae

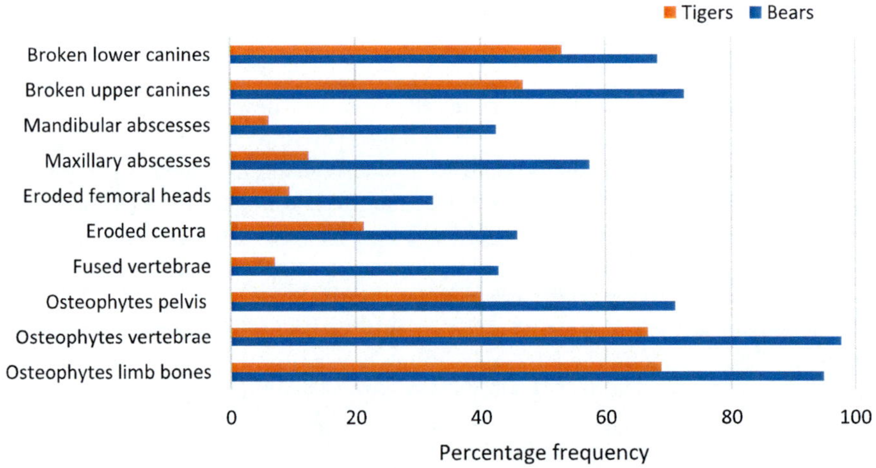

Fig. 18 A comparison of the prevalence of dental and skeletal pathologies in tigers (*Panthera tigris*) ($n = 32$), more than 3 years old, and bears, family Ursidae ($n = 44$), more than 14 years old from zoos

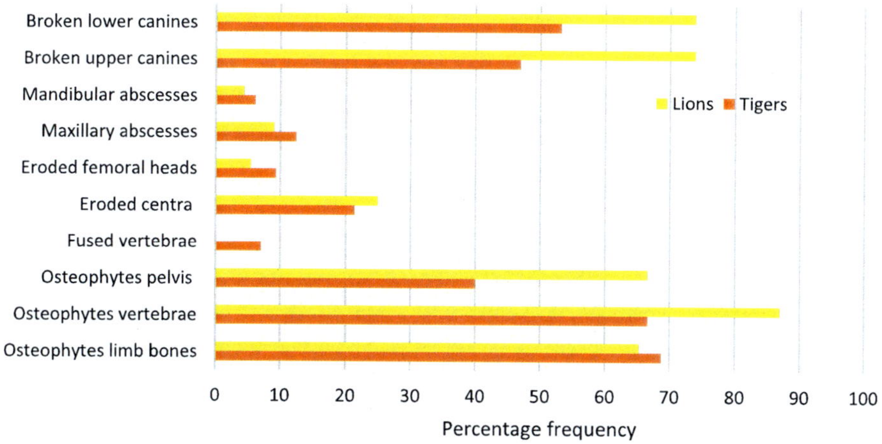

Fig. 19 A comparison of the prevalence of dental and skeletal pathologies in tigers (*Panthera tigris*) ($n = 32$), more than 3 years old, and lions (*P. leo*) ($n = 28$), more than 5 years old from zoos

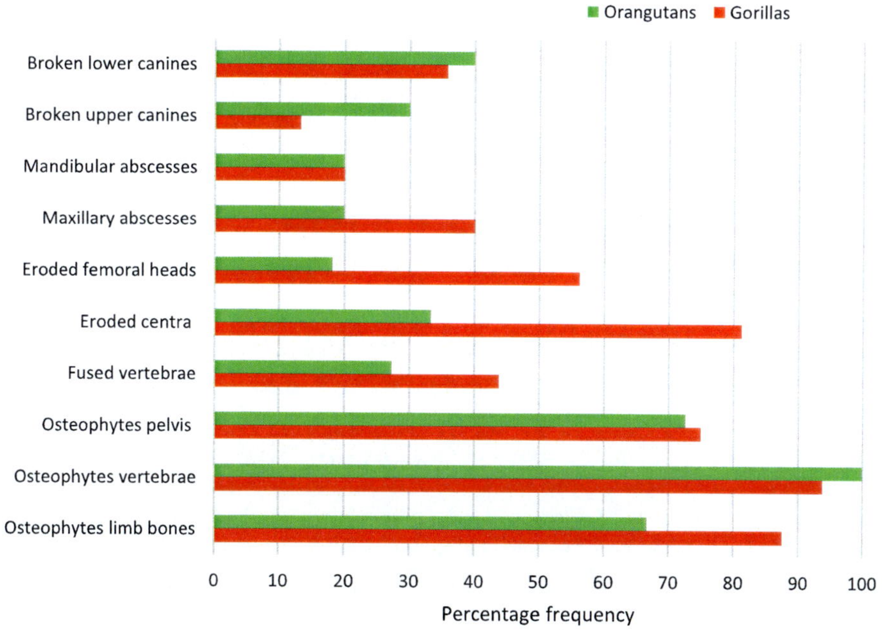

Fig. 20 A comparison of the prevalence of dental and skeletal pathologies in gorillas (*Gorilla* spp.) ($n = 16$), more than 14 years old, and orangutans (*Pongo* spp.) ($n = 12$), more than 17 years old from zoos

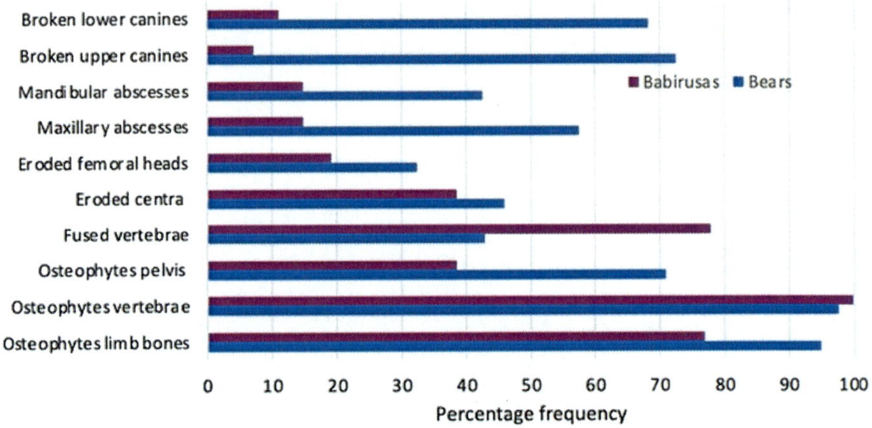

Fig. 21 A comparison of the prevalence of dental and skeletal pathologies in babirusas (*Babyrousa celebensis*) ($n = 27$), more than 8 years old, and bears, family Ursidae ($n = 44$), more than 14 years old from zoos

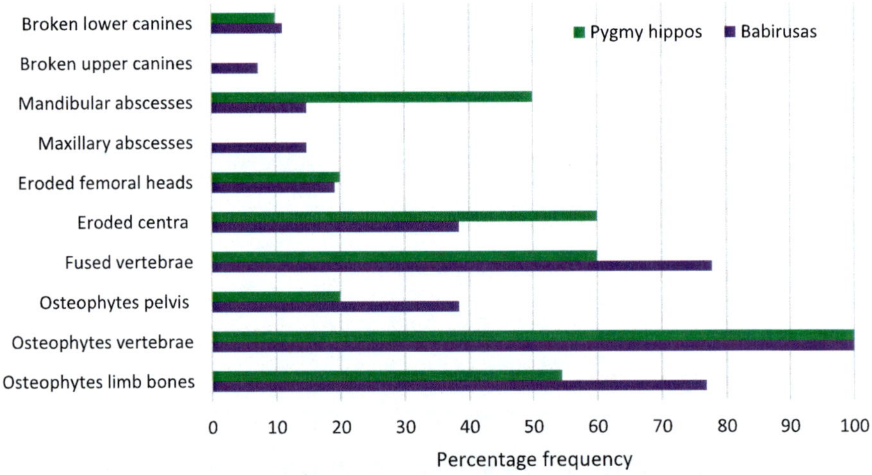

Fig. 22 A comparison of the prevalence of dental and skeletal pathologies in babirusas (*Babyrousa celebensis*) ($n = 27$), more than 8 years old, and pygmy hippos (*Choeropsis l. liberiensis*) ($n = 11$), more than 15 years old from zoos

(77.8%) than bears (42.9%). However, dental problems were much less prevalent in babirusas compared with bears (Fig. 21).

Pygmy hippos more than 15 years old showed a different pattern of skeletal pathologies to those of babirusas (Fig. 22). Although all individuals had osteophytes on their vertebrae, in pygmy hippos the prevalence of osteophytes on limb bones (54.5% pygmy hippos, 76.9% babirusas) and the pelvis (20% pygmy hippos, 38.5% babirusas) was lower, as was the prevalence of fusion of vertebrae (60% pygmy hippos, 77.8% babirusas), but prevalence of erosion of femoral heads was similar (c.20%). Dental problems in pygmy hippos were confined to the lower jaw with 50% of individuals with abscesses.

4 Discussion

4.1 Comparison of skeletal and dental pathologies with previous studies

Compared with previous studies, it would appear that dental and skeletal pathologies are widespread in older individuals of a range of large mammal species, although prevalence and minimum ages at which these first occur may vary considerably between some species. However, they are not restricted to large species (Fig. 14c) and are probably overlooked in smaller species, which may have long life spans. The high prevalence reported here compared with previous studies are in part due to the different way of calculating the percentage of animals, including the youngest of those that first showed a pathology in this study compared with all available skeletons. However, in practice very few younger adult skeletons were available in the collections of National Museums Scotland, so percentages would only be slightly lower, if calculated in the same way as other studies.

Amongst 44 bear skeletons of seven species examined so far, all had been suffering from some form of dental or skeletal pathology. From the range of ages of the bears, their skeletons appear to show a progressive deterioration from about 15 years of age, so that by their late teens spondyloarthropathy and osteoarthroses are evident in c. 98% of bear skeletons. These pathologies are highly developed in animals greater than 25 years old. In human terms, these would be regarded as very painful, except where fusion of the vertebrae has occurred. DISH was not distinguished from SA, so it is unclear how debilitating vertebral pathologies were. In humans, DISH is regarded as mostly asymptomatic, but studies have linked it to a variety of medical issues, such as obesity, hypertension, diabetes mellitus, hyperinsulinemia, dyslipidemia, and hyperuricaemia (Nascimento et al. 2014). It may also cause dysphagia, if it occurs in the cervical vertebrae, and affect locomotion (Mader et al. 2013). Dysphagia as a result of DISH has been recorded in a spider monkey (*Ateles geoffroyi*) (Ratliff et al. 2020).

The bears' teeth also commonly show broken or open-tipped canines from erosion against hard surfaces, often resulting in abscesses in the jaws, a problem which would again be regarded as painful for humans. However, it should be noted that dental infections may cause secondary infections, such as endocarditis (heart infection), resulting in death, and may also be causing reactive arthritis in the skeleton (Rothschild et al. 1993). Given the limited sample size available (albeit greater than in previous studies), it is vital that zoos record these pathologies during life (and their impact on behaviours) and during post-mortem examinations, and also work with museums to preserve as many complete skeletons as possible. In particular, skeletons of younger bears should be examined for dental and skeletal pathologies to refine data on when these first appear.

There are few data from wild bears for comparison. For dental problems, there are data for two populations of the American black bear (*Ursus americanus*) from northern Wisconsin ($n = 95$) from 1974 to 1975 and from the northern Lower Peninsula in Michigan ($n = 35$) from 1977 to 1980 (Manville 1990). Whilst dental caries was common in Wisconsin (10%) and Michigan (20%) and developed in animals from 3.5 years or older to the maximum-recorded age of 15.5 years, periodontal disease was rare in Wisconsin bears (1.1%), but common in Michigan bears (37%) (cf. 4% in 618 black bears from North America). It was suggested that low levels of selenium in the diet of these bears were a cause or contributing factor to the higher prevalence of dental caries in Michigan. Whilst the prevalence of dental caries in captive bears was similar (41.2%), the ages of the captive sample exceeded the maximum age of the wild-caught bears, and we would expect to see increasing prevalence with age. It is likely that captive bears had diets with more sugar than the wild bears.

In a study of 243 adult bear skeletons of several species (including the giant panda, *Ailuropoda melanoleuca*), 17.3% had evidence of skeletal pathologies (Rothschild et al. 1993). This compares with 96.3% of the captive bear sample. For the affected bears, the affected joints included shoulders (7%), elbows (14%), wrists (29%), proximal interphalangeal joints (12%), hips (7%), knees (24%), ankles (17%), and metatarsal phalangeal joints (17%). Spondyloarthropathy was found in 86% of the affected bears, but the nature of the new bone between the vertebrae was apparently different to those observed in captive bears at National Museums Scotland. Rothschild et al. (1993) suggested that the reactive arthritis seen in these wild bears may be the result of a sexually transmitted disease. Kompanje et al. (2000) found similar skeletal pathologies in the skeletons of three species of Indo-Malayan bears, *Melursus* spp. from Dutch zoos. They concluded that whilst some of the pathologies resulted from degeneration from ageing, including osteoarthrosis of the limbs and vertebrae (as well as vertebral erosion), the development of syndesmophytes, causing vertebral fusion, resulted from a combination of genetic and environmental factors. For example, some pathologies, such as Reiter's syndrome or reactive arthritis, are triggered by infections (e.g., *Campylobacter*, *Chlamydia*, *Mycoplasma*, *Salmonella*, and *Shigella*). In humans, these infections result from sexually transmitted diseases, gastroenteritis, and infections of roundworms, *Ascaris* (Rothschild et al. 1993; Kompanje et al. 2000). In this study, many of the spondyloarthropathies in bears were suggestive of DISH with no discarthrosis (Fig. 13) and no loss of intervertebral discs when the vertebrae fuse (Fig. 14b).

Taking bears as a baseline for comparisons, it is interesting to see varying prevalence and ages of first occurrence amongst different large mammals. Big cats begin to develop skeletal pathologies from only four years old, whilst in babirusas these occur from nine years old. In part, this may reflect their overall shorter longevities of about 20 years or so, whereas bears may live several decades. Broken canines were frequent amongst bears and big cats, especially lions, which suggests that care should be taken in designing captive environments, which may cause these breakages.

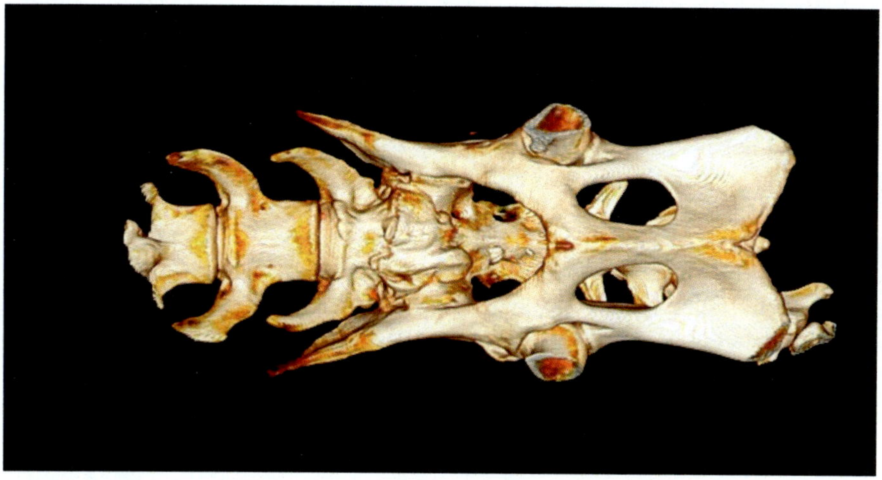

Fig. 23 CT scan of ventral view of pelvis and lumbar vertebrae of an 18-year-old Amur tiger (*Panthera t. tigris*), showing spondyloarthropathy, probably DISH, resulting in vertebral fusion. © Jonathan Cracknell

Closely related species showed mostly similar prevalence of dental and skeletal pathologies. Lions and tigers showed little evidence of fusion of vertebrae compared with the other taxa. However, vertebral fusion has been reported in an 18-year-old Amur tiger (Fig. 23) (J. Cracknell pers. comm.), and it has been recorded in a snow leopard (*Panthera uncia*) at National Museums Scotland. This may reflect the very much more flexible vertebral column of the cats, which needs to move, in order to assist locomotion, especially when running and climbing (Kitchener et al. 2010). Even so, the prevalence of skeletal pathologies reported here is far higher than the 3.7% of 386 large cat skeletons reported with spondyloarthropathy, 2.1% with OA, and 2.1% with CPPD by Rothschild et al. (1998), with no difference in disease frequency between wild and captive animals. Kolmstetter et al. (2000) examined 37 big cats from all five species of *Panthera* from Knoxville Zoo and found degenerative spinal disease (spondyloarthropathy) in eight (21.6%), including three lions (23.1% of lions), four tigers (25%), and one leopard (25%). The age of onset was 10–19 years (median 18 years), which was older than that recorded in specimens (four years and older) at National Museums Scotland. These spinal pathologies caused a variety of problems for these cats, including decreased activity and increased rear limb muscle atrophy, chronic intermittent paresis, and ataxia (Kolmstetter et al. 2000). Acute or chronic damage was recorded in the spinal cords of five out of six animals, as a result of disc protrusion.

Compared with orangutans, gorillas showed a higher prevalence of erosion of femoral heads, osteophytes on the limb bones and pelvis, and fusion of vertebrae, which may reflect greater wear and tear, resulting from their terrestrial existence. Livingstone et al. (2020) examined DISH in 11 captive gorillas of both sexes and found that it occurred mostly in the lumbar and cervical vertebrae, whereas in humans it occurs primarily in the thoracic vertebrae. In most cases, DISH was extensive and caused fusion of vertebrae. One possible factor is that all eleven gorillas with DISH had been hand-reared, either as wild-caught infants or born in captivity, whereas a captive-bred, mother-reared male was unaffected at the age of 31 years when he died (Livingstone et al. 2020). Further research is underway to examine DISH in other great apes at National Museums Scotland. In a sample of 99 wild western lowland gorillas, spondyloarthropathy was reported in 21% of individuals (60% of males), fractures in 8%, and severe OA in 8% (Rothschild and Woods 1989; Rothschild and Rühli 2005a, b; Rothschild 2005). Overall, 78% of males had some kind of skeletal pathology. Spondyloarthropathy was reported in 17% of wild orangutans, but osteoarthritis was not reported in contrast to the captive animals in this study (Rothschild 2005; Lowenstine et al. 2016). Prevalence of skeletal pathologies in the captive sample at National Museums Scotland is much higher than recorded in these previous studies.

It is unclear why there are differences in dental problems between gorillas and orangutans, but it is possible that this may reflect either differences in their diets and/or methods of feeding in captivity. The prevalence of abscesses in captive orangutans was similar to that in a sample of 123 wild Bornean orangutans (*Pongo pygmaeus*), where 26% had periodontal abscesses and 23% showed periodontal bone loss, but only 3% suffered tooth loss (Lowenstine et al. 2016).

Pygmy hippos showed a lower prevalence of osteophytes on their limb bones compared with babirusas (and all other taxa), but this still exceeded 50% of animals. This may be because they have access to heated indoor pools, so that for long periods their weight is supported by water rather than their limbs, resulting in less wear and tear. Musculoskeletal and neuromuscular issues are a common cause of mortality (15.9%) in geriatric (30+ years old) pygmy hippos (Flacke et al. 2016). Animals were mostly euthanised humanely because they were suffering from arthritis (sic.), degenerative joint disease, ankylosing spondylitis, lameness, impaired mobility, posterior (hind limb) paresis, and inability to rise/stand (Flacke et al. 2016). Babirusas and pygmy hippos showed few dental problems. Babirusas' upper canines are non-functional, and their lower canines are adapted to a high degree of wear through their use in foraging (Macdonald 2017). Breakages matter less to babirusas because their canines continue to grow throughout life. The few abscesses in babirusas were associated with the premolars and molars, whilst abscesses occurred in 50% of pygmy hippo mandibles, but these were absent from their maxillae. Pygmy hippos frequently suffer dental problems, particularly their canines, which grow continuously, but which may not get sufficient wear in zoos.

4.2 Causes of Pathologies

Nunn et al. (2007) reviewed the prevalence and causes of arthritis in 34 species of primate and 100 species of carnivoran. They found that 5.6% of primates and 3.6% of carnivorans had spondyloarthropathy, with the highest levels in great apes (22%) and bears (27%). Amongst the causal factors considered were social contact, diet, body mass, and life history traits, such as longevity, terrestriality, and mating promiscuity, but only one factor appeared to be significant. This was body mass, which might be important in that larger mammals are longer-lived and use more resources, exposing themselves to infections that might cause SA, or heavier animals are more likely to suffer damaging stresses on joints that cause SA (Nunn et al. 2007). However, no firm conclusions could be drawn. It is clear that we need much more information about the prevalence and causal factors of skeletal and dental pathologies of wild and captive mammals. National Museums Scotland continues to collect complete skeletons of captive vertebrates which are available for collaborative research on dental and skeletal pathologies. But what are the principal causes of these pathologies?

Some forms of reactive arthritis are caused by infections, such as Reiter's syndrome, and it has been suggested that these could be treated with antibiotics. For example, a juvenile western lowland gorilla suffering from reactive arthritis was treated using long-term sulfasalazine therapy (Neiffer et al. 2000; Brown et al. 1971; Raphael et al. 1995). Pathologies that have developed owing to nutritional deficiencies or environmental issues may also be addressed.

Most cases of osteoarthrosis and probably spondylosis are caused by ageing or wear and tear (Kompanje 1999; Kompanje et al. 2000). The skeletons of older animals may just wear out because captive vertebrates often live much longer than wild counterparts and they are often obese, thereby increasing forces on joints. This may be mitigated by appropriate environmental enrichment for zoo animals, which mimics the behaviours and activity levels of wild animals and reduces body weight and/or fat levels. Recently, Law and Kitchener (2020) reviewed the use of the tiger feeding pole in mainly British zoos. This enrichment device was developed by Graham Law and his colleagues at Glasgow Zoo in the 1990s (Law et al. 1997), and mimics natural hunting by stimulating a tiger (or other cat) to climb a telegraph pole to get access to meat. The forelimb musculature used to climb the pole is the same as that used by big cats to capture and bring down large prey (Anton and Turner 1997). The sudden burst of activity and associated behaviours closely mirror those experienced by a wild tiger or lion hunting in the wild (Law and Kitchener 2020).

A close examination of tigers that have used a feeding pole and those that have not indicates significant differences. Feeding-pole tigers have almost no subcutaneous fat and deposited fat around the internal organs and show well-developed musculature, but non-feeding-pole tigers are often obese and show poor muscle development. However, there were also significant differences between their skeletons. Although there was no significant difference in the ages of the sample of feeding-pole and non-feeding-pole tigers, there were significant differences in the development of osteophytes on their limbs and vertebrae; non-feeding-pole tigers had arthrosis scores more than four times greater than feeding-pole tigers (Law and Kitchener 2020).

These results are based on small samples ($n = 16$), but they suggest that activity levels in captivity may be important in determining whether osteoarthrosis and spondyloarthropathy develop (or else they may have been suffering from bacterial infections that triggered reactive arthritis). Simply by being fitter, with lower levels of subcutaneous and deposited fat, perhaps the loading on limbs and vertebrae is lower, so that osteoarthrosis and spondyloarthropathy are less likely to develop. This observation also supports the use of appropriate enrichment methods to stimulate appropriate activity levels, as well as behaviours, of captive mammals.

Enrichment may not only be good for the mental health of captive animals, but it may also be essential for their long-term physical health. Further study of the limb bones in comparison with those from wild big cats may indicate whether feeding-pole tigers have skeletons that are more similar to those of wild tigers. The study of skeletal material from captive animals provides us with an important lifetime record of the behaviour and nutrition of an individual, which far exceeds what is possible through direct behavioural observations of living animals. Tooth and bone can be analysed for stable isotopes and elements that can provide considerable information about diet and contaminants. Therefore, it is essential to keep good records of captive husbandry, including diets and veterinary treatments, in order to assess which factors may be important in determining appropriate skeletal development and health. The ultimate goal of environmental enrichment and other care activities will be achieved when it is no longer possible to distinguish between the skeletons of wild and captive animals. We are still a long way from this, but there are positive signs that enrichment and best-practice animal care are moving in the right direction. Clearly further data are needed, particularly for animals where we have lifetime records of their husbandry and behaviour, in order to test this hypothesis.

Although this study of zoo mammal skeletons has demonstrated high prevalence of dental and skeletal pathologies, the results should be treated with some degree of caution. Firstly, sample sizes are generally small (although greater than previous analyses), so that we cannot be certain that this prevalence would be found in much larger samples of similar age ranges. Secondly, severity of these pathologies needs to be assessed and their potential for impairment of normal behaviour by causing significant pain, or indeed whether pain is perceived at all. A possible problem is that wild mammals may, even in zoos, try to hide any pain or other discomfort that they may be suffering, so that they are not at a disadvantage in a social group. Even when not social, they may change their behaviour, in order to avoid pain and discomfort. Whatever the case, it is likely that their normal behavioural repertoire will have been affected.

There is a clear need to develop a series of sensitive behavioural indicators to see if normal behaviour has been affected. For example, the use of a wobble tree, which encourages bears to stand bipedally to obtain food items (Law and Kitchener 2002; Law et al. 2008), might allow an assessment of how badly affected hip joints and vertebral columns of individuals are. Long-term studies on individuals might also show how behaviours change over a lifetime, which can then be correlated with observed skeletal changes. A quick way of assessing the degree of behavioural change may be to give the individuals specific anti-inflammatories or painkillers and see if their behaviour has been transformed.

There is a real problem that data, such as those presented here, can be seen as a panacea to get over the problem of deciding whether to carry out humane euthanasia on management grounds. We are not suggesting that all bears should be culled at 16 years or more, or all babirusas at nine years old and greater. Clearly, there is individual variation in the prevalence and progress of these pathologies. However, we are alerting the zoo and veterinary communities to the fact that these problems may begin at a much younger age than was previously thought and that there is an obligation to develop veterinary and animal care and welfare protocols. Whilst we cannot necessarily prevent them, we can do our utmost to monitor and assess ageing animals, and to ensure that dental problems are treated when diagnosed. There can be no excuse for allowing an animal to suffer unnecessarily when it should either be treated appropriately or euthanased humanely to end its suffering.

In the meantime, what can be done for aged mammals in zoos? It is unacceptable that any zoo mammal should be left to suffer long-term dental problems. Annual checks on teeth would allow any problems to be addressed as they arise; training is often used to assist keepers and veterinarians to examine teeth without the need for anaesthetics. Changes in diets to increase cleaning of teeth, coupled with descaling and appropriate use of antibiotics, could all help minimise dental problems, as well as possibly having an impact on skeletal pathologies.

Dealing with skeletal pathologies is more difficult because it may be uncertain when these arise and whether they are painful. Detection of the development of spondyloarthropathy, including spondylosis and DISH, may also be challenging, because of the difficulty in carrying out radiography on large-bodied often obese animals with such large girths (Aminkov et al. 2018), although CT scanning would overcome this difficulty. However, by developing sensitive behavioural indicators, it may be possible to assess which movements are painful to an animal and what it avoids doing even for its most favourite food. For example, the "wobble tree" may offer a way of assessing the degeneration of the hip joints and vertebral columns of bears (Law and Kitchener 2002). It would also be possible to administer short-term anti-inflammatories and/or painkillers to see if an animal's behaviour and movements improve. The availability of cheap video cameras, such as GoPro cameras, allow the routine recording of normal locomotory and other behaviours that can be archived for many years for later qualitative comparison, and even quantitative gait analyses to see when skeletal deterioration can be detected and could be impacting quality of life.

At the end of the day, hard decisions may have to be made as to whether to euthanase the animal or not, but this must be in the best interests of an animal, which should not be kept alive unnecessarily for the social group or the breeding programme. From the data we have seen so far, this should probably be done much sooner rather than much later, in order to avoid large mammals suffering a great deal of pain in their old age.

5 Conclusion

Osteoarthrosis and spondyloarthropathies, including reactive arthritis and spondylosis, are common in zoo mammals and appear to increase with age in bears, big cats, great apes, babirusas, and pygmy hippos. Although we cannot be sure when or if they cause pain, zoo managers and veterinarians should be aware that these problems may arise at a much younger age than may be evident from their behaviour. There is a clear need to monitor more closely the skeletal and dental health of mammals in zoological collections and to develop behavioural or other indicators to determine whether the behavioural repertoire of the older mammals is negatively affected by the development of these pathologies. Dental problems should be detected and treated, but skeletal pathologies may only be alleviated by the use of anti-inflammatory drugs, or euthanasia. The decision as to how to treat an individual can only be made on a case-by-case basis, which considers the welfare needs of the individual and communicates clearly and effectively the necessary outcomes to zoo staff and the wider public. It is vitally important that zoos, pathologists, and museums work closely together to ensure that more skeletal specimens are preserved to add to our understanding of the prevalence, severity, progress, and causes of skeletal and dental pathologies in a wider range of mammals and other vertebrates as one aspect of the challenges of old age.

Acknowledgements I am most grateful to Brian Livingstone for his many helpful comments on an earlier version of this chapter. I also thank Sharon Redrobe for her comments and Jonathan Cracknell for permission to use Fig. 23. I am also very grateful to Alan Lothian, Georg Hantke, Stephen Rodger, Ruth Pollitt, Drew Bain, Phil Howard, and Peter Summers for preparing the many skeletons at NMS and to all zoos, who have kindly donated many specimens over the last 30 years. I am grateful to Morgane Tidière for permission to reproduce Fig. 1 and providing a high-resolution copy for this chapter.

References

Adkesson MJ, Langan JN (2007) Metabolic bone disease in juvenile Humboldt penguins (*Spheniscus humboldti*): Investigation of ionized calcium, parathyroid hormone, and vitamin D as diagnostic parameters. J Zoo Wildl Med 38(1):85–92

Aminkov B, Nanev P, Aminkov K (2018) Radiographic findings in joint disease in three brown bears. Tradit Modernity Vet Med 3(4):3–6

Anton M, Turner A (1997) Big cats and their fossil relatives. Columbia University Press, New York

Ballou JD, Gilpin M, Foose TJ (eds) (1995) Population management for survival and recovery. Columbia University Press, New York

Barongi R, Fisken FA, Parker M, Gusset M (eds) (2015) Committing to conservation: The World Zoo and Aquarium Conservation Strategy. WAZA Executive Office, Gland. 69 pp

Blunt W (1976) The ark in the park. Hamish Hamilton, London

Brothwell DR (1981) Digging up bones. British Museum (Natural History). Oxford University Press, London

Brouwer K, Jones ML, King CE, Schifter H (1992) Longevity and breeding records of storks Ciconiidae in captivity. Int Zoo Yearb 31:131–139

Brouwer K, Schifter H, Jones ML (1994) Longevity and breeding records of ibises and spoonbills Threskiornithidae: in captivity. Int Zoo Yearb 33:94–102

Brouwer K, Jones ML, King CE, Schifter H (2000) Longevity records for Psittaciformes in captivity. Int Zoo Yearb 37:299–316

Brown TM, Clark HW, Bailey JS, Gray CW (1971) A mechanistic approach to treatment of rheumatoid type arthritis naturally occurring in a gorilla. Trans Am Clin Climatol Assoc 82:227

Browning H (2018) No room at the zoo: Management euthanasia and animal welfare. J Agric Environ Ethics 31(4):483–498
Byers O, Lees C, Wilcken J, Schwitzer C (2013) The One Plan Approach: The philosophy and implementation of CBSG's approach to integrated species conservation planning. WAZA Mag 14:2–5
Canfield PJ, Spencer AJ (1993) Secondary degenerative arthropathy (osteoarthritis) of the hip joints in ageing, free-living koalas. Aust Vet J 70(10):394–395
Chesney RW, Hedberg G (2010) Metabolic bone disease in lion cubs at the London Zoo in 1889: the original animal model of rickets. J Biomed Sci 17(1):1–4
Cole J, Fraser D (2018) Zoo animal welfare: The human dimension. J Appl Anim Welf Sci 21(suppl):49–58
Colman RJ, Binkley N (2002) Skeletal aging in macaque monkeys. In: Erwin JM, Hof PR (eds) Aging in nonhuman primates. Karger, Basel, pp 32–47
Colman RJ, Kemnitz JW, Lane MA, Abbott DH, Binkley N (1999) Skeletal effects of aging and menopausal status in female rhesus macaques. J Clin Endocrinol Metabol 84:144–148
Conde DA, Flesness N, Colchero F, Jones OR, Scheuerlein A (2011) Zoos can lead the way with *ex situ* conservation. WAZA Mag 12:26–29
Coni N, Davison W, Webster S (1992) Ageing. The facts, 2nd edn. Oxford University Press, Oxford
Du Boulay GH, Crawford MA (1968) Nutritional bone disease in captive primates. Symp Zool Soc Lond 21:223–236
Duckworth WLH (1911) On the natural repair of fractures, as seen in the skeletons of anthropoid apes. J Anat Physiol 46(1):81–85
Erwin JM, Hof PR, Ely JJ, Perl DP (2002) One gerontology: Advancing understanding of aging through studies of great apes and other primates. In: Erwin JM, Hof PR (eds) Aging in nonhuman primates. Karger, Basel, pp 1–21
Fa JE, Funk SM, O'Connell D (2011) Zoo conservation biology. Cambridge University Press, Cambridge
Farrell M, Rando C, Garrod B (2015) Lessons from the past: Metabolic bone disease in historical captive primates. Int J Primatol 36(2):398–411
Fernandez EJ, Martin AL (2021) Animal training, environmental enrichment and animal welfare: A history of behavior analysis in zoos. J Zool Bot Gardens 2(4):531–543
Flacke GL, Tkalčić S, Steck B, Warren K, Martin GB (2016) A retrospective analysis of mortality in captive pygmy hippopotamus (*Choeropsis liberiensis*) from 1912 to 2014. Zoo Biol 35(6):556–569
Föllmi J, Steiger A, Walzer C, Robert N, Geissbühler U, Doherr MG, Wenker C (2007) A scoring system to evaluate physical condition and quality of life in geriatric zoo mammals. Anim Welfare 16(3):309–318
Fox H (1939) Chronic arthritis in wild mammals. Being a description of lesions found in the collections of several museums and from a pathological service. Trans Am Philos Soc 31(2):73–148
Freedman MT, Bush M, Novak GR, Heller RM, James AE (1976) Nutritional and metabolic bone disease in a zoological population. Skeletal Radiol 1(2):87–96
Gardner DL (1983) The nature and causes of osteoarthrosis. Br Med J (Clin Res Ed) 286(6363):418–424
Geissmann T, Geschke K, Blanchard BJ (2009) Longevity in gibbons (Hylobatidae). Gibbon J 5:81–92
Giel JM, Zinn SA, Capacchione C, Bushmich SL (2000) Canine geriatric veterinary medicine: Attitudes and observations of small animal practitioners in the State of Connecticut. J Vet Med Educ 27(3):21–24
Greer M, Greer JK, Gillingham J (1977) Osteoarthritis in selected wild mammals. Proc Okla Acad Sci 57:39–43
Guatelli-Steinberg D, Skinner M (2000) Prevalence and etiology of linear enamel hypoplasia in monkeys and apes from Asia and Africa. Folia Primatol 71(3):115–132

Guatelli-Steinberg D, Ferrell RJ, Spence J (2012) Linear enamel hypoplasia as an indicator of physiological stress in great apes: Reviewing the evidence in light of enamel growth variation. Am J Phys Anthropol 148(2):191–204
Haberstroh LI, Ullrey DE, Sikarski JG, Richter NA, Colmery BH, Myers TD (1984) Diet and oral health in captive Amur tigers (*Panthera tigris altaica*). J Zoo Anim Med 15(4):142–146
Halstead LB (1974) Vertebrate hard tissues. Wykeham, London
Hatt J-M, Müller DWH, Bingaman Lackey L, Clauss M (2011) Life expectancy in zoo mammals: What a zoo veterinarian should know. In: Proceedings of the AAZV conference, Kansas City, Missouri, USA, 23–28 Oct 2011, pp 181–183
Herndon JG, Moss MB, Rosene DL, Killiany RJ (1997) Patterns of cognitive decline in aged rhesus monkeys. Behav Brain Res 87:25–34
Herndon JG, Tigges J, Anderson DC, Klumpp SA, McClure HM (1999) Brain weight throughout the life of the chimpanzee. J Comp Neurol 409:567–572
Hillson S (2005) Teeth, 2nd edn. Cambridge University Press, Cambridge
Hutton CW (1989) Osteoarthritis: The cause not result of joint failure? Ann Rheum Dis 48(11):958
Jones ML (1962) Mammals in captivity—primate longevity. Lab Primate Newsl 1(3):3–13
Jones ML (1968) Longevity of primates in captivity. Int Zoo Yearb 8:183–192
Jones ML (1977) Longevity of mammals in captivity. Int Zoo News 159:16–19
Jones ML (1993) Longevity of ungulates in captivity. Int Zoo Yearb 32:159–169
Kitchener AC (2004) The problems of old bears in zoos. Int Zoo News 51(5):282–293
Kitchener AC, Macdonald AA (2002) The longevity legacy: the problem of old animals in zoos. Adv Ethol 37:7–10
Kitchener A, Macdonald A (2005) The longevity legacy the problem of old mammals in zoos. In: Hiddinga B (ed) Proceedings of the EAZA Conference 2004, Kolmarden. EAZA Executive Office, Amsterdam, pp 131–137
Kitchener AC, Merryweather J, Allchurch T (2000) The effect of captivity on the flight musculoskeletal system of fruit bats (*Pteropus* spp.). In: Rietkerk F, Hiddinga B, Brouwer K, Smits S (eds) EEP yearbook 1998/99 including the proceedings of the 16th EAZA conference, Basel 7-12 September 1999. EAZA Executive Office, Amsterdam, pp 553–555
Kitchener AC, Kolter L, Brownstein D (2001) Problems with old bears in zoos. In: Hiddinga B, Brouwer K (eds) EEP Yearbook 1999/2000 including proceedings of the 17th EAZA conference, Aalborg, 19–24 Sept 2000. EAZA Executive Office, Amsterdam, pp 625–628
Kitchener AC, Van Valkenburgh B, Yamaguchi N (2010) Felid form and function. In: Macdonald DW, Loveridge AJ (eds) Biology and conservation of wild felids. Oxford University Press, Oxford, pp 83–106
Kolmstetter C, Munson L, Ramsay EC (2000) Degenerative spinal disease in large felids. J Zoo Wildl Med 31(1):15–19
Kolter L (1995) European collection planning for bears. In: Rietkerk F, Brouwer K, Smits S (eds) EEP yearbook 1994/95 including the 12th EEP conference, Poznan 30 June–2 July 1995. EAZA/EEP Executive Office, Amsterdam
Kolter L, Sergiel A, Huber D, Silver S (2021) *Ex-situ* conservation of bears: Roles, status and management. In: Penteriani V, Melletti M (eds) Bears of the world. Ecology, conservation and management. Cambridge University Press, Cambridge, pp 338–348
Kompanje EJO (1999) Considerations on the comparative pathology of the vertebrae in Mysticeti and Odontoceti; evidence for the occurrence of discarthrosis, zygarthrosis, infectious spondylitis and spondyloarthritis. Zoologische Mededelingen 73:99–130
Kompanje EJO, Klaver PSJ, de Vries GT (2000) Spondyloarthropathy and osteoarthrosis in three Indomalayan bears: *Ursus ursinus* Cuvier, 1823, *Ursus thibetanus* Raffles, 1821, and *Ursus malayanus* Shaw & Nodder, 1791 (Mammalia: Carnivora: Ursidae). Contrib Zool 69(4):259–269
Krebs BL, Marrin D, Phelps A, Krol L, Watters JV (2018) Managing aged animals in zoos to promote positive welfare: A review and future directions. Animals 8(7):116
Lacreuse A, Herndon JG, Killiany RJ, Rosene DL, Moss MB (1999) Spatial cognition in rhesus monkeys: Male superiority declines with age. Horm Behav 36:70–76

Lane MA (2000) Nonhuman primate models in biogerontology. Exp Gerontol 35(5):533–541
Law G, Kitchener A (2002) Simple enrichment techniques for bears, bats and elephants—untried and untested. Int Zoo News 49(1):4–12
Law G, Kitchener AC (2020) Twenty years of the tiger feeding pole: Review and recommendations. Int Zoo Yearb 54:174–190
Law G, MacDonald A, Reid A (1997) Dispelling some common misconceptions about the keeping of felids in captivity. Int Zoo Yearb 35:197–207
Law G, Kitchener AC, Van Etten LE (2008) Zoo animal enrichment: Where next? Int Zoo News 55(4):212–219
Livingstone B, Kitchener AC, Hull G, Schwarz T, Vijayanathan S, Allen MJ, Liptovszky M, Portela Miguez R (2020) Diffuse Idiopathic Skeletal Hyperostosis in captive gorillas (*Gorilla* spp.): Appearances and diagnosis. J Zoo Wildl Med 51(3):578–590
Lowenstine LJ, McManamon R, Terio KA (2016) Comparative pathology of aging great apes: Bonobos, chimpanzees, gorillas, and orangutans. Vet Pathol 53(2):250–276
Macdonald AA (2017) Erosion of canine teeth in babirusa (genus *Babyrousa*). C R Biol 340(5):271–278
Mader R, Verlaan JJ, Buskila D (2013) Diffuse idiopathic skeletal hyperostosis: Clinical features and pathogenic mechanisms. Nat Rev Rheumatol 9(12):741
Mallinson JJ (1971) The breeding and maintenance of marmosets at Jersey Zoo. Int Zoo Yearb 11:79–83
Manville RH (1957) Longevity of captive mammals. J Mammal 38(2):279–280
Manville AM (1990) Variability of dental diseases in two populations of Great Lakes black bears. Bears: Their Biol Manag 8:129–134
McKinney B (1970) Hyaline arteriosclerosis in wild animals. J Comp Pathol 80:275–279
McWilliams DA, Leeson S (2001) Metabolic bone disease in lizards: Prevalence and potential for monitoring bone health. Nutr Advisory Group 19:120
Mohr U, Carlton WW, Dungworth DL, Benjamin SA, Capen CC, Hahn FF (eds) (2001) Pathobiology of the aging dog. Iowa State University Press, Ames
Morbeck ME, Galloway A, Sumner DR (2002) Getting old at Gombe: Skeletal aging in wild-ranging chimpanzees. In: Erwin JM, Hof PR (eds) Aging in nonhuman primates. Karger, Basel, pp 48–62
Nakai M (2003) Bone and joint disorders in wild Japanese macaques from Nagano Prefecture, Japan. Int J Primatol 24(1):179–195
Nascimento FA, Gatto LAM, Lages RO, Neto HM, Demartini Z (2014) Diffuse idiopathic skeletal hyperostosis: A review. Surg Neurol Int 5(Suppl 3):S122
Neiffer DL, Rothschild BM, Marks SK, Urvater JA, Watkins DI (2000) Management of reactive arthritis in a juvenile gorilla (*Gorilla gorilla gorilla*) with long-term sulfasalazine therapy. J Zoo Wildl Med 31(4):539–551
Nichols KA, Zihlman AL (2002) Skeletal and dental evidence of aging in captive western lowland gorillas: A preliminary report. In: Erwin JM, Hof PR (eds) Aging in nonhuman primates. Karger, Basel, pp 22–31
Nowak R (1991) Walker's mammals of the world, 5th edn. John Hopkins University Press, Baltimore
Nowak RM (1999) Walker's mammals of the world, 6th edn. John Hopkins University Press, Baltimore
Nunn CL, Rothschild B, Gittleman JL (2007) Why are some species more commonly afflicted by arthritis than others? A comparative study of spondyloarthropathy in primates and carnivores. J Evol Biol 20(2):460–470
Obregon EB, Ramirez O (1997) Ageing and testicular function in *Octodon degus*. Andrologia 29(6):319–326
Packer J, Ballantyne R (2010) The role of zoos and aquariums in education for a sustainable future. New Dir Adult Contin Educ 2010(127):25–34
Packer C, Tatar M, Collins A (1998) Reproductive cessation in female mammals. Nature 392:807–811

Phelan JP (2001) Genetic variability and rodent models of human ageing. Exp Gerontol 27(2):147–159

Rabb GB (1960) Longevity records for mammals at the Chicago Zoological Park. J Mammal 41(1):113–114

Ramsay MA, Stirling I (1988) Reproductive biology and ecology of female polar bears (*Ursus maritimus*). J Zool 214(4):601–633

Raphael BL, Calle PP, Haramati N, Watkins DI, Stetter MD, Cook RA (1995) Reactive arthritis subsequent to *Shigella flexneri* enteritis in two juvenile lowland gorillas (*Gorilla gorilla gorilla*). J Zoo Wildl Med 26(1):132–138

Ratliff C, Waller KR, Steinberg H, Clyde VL (2020) Diffuse Idiopathic Skeletal Hyperostosis with secondary dysphagia in a black-handed spider monkey (*Ateles geoffroyi*). J Zoo Wildl Med 51(2):455–458

Resnick D, Niwayama G (1976) Radiographic and pathologic features of spinal involvement in diffuse idiopathic skeletal hyperostosis (DISH). Radiology 119:559–568

Roe K, McConney A, Mansfield CF (2014) The role of zoos in modern society—A comparison of zoos' reported priorities and what visitors believe they should be. Anthrozoös 27(4):529–541

Roller M, Müller DW, Bertelsen MF, Bingaman Lackey L, Hatt JM, Clauss M (2021) The historical development of juvenile mortality and adult longevity in zoo-kept carnivores. Zoo Biol 1–8. https://doi.org/10.1002/zoo.21639

Rothschild BM (1993) Arthritis of the spondyloarthropathy variety in *Callithrix jacchus*. J Med Primatol 22(5):313–316

Rothschild BM (2005) Primate spondyloarthropathy. Curr Rheumatol Rep 7(3):173–181

Rothschild B (2009) Scientifically rigorous reptile and amphibian osseous pathology: Lessons for forensic herpetology from comparative and paleo-pathology. Appl Herpetol 6(1):47–79

Rothschild B (2010) Macroscopic recognition of nontraumatic osseous pathology in the postcranial skeletons of crocodilians and lizards. J Herpetol 44(1):13–20

Rothschild B (2015) Spondyloarthropathy, spondyloarthritis, axial spondyloarthritis: A rose by any other name. J Rheum Dis Treat 1:e010

Rothschild B (2021) Survey of post-cranial skeletal pathology in snakes. J Comp Pathol 183:39–44

Rothschild BM, Panza RK (2005) Epidemiologic assessment of trauma-independent skeletal pathology in non-passerine birds from museum collections. Avian Pathol 34(3):212–219

Rothschild BM, Panza R (2006) Osteoarthritis is for the birds. Clin Rheumatol 25(5):645–647

Rothschild BM, Rühli FJ (2005a) Comparison of arthritis characteristics in lowland *Gorilla gorilla* and mountain *Gorilla beringei*. Am J Primatol 66(3):205–218

Rothschild BM, Rühli FJ (2005b) Etiology of reactive arthritis in *Pan paniscus, P. troglodytes troglodytes,* and *P. troglodytes schweinfurthii*. Am J Primatol 66(3):219–231

Rothschild BM, Rühli FR (2007) Comparative frequency of osseous macroscopic pathology and first report of gout in captive and wild-caught ratites. J Vet Med Ser A 54(5):265–269

Rothschild BM, Woods RJ (1989) Spondyloarthropathy in gorillas. Semin Arthritis Rheum 18(4):267–276

Rothschild BM, Woods RJ (1991) Reactive erosive arthritis in chimpanzees. Am J Primatol 25(1):49–56

Rothschild BM, Woods RJ (1992a) Spondyloarthropathy as an Old World phenomenon. Semin Arthritis Rheum 21(5):306–316

Rothschild BM, Woods RJ (1992b) Osteoarthritis, calcium pyrophosphate deposition disease, and osseous infection in Old World primates. Am J Phys Anthropol 87(3):341–347

Rothschild BM, Woods RJ (1992c) Erosive arthritis and spondyloarthropathy in Old World primates. Am J Phys Anthropol 88(3):389–400

Rothschild BM, Woods RJ (1993) Arthritis in New World monkeys: Osteoarthritis, calcium pyrophosphate deposition disease, and spondyloarthropathy. Int J Primatol 14(1):61–78

Rothschild BM, Woods RJ (1996) Inflammatory arthritis in Pongo. J Med Primatol 25(6):414–418

Rothschild BM, Wang X, Cifelli R (1993) Spondyloarthropathy in Ursidae: A sexually transmitted disease? Res Explor 9(3):382–384

Rothschild BM, Rothschild C, Woods RJ (1998) Inflammatory arthritis in large cats: An expanded spectrum of spondyloarthropathy. J Zoo Wildl Med 29(3):279–284

Rothschild BM, Hong N, Turnquist JE (1999) Skeletal survey of Cayo Santiago rhesus macaques: Osteoarthritis and articular plate excrescences. Semin Arthritis Rheum 29(2):100–111

Rothschild BM, Prothero DR, Rothschild C (2001a) Origins of spondyloarthropathy in Perissodactyla. Clin Exp Rheumatol 19(6):628–632

Rothschild BM, Rothschild C, Woods RJ (2001b) Inflammatory arthritis in canids: Spondyloarthropathy. J Zoo Wildl Med 32(1):58–64

Rothschild BM, Mitchell EDM, Michael J, Early GA (2005) What causes lesions in sperm whale bones? Science 308(5722):631–633

Schultz AH (1935) Eruption and decay of the permanent teeth in primates. Am J Phys Anthropol 19(4):489–581

Schultz AH (1939) Notes on diseases and healed fractures of wild apes and their bearing on the antiquity of pathological conditions in man. Bull Hist Med 7(6):571–582

Schultz AH (1940) Growth and development of the chimpanzee. Contrib Embryol 28:1–63

Schultz AH (1941) Growth and development of the orang-utan. Contrib Embryol 29:57–110

Schultz AH (1942) Growth and development of the proboscis monkey. Bull Mus Comp Zool Harv 89:279–314

Schultz AH (1944) Age changes and variability in gibbons. A morphological study on a population sample of a man-like ape. Am J Phys Anthropol 2(1):1–129

Schultz AH (1956) The occurrence and frequency of pathological and teratological conditions and of twinning among non-human primates. Primatologia 1:965–1014

Schultz AH (1960) Age changes and variability in the skulls and teeth of the Central American monkeys *Alouatta, Cebus* and *Ateles*. Proc Zool Soc Lond 133(3):337–390

Sclater PL (1879) List of the vertebrated animals now or lately living in the gardens of the Zoological Society of London, 7th edn. Longmans, Green, Reader and Dyer, London

Sclater PL (1896) List of the vertebrated animals now or lately living in the gardens of the Zoological Society of London, 9th edn. Longmans, Green, London

Smucny DA, Allison DB, Ingram DK, Roth GS, Kemnitz JW, Kohama SG, Lane MA (2001) Changes in blood chemistry and hematology variables during aging in captive rhesus macaques. J Med Primatol 30(3):161–173

Snyder RL, Moore SC (1968) Longevity of captive mammals in Philadelphia Zoo. Int Zoo Yearb 8:175–183

Stott K (1948) Notes on the longevity of captive birds. Auk 65(3):402–405

Tidière M, Gaillard JM, Berger V, Müller DW, Lackey LB, Gimenez O, Clauss M, Lemaître JF (2016) Comparative analyses of longevity and senescence reveal variable survival benefits of living in zoos across mammals. Sci Rep 6(1):1–7

Ward SJ, Sherwen S, Clark FE (2018) Advances in applied zoo animal welfare science. J Appl Anim Welf Sci 21(suppl):23–33

Weigl R (2005) Longevity of mammals in captivity; from the living collections of the world. Schweizerbart, Stuttgart

Willis GP, Kapustin N, Warrick JM, Miller LL, Stookey GK, Hopkins DT, Doan EJ, Ross SR (1999) Preventing dental calculus formation in lemurs (*Lemur catta, Eulemur fulvus collaris*) and baboons (*Papio cynocephalus*). J Zoo Wildl Med 30(3):377–382

Wolfensohn S, Shotton J, Bowley H, Davies S, Thompson S, Justice WS (2018) Assessment of welfare in zoo animals: Towards optimum quality of life. Animals 8(7):110

Wolstenholme GEW, O'Connor M (1959) CIBA Foundation colloquia on ageing. Vol. 5. The lifespan of animals. Churchill, London

Mourning-Like Behaviour in a Malayan Sun Bear

Friederike Schmitz and Simone Schehka

Klaus and Josephine. Klaus and Josephine (Josy) were two Malayan sun bears (*Helarctos malayanus*) who met in 2006 when Josy arrived at Allwetterzoo Münster, Germany. They soon became close companions and can be seen in this photograph (taken in January 2012) interacting with each other and their enrichment. Over their years together, a strong relationship and bond was created between them. Play behaviour between them consisted of social play, wrestle and play fighting, knocking each other over, making faces, or biting into the skin folds (but without injuries; it was rather teasing), sometimes accompanied by playful roars. They both enjoyed food

F. Schmitz · S. Schehka (✉)
Allwetterzoo Münster, Germany
e-mail: schehka@allwetterzoo.de

S. Brando, S. Chapman (eds.), *Optimal Wellbeing of Ageing Wild Animals in Human Care*, https://doi.org/10.1007/978-3-031-30659-4_13

very much (typically bear!). Especially Josy liked to eat, and with an amazing speed, so her caregivers had to keep an eye on her weight. As Klaus got older, he ate much slower, and he also needed a few extra calories to maintain his weight. Some of their preferred food was fruits (grapes, melon, and berries), nuts, raisins, honey, eggs, whole carcasses, or meat (they liked to rub themselves with the meat and spread the smell in their fur). One of Josy's favourites was fish, and she always got very impatient when the smelled it. Both bears had very special personalities, and it was a privilege for the caregivers to work with them. Josy was an extremely smart, curious, creative, and very self-confident bear. She was very attentive to what was happening around her, and she always had a positive connection with her caregivers. However, patience and sharing food were not her strengths. When Klaus was younger, he used to be more temperamental, also towards the keepers, but as he got older, he got calmer. Especially his stoic patience towards Josy was amazing. For example, when Josy wanted food from him, she sat down in front of him and complained loudly. Klaus just kept eating, but in the end, Josy mostly got what she wanted and was able to steal some leftovers. To encourage natural behaviour, the bears were presented with different visual, manipulation, scent, and food-based enrichment, and it was frequently changed to maintain the bears' interest. Klaus passed away in 2017 at over 30 years old, following a deterioration in his health. Especially after Klaus's death, when Josy was alone for a while, she was given new things to keep her stimulated. Josy used to enjoy a yoghurt and bran mix and had to lick herself clean afterwards because she greedily spilled the yoghurt. Josy and a male Frodo, who arrived to join her in 2018, have both since been transferred to Tierpark Berlin, Germany in 2022.

Abstract

The behavioural response of animals to death, also called mourning-like behaviours, is an area of science which is underrepresented in the literature. Grief is the internalised psychological processing of a loss, whilst mourning is the external expression. In a survey by McGrath et al. (Anim Welf 22, 2013), "*90% of the general public surveyed believed that some or all animals can experience grief, with 23% believing that all animals can grieve*" (p. 33). Many caregiving staff have reported anecdotally that they believe they have witnessed changes in animal's behaviour when other animals have died, either suddenly or following a longer illness or deterioration. Within this chapter, we refer to the literature with regard to animal behavioural responses to death and also describe an example of behaviour we observed following the death of a companion Malayan sun bear (*Helarctos malayanus*). The appreciation that animals may have feelings of grief and show mourning behaviour is an important factor in their care in captivity and should be considered when managing older animals in zoos.

Keywords

Mourning · Grief · Ageing · Zoo · Behaviour · Companion · Animal–animal relationship · Bond · Death · Sun bear · Animal emotions

1 Introduction

Negative emotions such as grief or sadness are adaptive human reactions to stressful life events like the loss of a bonded partner. They are accompanied by a physiological arousal (Huron 2018). In animals, grief might be described similarly: as a substantial change in the behaviour of the surviving other(s), with possible affect display (King 2016). Grief studies in humans often rely on the verbal expression of emotions. We cannot ask animals what they feel after the conspecific's death. However, we can observe their behaviour in these situations, or, in the words of Bekoff (2007, p. 1): "*Many animals display their feelings openly... And when we pay attention, what we see outside tells us lots about what's happening inside an individual's head and heart.*"

2 Mourning-Like Behaviour in Animals

A variety of death-related responses have been described, particularly in socially complex mammals such a as non-human primates (Van Leeuwen et al. 2016; Watson and Matsuzawa 2018), elephants (Douglas-Hamilton et al. 2006; Sharma et al. 2020), and cetaceans (Bearzi et al. 2017; Reggente et al. 2016). Death or mourning-related behaviours have rarely been observed in wild carnivores (Appleby et al. 2013; Bekoff 2007; Boyd and Pletscher 1993 (as cited in Uccheddu et al. 2022)), especially not in species such as bears (Karamanlidis and Panagiotopoulos 2021). This is not surprising because highly social animals should generally be more receptive to grief-like behaviour (King 2016). However, the example we describe also indicates that a presumably solitary species such as the sun bear (*Helarctos malayanus*) can exhibit emotional arousal to the death of a close conspecific. The individual death response can hardly be predicted and is strongly influenced by "ontogeny and personality as well as the nature of the relationship with the deceased" (King 2016, p. 3).

3 Malayan Sun Bears Klaus and Josy

Called a "forgotten" species, the Malayan sun bear is the least studied of the bears (Crudge et al. 2019; Meijaard 1999). This small tropical carnivore, found in Southeast Asia, is extremely reclusive and little is known about their social life. In the wild, sun bears tend to remain solitary (Scotson et al. 2017). However, in human care, they often tolerate conspecifics and are kept in social pairs or groups. Bears communicate with each other through sounds, postures, and smells (Pruitt and Burghardt 1977). Sun bears even mimic their playmates' facial expressions, indicating a high social sensitivity (Taylor et al. 2019).

In 2006, a young female bear named Josy arrived at Allwetterzoo Münster, Germany, and met the older male Klaus. Quickly, the two bears became inseparable. They had a strong bond, with regular socio-positive interactions, like playful wrestling and mating behaviour. Part of their evening ritual was that Klaus worked on his bed enrichment, e.g. jute bags, whilst Josy waited patiently until their shared bed was ready to be slept in. In March 2016, Klaus, meanwhile in the advanced age of almost 30 years, started to show age-related signs of decreasing general condition. Therefore, a specific age-related quality-of-life assessment was initiated. Over the next few months, he was closely monitored by a team of caregivers, veterinarians, and curators. In May 2017, Klaus's health deteriorated drastically, and to prevent prolonged suffering, the decision was made to euthanise him.

To facilitate Josy processing the loss of her long-term partner following Klaus's euthanasia, she was given the opportunity to spend some time with Klaus's body. The behavioural responses were observed and filmed by the animal welfare team of the Allwetterzoo Münster. Hereinafter, descriptions of different behaviours are portrayed.

Klaus's body was lying on the ground in the indoor exhibit. The gate opened and Josy came in, quickly approaching her mate without hesitation. The access to the outdoor stayed open, so Josy could leave whenever she wanted to. Immediately, she started investigating Klaus and gently sniffing his head, followed by the torso, anogenital region, and paws. Then, Josy checked the surroundings, taking in all the unfamiliar smells. When she returned to his body, she started sniffing more intensely this time, followed by light nudges. The only sounds were short forceful puffs of air from her nose. The female kept touching Klaus's body with her snout and paws, carefully examining every centimetre of his body. After about ten minutes, her behaviour changed abruptly. The inspecting became more forceful, and the otherwise quiet Josy began to vocalise: first a light moan and then a stronger roaring. These were similar sounds she used, when she seems to be upset (e.g. when she wanted some of Klaus's food and he was not willing to share). Josy bit his lips and pulled his head with her teeth and claws closer to her. Then, she barked which was an unexpected loud noise. Her movements became more rapid and increasingly energetic. She kept vigorously scratching and biting Klaus's body, without damaging the skin. The behaviour increased, far beyond simple exploration with Josy showing all signs of distress, frustration, and anger. She kept whining and yelling and pulled Klaus's body more intensively. Then her movements slowed down and she stopped vocalising. After more than twenty minutes, the team decided to separate Josy from Klaus to remove the body. Later, when she returned to the empty enclosure, she wandered around and then started a clucking noise. This noise is a typical friendly contact call in sun bears (personal observation).

The days after Klaus's death, Josy was very attentive to her caregivers, more than usual. Then, about 3 weeks later, she suddenly started swimming in the moat of the outdoor exhibit for a couple of days. Up to this point, Josy has only shown this behaviour once at the Allwetterzoo Münster, almost 11 years ago, shortly after arrival, when she examined the unfamiliar outdoor enclosure for the very first time.

4 Discussion

Josy's reaction to Klaus's dead body consisted mainly of explorative behaviour and signs of emotional affect and distress. However, does this mean that Josy was sad about her loss, or grieving? Charles Darwin was one of the first scientists who wrote about the existence and nature of emotions in animals. In his book, *The Expression of the Emotions in Man and Animals* (1872), Darwin notes the universal nature, with both humans and animals expressing the same state of mind by the same movements. Of course, any theory about animal emotions cannot be tested as in humans, but modern neurosciences studies show similarities in how brains in other animals function and that all animals have many neural processes in common with humans (e.g. Berns 2018).

Other examples of death responses in animals have been reported in zoos. These have included Asian elephants (*Elephas maximus*) being given access to the body of a calf that was stillborn and the team seeing and hearing the herd's response. This response included gentle touching of the calf's body, roaring, and rumbling noises. The herd's general behaviour appeared subdued in the days following the incident. Another anecdotal report involved an aged chimpanzee (*Pan troglodytes*) that lost her long-term companion suddenly. She was given access to his body and was seen pushing a long stick down the back of this throat as if trying to stimulate his breathing. She appeared to lose interest and walked away leaving the stick in place. Great apes and primates have been known to carry their stillborn infants around for many days before dropping them (S. Chapman, pers comms).

Until today, descriptions of death-associated behaviour have largely been based on opportunistic observations in the wild with limited published literature on the subject. One example of research proposed that nonhuman primates were capable of an implicit awareness of death following reports of how the animal behaved after the death of their conspecifics (Gonçalves and Carvalho 2019). Reports of animals' responses to death in captivity are also rare and restricted to few taxa (e.g. Anderson et al. 2010; De Marco et al. 2020; Jakucińska et al. 2019; Rutherford and Murray 2021; Uccheddu et al. 2022; Walker et al. 2016). Together with approaches in domains covered in this book, this data is valuable in changing how we can expand and improve animal care programmes by supporting animals in these last phases of their life. This might help us better understand an animal's emotions and his or her welfare needs, leading to significant improvements in quality of life under human care.

References

Anderson JR, Gillies A, Lock LC (2010) Pan thanatology. Curr Biol 20(8):R349–R351. https://doi.org/10.1016/j.cub.2010.02.010

Appleby R, Smith B, Jones D (2013) Observations of a free-ranging adult female dingo (*Canis dingo*) and littermates' responses to the death of a pup. Behav Process 96:42–46. https://doi.org/10.1016/j.beproc.2013.02.016

Bearzi G, Eddy L, Piwetz S, Reggente MAL, Cozzi B (2017) Cetacean behavior toward the dead and dying. In: Vonk J, Shackelford T (eds) Encyclopaedia of animal cognition and behavior. Springer, Cham, pp 1–8. https://doi.org/10.1007/978-3-319-47829-6_2023-1

Bekoff M (2007) The emotional lives of animals: a leading scientist explores animal joy, sorrow, and empathy—and why they matter. New World Library, Novato, CA

Berns G (2018) What it's like to be a dog: and other adventures in animal neuroscience. Basic, New York

Boyd D, Pletscher D (1993) Evidence of wolves, Canis lupus, burying dead wolf pups. Can. Field Nat. 107:230–231

Charles D (1872) The expression of the emotions in man and animals. John Murray, London

Crudge B, Lees C, Hunt M, Steinmetz R, Fredriksson G, Garshelis D (eds) (2019) Sun bears: Global status review & conservation action plan, 2019-2028. IUCN SSC Bear Specialist Group/IUCN SSC Conservation Planning Specialist Group/Free the Bears/Traffic

De Marco A, Cozzolino R, Thierry B (2020) Responses to a dead companion in a captive group of tufted capuchins (*Sapajus apella*). Primates 61(1):111–117. https://doi.org/10.1007/s10329-019-00743-y

Douglas-Hamilton I, Bhalla S, Wittemyer G, Vollrath F (2006) Behavioural reactions of elephants towards a dying and deceased matriarch. Appl Anim Behav Sci 100(1–2):87–102. https://doi.org/10.1016/j.applanim.2006.04.014

Gonçalves A, Carvalho S (2019) Death among primates: a critical review of non-human primate interactions towards their dead and dying. Biol Rev 94(4):1502–1529. https://doi.org/10.1111/brv.12512

Huron D (2018) On the functions of sadness and grief. In: Lench HC (ed) The function of emotions: when and why emotions help us. Springer, Cham, pp 59–91. https://doi.org/10.1007/978-3-319-77619-4_5

Jakucińska A, Trojan M, Sikorska J, Farley D (2019) Reaction to the death of the oldest female in a group of chimpanzees at the Municipal Zoological Garden, Warsaw. Primates 61(1):103–109. https://doi.org/10.1007/s10329-019-00772-7

Karamanlidis AA, Panagiotopoulos N (2021) Burying of dead cubs by a brown bear in Greece: Food caching or 'grief' behavior? Ursus 2021(32e7):1–5. https://doi.org/10.2192/ursus-d-20-00015.2

King BJ (2016) Animal mourning: Précis of How animals grieve (King 2013). Anim Sentience 4(1). https://doi.org/10.51291/2377-7478.1010

McGrath N, Walker J, Nilsson D, Phillips C (2013) Public attitudes towards grief in animals. Anim Welf 22:33–47. https://doi.org/10.7120/09627286.22.1.033

Meijaard E (1999) Ursus (Helarctos) malayanus, the neglected Malayan sun bear. Nederlandse Commissie voor Internationale Natuurbescherming Mededelingen 34:1–62

Pruitt CH, Burghardt GM (1977) Communication in terrestrial carnivores: Mustelidae, Procyonidae, and Ursidae. In: Seboek TA (ed) How animals communicate. Indiana University Press, Bloomington, IN, pp 767–793

Reggente MAL, Alves F, Nicolau C, Freitas L, Cagnazzi D, Baird RW, Galli P (2016) Nurturant behavior toward dead conspecifics in free-ranging mammals: new records for odontocetes and a general review. J Mammal 97(5):1428–1434. https://doi.org/10.1093/jmammal/gyw089

Rutherford L, Murray LE (2021) Personality and behavioral changes in Asian elephants (*Elephas maximus*) following the death of herd members. Integr Zool 16(2):170–188. https://doi.org/10.1111/1749-4877.12476

Scotson L, Fredriksson G, Augeri D, Cheah C, Ngoprasert D, Wai-Ming W (2017) Helarctos malayanus (errata version published in 2018). The IUCN Red List of Threatened Species 2017. e.T9760A123798233. https://doi.org/10.13140/RG.2.2.13343.66721

Sharma N, Pokharel SS, Kohshima S, Sukumar R (2020) Behavioural responses of free-ranging Asian elephants (*Elephas maximus*) towards dying and dead conspecifics. Primates 61(1):129–138. https://doi.org/10.1007/s10329-019-00739-8

Taylor D, Hartmann D, Dezecache G, Te Wong S, Davila-Ross M (2019) Facial complexity in sun bears: exact facial mimicry and social sensitivity. Sci Rep 9(1). https://doi.org/10.1038/s41598-019-39932-6

Uccheddu S, Ronconi L, Albertini M, Coren S, Da Graça Pereira G, De Cataldo L, Haverbeke A, Mills DS, Pierantoni L, Riemer S, Testoni I, Pirrone F (2022) Domestic dogs (Canis familiaris) grieve over the loss of a conspecific. Sci Rep 12(1). https://doi.org/10.1038/s41598-022-05669-y

Van Leeuwen EJ, Mulenga IC, Bodamer MD, Cronin KA (2016) Chimpanzees' responses to the dead body of a 9-year-old group member. Am J Primatol 78(9):914–922. https://doi.org/10.1002/ajp.22560

Walker J, Waran N, Phillips C (2016) Owners' perceptions of their animal's behavioural response to the loss of an animal companion. Animals 6(11):68. https://doi.org/10.3390/ani6110068

Watson CFI, Matsuzawa T (2018) Behaviour of nonhuman primate mothers toward their dead infants: uncovering mechanisms. Philos Trans Roy Soc B Biol Sci 373(1754):20170261. https://doi.org/10.1098/rstb.2017.0261

Caring for Elderly Wild Animals: The Human Experience

Sabrina Brando, Mickey Gjerris, Nicola Field, and Lynette Hart

You've left this world no longer a number, a letter or bear just the same,

For you will always be remembered as a bear with a name, from Spirit of Hope

S. Brando (✉)
AnimalConcepts, Teulada, Spain
e-mail: sbrando@animalconcepts.eu

M. Gjerris
University of Copenhagen, Copenhagen, Denmark

S. Brando, S. Chapman (eds.), *Optimal Wellbeing of Ageing Wild Animals in Human Care*, https://doi.org/10.1007/978-3-031-30659-4_14

Jade. Jade was a wild-born bald eagle (*Haliaeetus leucocephalus*), estimated to have hatched in 1982. He was brought to the Toronto Zoo after he was injured when his nest was blown out of a tree near Red Lake in Northern Ontario. Jade spent most of his time at the Toronto Zoo on display to the public in the Canadian Domain, along with a female bald eagle named Gronk. After his companion passed away, Jade remained a few years in the Domain and then retired to the Outreach & Discovery area. During his time with the Outreach team, he was calm and content. He generally liked to keep his distance from the keepers, but would occasionally approach keepers to gently take offered food. Given his injured wing, he could not fly, but would still navigate well around his enclosure, making use of his various perching. Compared to the other eagles, Jade was always the kindest one to work with, and great for younger/new/not as confident keepers to get hands-on experience when handling the eagles. He was a great teacher! Caregivers would offer him food in foraging toys, or in his pool to encourage fishing! He also liked tearing up heads of lettuce when they were offered to him. Migizi (a word to describe the bald eagle by the Anishnaabek) is a symbol of immense honour to many Indigenous people on Turtle Island (now known as North America). It is said that because the eagle flies the highest, they are the closest to Creator, taking offerings and prayers directly to Creator. Jade was approximately 40 years of age and spent his entire life at the Toronto Zoo, until passing away naturally in December 2022. A ceremony was held at the zoo, led by Indigenous women on staff, to honour Jade's spirit and prepare him for his journey to the Spirit World. His body was smudged with sacred medicines (tobacco, sage, cedar, and sweetgrass) and a prayer said, thanking him for his life and the important role that he played at Toronto Zoo. A cedar bath was prepared and Jade was gently cleansed with the medicine-infused water. The cedar bath symbolises the transition from one stage of life to the next, and a common traditional practice when an individual passes, to wash their bodies of spiritual and physical negativities prior to burial. To close out the ceremony, Jade was wrapped in red cloth with tobacco and sage offered by ceremony observers, an eagle honour song was sung, and a plate with Jade's favourite foods, such as fish and rodents, to feast his spirit was made. All those who worked with Jade were grateful for the opportunity to build trust with him over his 40 years in the care of the Toronto Zoo.

Abstract

For many who work closely with animals, it is inevitable that an emotional connection is forged between animals and their human caretakers, especially for those who perceive their work in animal care and welfare as a calling. There is often a deep emotional investment in the animal's life and wellbeing, and the experience of caring for that animal comes with a range of positive feelings such as joy, love, and trust—and, inevitably, a range of negative feelings such as sorrow, pain, and grief when the animal reaches the end of her or his life. It is essential to recognise the importance of the human experience as it pertains to the relationships and bonds between animals and those who care for them, and the impacts these experiences can have on human wellbeing. The human experience can be especially pertinent when it comes to making decisions surrounding qual-

N. Field
Global Animal Welfare, Buckinghamshire, UK
e-mail: nfield@globalanimalwelfare.org.uk

L. Hart
University of California Davis, Davis, CA, USA
e-mail: lahart@ucdavis.edu

ity of life and euthanasia. Those working with animals must feel sufficiently supported in their emotional experiences including through opportunities to remember and mourn the animals they have lost. Bonds are to be celebrated and nurtured as foundational to the human–animal care experience; to experience grief is to have experienced profound love and care. Fostering a culture of care and respect allows for humans to be at their best for the animals they care for, feeling involved in all management decisions and safe to express their emotional experiences with animals. This can be achieved through providing safe spaces for staff to share in their mutual experiences, providing opportunities to celebrate animals they have lost and pay their respects such as through memorial services, and supporting the mental wellbeing of those who may be struggling with the loss of a beloved animal.

Keywords
Ageing · Death · Zoos · Sanctuaries · Human wellbeing · Grief · Dying · Human–animal relationships · Celebration · Remembrance

1 Experiencing the Animals' Ageing Process

Human fascination with animals goes back many thousands of years, with the oldest cave paintings showing more images of animals than peoples, such as in the Chauvet cave in southwestern France (Kalof 2007). Humans began to domesticate plants and animals approximately 10,000 years ago in developing agricultural communities, and there is archaeological evidence of domestication in many scenes that represent husbandry or remains of objects linked with husbandry (e.g. yokes) (Vigne 2011). Examples of human–animal relationships dating back to the domestication of the wolf (*Canis lupus*) several thousand years ago have also been found (Serpell 1996). Wild animals in human care have a long history, e.g. a public display of animals in 1480 BC by Queen Hatshepsut in Egypt, the Garden of Intelligence in 1060 in China by Emperor Wen-Wang housing many fishes, deer, and birds, and King Henry I who in 1200 set up a menagerie including tigers and lions (Hoage and Deiss 1996). Zoo Schönbrunn in Vienna, Austria, was built in 1752 and is considered the oldest continuously operating zoo in the world.

Many animals have held a special place in people's lives, and still do today, as apparent through human activities and rituals during the animals' lives and deaths, some commemorated in animal cemeteries dating back to the early nineteenth century in New York, London, and Paris (Kean 2013). Mary Ann, the first elephant at the Baltimore Zoo, received an official burial in a Maryland graveyard when she died in 1941. Countless published and unpublished stories of animals who touched people's hearts are available, including many wild animals in human care in zoos, aquariums, sanctuaries, and the like, henceforth zoos. Animal care and wellbeing programmes continue to evolve employing the lifespan approach (Brando and

Buchanan-Smith 2018), and contemporary facilities today have specific programmes in place for animals as they age.

Caring for animals through daily actions such as feeding, animal training, and behaviour observations gives animal caregivers and the animals many opportunities to develop a positive relationship and over time a bond (Hosey and Melfi 2012). Whilst specific research on caring for elderly animals in zoos from a human perspective is sparse, animal care staff often develop a shared relationship with the animals, just as with companion dogs and cats. Each partner in the relationship responds in special ways to the other and, over time, they come to know the mannerisms, typical activities, and behaviours of each other more specifically. As the animal ages, the number of routinely shared activities increases. As animals often require treatments for age-related conditions; more behavioural observations, as well as communication around certain iconic animals towards the public at the zoo, may increase. The caregiver may become more attached to the animal and may suffer a serious sense of loss when the animal dies.

When caring for a farm, exotic, or laboratory animal, the relationship might be more difficult to characterise than with a companion dog or cat. Visible and tangible responses of the animal to the person may be more subtle, or not evident. Hank Davis was especially interested in this phenomenon and set up experiments giving animals a choice of selecting either their regular caregiver or a strange person, and not surprisingly animals in a wide range of species preferred their familiar caregiver. Rats, sheep, and rabbits showed preferences for familiar humans (Davis et al. 1977, 1998; Davis and Gibson 2000); llamas responded differently to familiar and unfamiliar humans (Taylor and Davis 1996); cattle differentiated between and preferred individual humans (Taylor and Davis 1998); and two species of seals were also found to be more vigilant with unfamiliar humans (Taylor et al. 1998). These results demonstrate that the relationships humans developed with the farm, exotic, or laboratory animals they cared for were reciprocal and involved both the person and the animal. In addition to these many published papers, the authors compiled related information in an edited book highlighting the inevitable bond that scientists have with the animals they work with (Davis and Balfour 1992).

More recent studies also show that animals have a lower level of fear towards familiar keepers (Martin and Melfi 2016), and longer familiarity with a caregiver can positively support animals in novel situations (Liehrmann et al. 2021). Using playback experiments, gorillas at Zoo Atlanta responded to the voices of unfamiliar and familiar-negative individuals with a larger number of distress behaviours, higher gazing frequency, longer gazing duration, and shorter latency, than to those of familiar-positive individuals (Salmi et al. 2022). So, the familiarity of the person brings comfort to the animal, and the animal develops a preference for that person over a stranger. Building good relationships is important for most animal caregivers, so that they know the animals experience a good life without fear or anxiety and know that the animals are enjoying themselves and find joy in interacting with or being near them.

1.1 Self-Object: You Are a Good Person

Why is an animal so important to a person? The deep connections that people develop with certain animals are everyday occurrences. One explanation of an animal's importance to a person is that we learn who we are from our interactions with others—the reflection they give back to us indicating that we are valued and that we matter to them. The essential communication we receive from others is how we learn of our value and goodness; a healthy sense of self, with self-esteem, wellbeing, and cohesion, is supported and maintained by responses from others that are calming, soothing, and affirming. Kohut (1971, 1984) proposes the psychoanalytic idea of the concept of "self-objects", or experiences that are not experienced as separate and independent of the self: persons, objects, or activities which are seen to "complete" the self.

Another person can be a self-object for us, becoming an empathy-based partner, by (1) reflecting or mirroring back to us as a partner who affirms that we are appreciated, (2) serving as an ideal perfect partner that has the traits we would like to have, or (3) serving as an identical twin partner that agrees with us. The concept of a self-object, which tells us who we are, is first experienced with a primary caregiver, such as a parent, who reflects to us our immense value and goodness. As we grow older, others can provide this type of feedback. An animal also has the capacity to communicate this information to us, as reported by Sue-Ellen Brown (2007, 2011), who in a survey of pet owners found that a variety of companion animal species served as self-objects for their human caregivers.

People believe that animals are good judges of character, so interactions with an animal, including wild animals, can be inspiring and highly rewarding. In fact, even a brief experience with a strange wild animal can be especially important and affirming to a person; tourists in one study drove up Mount Evans in Colorado just to experience feeding junk food to wild mountain sheep (Lott 1988). They were seeking to have the wild mountain sheep take food from their hands, believing that this demonstrated that the animal trusted them. The tourists concluded that animals are better judges of character than people and thought that it reflected favourably on them if they were trusted by a wild animal.

Contemporary animal caregivers in zoos are often attentive in creating positive conditions for the animals and building positive relationships with the animals in their care, sometimes for many years. The animals respond in special ways to the caregiver, indicative of their mutual relationship. Just as with companion animals, these animals can come to assume a role of supportive self-object for the person. Their poor health and even suffering when ageing are often distressing to the caregiver, and their deaths can leave a sad gap in the life of the caregiver. They too may suffer a serious sense of loss when the animal dies.

1.2 5H2M Scale for Assessing Quality of Life

Whatever the context, caring for an animal that is ageing and potentially suffering is a painful task for the caregiver. This can set up a difficult dilemma, with the caregiver witnessing the animal's discomfort and considering whether euthanasia should be recommended to relieve the suffering of the animal. It can become a challenging reality to accept or even to understand. One approach that can clarify the health and welfare status of the animal is to teach caregivers how to regularly assess the animal's quality of life. This provides a way to identify problem areas to work on, notice changes in the animal's health, and be aware of a decline in health.

The 5H2M Quality of Life Scale can be used by caregivers of any animal. Caregivers, using this scale, provide a daily score on each of seven variables: hurt, hunger, hydration, hygiene, happiness, mobility, and "more good days than bad" (Villalobos and Kaplan 2018). The method is easy to learn and empowers the caregiver by making it possible to track quality of life over time and strive to improve the lower scores wherever it is possible. The importance of having good record keeping is fundamental to monitoring elderly animals as well as discussions on the practical, medical, and moral issues when confronted with ageing animals, end-of-life care, and death. As caring for elderly zoo animals can bring pain or physical challenges for the animal which can be distressing to caregivers, having the social support as well as tools to help manage these challenges and bring comfort is very important. Educating animal care professionals on this method and other meaningful tools and frameworks creates an avenue for their greater peace of mind when the animal's decline is inexorable and cannot be reversed, leading to a decision of euthanasia when it is necessary.

1.3 Assuring the Animal's Comfort

Animal care staff should work to modify the animal's habitat to make it more comfortable and suitable to move around in, rest, and interact with social companions. As the animal ages, certain activities or routine tasks that were formerly easy may become challenging—or even impossible. Steps can be taken to improve the animal's comfort and quality of life (Hart et al. 2001); e.g. climbing or other types of mobility may no longer be feasible and this may mean that the animal needs more accessible shelter, feeding, and toileting areas, or modifications to their habitat that make movement between spaces easier. Changes may need to be made to food preparation and presentation if the animal's teeth have been lost or are less effective in chewing food; if living with conspecifics, food may need to be provided separately to ensure the older animal has access without competition. Diminished vision, hearing, or other senses may require modifications to aspects of the animal's care and the way they are approached. Some animals may suffer pain from arthritis or other age-related conditions, for which medications can provide relief. All these various changes and needs may require collaborating with architects, veterinarians, and other specialists to not only improve the environment but also simplify the extra care if and when required. Care staff already training the animals may train additional cues in case of loss of sight or hearing and modify the training area for easy access and treatments.

Sometimes elderly animals are moved from one facility to another for management reasons, as not all organisations have the opportunity, space, or expertise to care for an animal who is advancing in years. Sometimes elderly animals are moved within a facility to more appropriate areas to care for them, as well as being able to communicate with and about them better, which is better than isolation in the back of house. Care staff can have joy in telling stories about the animals to the visitors and the larger community and explain why older animals may look different in their fur or feathers, perhaps look a little scruffy, and/or rest and sleep a lot. Sharing stories about the individuals as well as planning different activities to enjoy together with the animal not only brings joy in the moment, but also new memories to cherish. Especially, caregivers benefit from spending time and effort caring for elderly animals to make them as comfortable as possible whilst knowing they might not be with them for very much longer.

1.4 Last Moments

When caring for animals (and humans), we never know when life will end, which is one of the reasons we want to do the best we can and support good wellbeing to provide the conditions for a good quality of life for the duration of the animal's life. As animals age, the likelihood that they will no longer be with us increases. The months and days before an animal's passing, especially when we know that the animal is not well and could die soon, may bring feelings of anticipatory grief, also referred to as anticipatory loss or preparatory grief. This is similar to the normal process of mourning, but it happens before the actual death, imagining what life will be like without this animal, and may bring feelings of fear, sadness, and/or depression. Care staff may enhance the enjoyment caring and being with the animal brings through creating extra time with the animal, doing special activities: an exceptional treat or meal, favourite toys, amongst many others, as well as finding support in each other, with friends and family, and sometimes with professional help.

We cannot forget the visiting public and the broader community. Many individuals, like year-pass holders and others who love the facility and animals, often come to see certain individuals they feel a connection with, and they are interested in knowing about the wellbeing of the animal, particularly when ill and/or ageing. Some organisations such as the Singapore Zoo shared with the public that they may have to euthanise Inuka, the elderly polar bear, if his health would not improve. Many people came to the zoo to say goodbye and expressed sadness that they may not be able to see him anymore. This could also be an option for other facilities where the public knows the animal well, including sponsoring and/or adopting the individual.

1.5 Death as a Basic Condition

As for man, his days are as grass: as a flower of the field, so he flourisheth. For the wind passeth over it, and it is gone; and the place thereof shall know it no more (Psalms 103:15–16). We all know it: "*Life is postponed death*", as the Danish theologian K.E. Løgstrup so bluntly framed it. And this obviously is not only true for humans but for all that exist. Everyone and everything have at some point not been here and will at some point not be here again. Whilst one is here, it can be seen as a local rebellion against the second law of thermodynamics, but entropy will take us all. Death is thus a basic condition for all living beings. Some of the beings that live are aware of this, and some are not, but the arrow of time points only in one direction. Trillions and fantasillions of living beings have died since life arose on planet Earth, and one could say that for a neutral observer, death is so common as to become banal: nothing to be concerned about as it is inevitable and the one thing that everyone shares.

Nonetheless, we care. Death and dying matters to us as humans to the extent that we find it "inhumane" if the death of a human being leaves others untouched. The degree to which we find that we should be touched depends on our closeness to the person who has died, but it is still considered proper behaviour to pay respects to the death of a stranger, at least in the way of driving slowly by a funeral procession. Life matters. And the closer we are to the person who has died, the more it matters. Losing ones we love is one of the hardest experiences a human being must have, and it leaves no one untouched. Loss is not something we "get over", but something that shapes us and which we carry forward with us. Loss leaves us changed, for better or worse.

To grieve is to be stranded with love that has nowhere to go. What we lose is a unique relationship with a human, a history that has been built over time (longer or shorter), but not only that. We lose a relationship with exactly that person: an individual that was in the world in exactly her or his own way. What we have shared with that person no longer is shared. When someone dies, something irreplaceable is lost. And thus, we grieve: grieve for the person who is dead, who will no longer be able to experience the everyday magic of simply being alive. We grieve that what we did together no longer happens. Thus, we grieve for ourselves, because we can no longer share experiences with this person, no longer build on our common history, and no longer express our gratitude that they are here with us. And we grieve because we are reminded that we and everyone we love will follow them into that good night before long. Death is the clouds that darken the sky of our future, as the French philosopher Camus wrote. Grieving is, thus, a very human phenomenon to the extent that people unable to grieve are offered help to do so. Grieving is recognition of what was, and what will be. It is not a phase to get over and through, but a process that enables us to continue our lives in a way where what was will continue to shape who we are without being decisive for what will come (Weller 2015).

1.6 Animal Deaths Matter as Well

As we grieve the humans that die, so we grieve our other companions—lives that have come close to us and, suddenly, are there no more. Anyone who has been close to a companion animal that died has experienced this whether they lost a dog, cat, horse, hamster, or other animal. They are not humans and might not leave us with the same degree of grief—but it is of the same kind. It is the loss of a common history and a particular existence in the world that has ended that we are grieving. To some, this is a misunderstanding. If we grieve the loss of an animal as we grieve the loss of a human, they claim that we have mixed the categories. We have forgotten the abyss between humans and animals. That humans are individuals, and animals are merely specimens of a species. Grieving the death of an animal is thus seen as an anthropomorphism—a humanisation of the animal—or, to put it more bluntly: "Get over it, it's just a dog".

But a dog is not "just a dog"—it is exactly *that* dog. Unless your relationship with a dog is purely professional in the sense that you only interact with her or him to have her or him provide some service—which may be the case with a guard dog, then person, in the sense that cannot be replaced in the way that you can buy a new phone and transfer your previous settings to it and simply discard the old one without missing something. Even with the guard dogs, it can very well be that they will only be able to provide the service you need, if you take them into consideration as more than simply a *guard* dog but pay heed to the *dog* itself. A dog is an individual person as a human is an individual person, in the sense that her or him differs from other dogs. Every dog has her or his own history that has been part of shaping her. Obviously, genetics play a significant role, but the epigenetic effects of everything from the time in womb, environmental exposures, and upbringing will have left the dog with exactly her or his way of being a dog (Heðinsdóttir et al. 2018). And as it is with dogs, so it is with many other animals: fish, spiders, lizards, and birds just to name a few (Ogden 2012). They are not just specimens, but each has their own way of being their creature.

Thus, it is not a sign of exaggerated sentimentalism or anthropomorphising to grieve because of the death of an animal that one has an emotional connection to. Having established that in the sphere of companion animals (broadly speaking), it only makes sense to widen this observation for other engagements with animals. There is someone and something to grieve every time an animal dies, as a unique experience of being in the world is lost. As with humans, we are not emotionally attached to all animals; therefore, we do not typically go around grieving the billions of animal lives that are lost every day, even though it can put a sombre tone to one's existence to live in recognition of the paradoxical nature of life: that each life is a limited resource and that, in many ways, the continuation of one life necessitates the death of another. One can obviously choose to reduce this to a scientific fact and simply accept it, but in the light of the understanding of life and death described above, it can also be seen as the fundamental tragedy of existence—which is the task of every rational creature to come to terms with.

It is thus meaningful to grieve the death of an animal, with the degree depending on the relation to her or him, simply because it is possible to have a relationship with an animal person that can be seen as an analogy to the relationships we have with human persons. But do humans also have an obligation to build such relationships with animals that are in their care?

1.7 Grieving Dead Animals

The claim in this chapter is not that we ought to grieve when an animal in our care dies; sorrow and grief are reactions that cannot be demanded into existence. So, the ethical demand does not oblige us to have certain feelings. The claim is rather that to be able to take the demand seriously, to be able to lift our responsibility, we need to enter a relationship with the animal where her or his individuality becomes visible to us; we need to get to know the animal as an individual, a 'person', and not just as a specimen of a species. Otherwise, we cannot act in an appropriate way. We need insight and understanding into exactly this animal to be able to use our imagination to care for them in the best possible way that we can. But if we do so, we will claim, feelings of sorrow and grief are appropriate reactions to the death of the animal. They follow the death of someone we knew, because someone important was here and now no longer is. Thus, the discussion is not whether we ought to feel sorrow or not, but that we ought to enter into a relationship with each animal in our care where we open ourselves to experiences of sorrow and grief when they die. Because we knew them and have lost someone important.

The ageing and, inevitably, the demise of the animal can bring a range of emotions such as a sense of loss and sadness to the caregiving and veterinary staff, regardless of whether euthanasia is involved or not (Hart et al. 1990), and/or guilt or anger if euthanasia decisions, or not making them in a timely fashion, is perceived. To the question, "how to mourn?"—there is no unequivocal answer. Thousands of people are involved in animal care in a range of contexts and the depth of feeling and the emotional involvement that human caregivers develop for the animals in their care should not be underestimated.

Grief and sorrow come in many shapes, coloured by culture, religion, personality, and other factors. Grief is unique and personal; as individuals, we experience bereavement differently. There is no uniform way to respond to loss. The most important thing is that the response is situated in the context with respect for the people involved with the animal. It should be an opportunity to remember and rejoice in what was and lament what is not here anymore: a genuine chance to express emotions, and not just an empty gesture. What that looks like can only be decided in each particular situation. Dealing with a dying animal's needs can be overwhelming. Many animal caregivers in zoos, sanctuaries, and other facilities housing wild animals, as well as other people in professions concerned with the care and wellbeing of animals, have loved and lost many animal friends. Many have experienced early in their career that navigating grief looked like attempting to hide feelings, often out of fear that an emotional response would be perceived as "unprofessional". Expression of grief or discontent with the process related to the last phase of an animal's life and subsequent death is also dependent on how supportive and communicative management is on these topics. Otherwise, caregivers may even feel they need to conceal their feelings from others (Hart and Mader 1995). Providing compassionate support can avert staff burnout and reduce stress and empathic strain for everyone involved. The significance of the animal's life and of each person's loss should be acknowledged: a step that assists healing for the people who knew the animal. Whilst working in an academic setting, Taylor and Davis (1993) created an annual memorial service to honour animals that had died in a teaching and research role at Guelph University. Later, Iliff (2002) advocated creating a 4th "R", for remembering animals that had died in research and teaching and provided a list of many institutions that already were holding annual memorial services.

Zoos, aquariums, sanctuaries, and other facilities have created occasions for remembrance, such as short stories, reports, videos, and memorials for the various animals, for animals of all ages, and not only when elderly. Some facilities have dedicated graveyards, statues, commemoration plaques, benches, sympathy cards, and other memorials, where staff and visitors can go to remember the animals. Facilities that communicate the death of an animal to the general public often receive sympathy cards or other tokens of support reflecting the fond memories of those who came to visit the animals. Sometimes regular visitors bring gifts, drawings, poems, or share their favourite photos.

At Animals Asia, every individual bear is remembered through evolved rituals; death is a social event for many species, and humans are no exception to this. The passing of these bears is an important event. Everyone involved with the bear in question, either for their direct care or those in the wider organisation, can acknowledge the bear by sharing personal reflections in an organisational obituary. In many cases, not only have these bears survived the unimaginable trauma of a lifetime on a bile farm but they have also been nurtured and received specialised care as geriatric and compromised individuals in their final days, weeks, months, and years. They have been subject to holistic care that considers their physical, behavioural, social, and psychological needs and preferences. A multidisciplinary and multi-cultural team of veterinary professionals, behavioural management experts, and husbandry caretakers have collectively considered the complex needs of each individual. Quality of life is constantly under review; the intensity and level of involvement from each bear's caretakers cannot be underestimated. The relationship with each bear is unique and complex. There is continuous monitoring of not only what is the norm for the species but also the norm for each individual.

There is awareness and recognition of the unique characteristics of every individual, and decisions about euthanasia involve all direct caretakers. For example, at Animals Asia, when euthanasia takes place, anyone on-site at the sanctuary can be present. Recognising human individuality there is no expectation to be present. For those who choose to attend, some stand quietly in the background, whilst others feel compelled to hold a paw during the bear's final moments. Importantly, this process is not rushed. The veterinary team will conduct a post-mortem once the bear has passed; however, every opportunity is given to reflect—and for those who wish to stay with the deceased bear for a while longer, the time and space to do so is readily available. Another example is when polar bear Inuka at the Singapore Zoo died he was surrounded by his current and former caregivers, with one of his caregivers saying her farewell with her hands over his paw.

Some facilities have funerals and other rituals, for example, at Animals Asia, every bear receives a funeral, at which all staff are welcome. Acknowledging how individual grief is, no one is obligated to attend a funeral. For the bereaved caretaking and extended animal team, the ritual provides an opportunity to mourn. The team, often alongside their companion animals, follows in procession as the bear is transported from the veterinary hospital to a forested graveyard beside the river. Gathered attendees surround the grave as the bear is carefully lowered. For those who do attend, a poem is read in English and Mandarin to honour the bear in death followed by a copy of the poem being placed in the grave. There is the option for favourite treats and toys to be placed in the grave with the bear. Sometimes, there are other messages from supporters, donors, and the wider team that are placed in the grave as well. In some faiths and cultures, the practice of throwing a handful of dirt into the grave symbolises the deceased returning to the earth from which they came; the bear ritual has also evolved to include this practice whereby all funeral attendees throw a handful of dirt onto the bear before the grave is filled. Some staff have been known to hand-craft wreaths from wildflowers growing nearby to be placed upon the grave once the burial is complete. A specially designed wooden marker in the image of the moon crescent, resembling the distinctive lemon-coloured V-shaped crescent on Asiatic black bear chests, marks every individual grave. A carved gravestone is further placed on the grave with the animal's name and date of death. Careful consideration is given to the location of each grave; where possible, bears are laid to rest next to former companions already buried in the ground, to acknowledge the bonds developed between their own species. This quiet and peaceful location surrounded by bamboo forest and the river running beside it provides a fitting location for the bears to rest in peace and for the humans to visit, spend time, and reflect. Every so often, flowers can be found carefully laid on graves, a reminder of their enduring legacy. The graveyard has become an important part of any visit to the sanctuary; whether the guests are local or international, along with marvelling at the living bears in the various habitats around the sanctuary, they are taken to visit the graveyard as part of their experience. Few have failed to be moved by the rows of these charismatic creatures laid to rest. As peaceful and as beautiful as the graveyard is, it also serves as a confronting reminder of why the sanctuary exists. It has been documented how funeral rituals can help people feel more in control when faced with loss (Romanoff and Terenzio 1998; Veldkamp 2009; Mitima-Verloop et al. 2021). Other facilities around the world also have different memorials and places where people can go to remember the animals. At Howletts Zoo in the UK, there is a memorial for Djoum the gorilla and for Torgamba, the Sumatran rhino, and at Jersey Zoo in the USA a statue memorial has been erected for Gina the orangutan. A bronze giraffe statue was erected in memory of Penny a little giraffe at Cheyenne Mountain Zoo in Colorado, USA, celebrating she was the 200th giraffe born at the zoo. At the Singapore Zoo, the staff and guests held a tribute ceremony and a temporary memorial wall situated at the Frozen Tundra was set up, where members of the public could pen their tributes and bid farewell to Inuka. The people from Singapore took to social media to share their tributes, including photos and artwork.

To the question "why do it?", there are several answers. Initially, it might seem tempting to avoid entering a relationship with the animals in one's care if it means having to experience grief and sorrow. They are not emotions most of us long for, or actively seek out. But if one accepts that animals are persons in the sense that they are individuals with individual needs, then it is a prerequisite for being able to care for them in an appropriate manner that one enters a "deep" relationship with the animal. At the same time, developing such a relationship with the animal is a way of showing her or him the respect that is due. As with humans, only engaging with animals as instruments to reach some external goals is to disrespect them as something important in themselves. For example, Animals Asia's sanctuary houses former bile-farmed bears in China as well as Vietnam. Each bear is an individual with their own story and character. Many of them have experienced unimaginable horrors, but thanks to the work of Animals Asia, their bodies and minds are expertly cared for to give them the best quality of life. Not one bear has died in vain; each individual was and is an ambassador for their species. Every piece of information documented about their physical and behavioural health has provided valuable insight and evidence into the impact of bear farming.

Some build memorials in the name of the animal beyond the facility grounds. When Karta, the Sumatran orangutan at the Adelaide Zoo, died in childbirth, the care staff and the zoo raised funds and constructed a guard post in Karta's native island of Sumatra. It serves as a hub for rangers, protecting the rainforest of Bukit Tigapuluh National Park in Sumatra, Indonesia, playing a vital role in the fight against illegal activities such as poaching and logging.

1.8 Acknowledge Significance of Loss

Fortunately, many organisations, leaders, and individuals today recognise that having feelings of grief is normal, and it is important to take time, have personal rituals, and reach out for help when needed, but there is still a lot of work to be done. This includes an organisational approach to communicating and caring for elderly animals, as well as facilitating support for animal care staff and others in the organisation who are affected by the loss of an animal. Whilst some instances of losing an animal can be sudden and unexpected, for many of the elderly animals, there is often a longer process of care and bearing witness to the ageing process. In a recent survey involving animal caregivers, curators, and veterinarians, all of the respondents indicated to be dissatisfied with the care and attention of elderly animals in the last phase of their life, including before, during, and after an animal's passing (Brando et al. 2023).

The organisation, be it a zoo or a sanctuary or something else, all within need to come together to create an end-of-life phase approach, including providing the best care possible for the animal and also the space and time for people from within the zoo and visitors to connect and say goodbye. What is needed is to have time and space to grieve and know where and who to contact. Having time to say goodbye and to be quiet together or alone can all be important expressions of grief. Support, time, and communication before, during, and after the passing of an animal are fundamental. For example, some people who are on holidays and/or days off might want to be informed of a deteriorating situation and be included in the decision-making process. Many of these aspects need to be discussed and planned for before these situations arise.

There will inevitably be emergencies and decisions that need to be taken fast with the animal's wellbeing in mind. In such situations, it is important to consider the human aspect of caring for elderly animals. People have different needs and preferences; some might not want to see the body or say goodbye whilst others might want to honour the animal's life with respectful body care after the animal has passed, take photos, or take a foot or paw print. People may want to take time to write a celebration letter, read it out, or combine it in a private or public booklet, or create other communications which can be shared or stored. Having places to remember the animals and/or marking a day to remember the death of loved animals does not need to be a public event. It might be the gesture of placing flowers on a grave, looking at photographs, raising a toast, or simply remembering and reflecting on the date in question. Talking and sharing together what we collectively would like beforehand can avoid many unnecessary sorrows from needs and preferences ignored.

Confronted by loss and dealing with the depth and breadth of grief can impact physical and psychological wellbeing. Those professionally involved in caring for wildlife under such conditions have an obligation to those still living. Sometimes, finding the strength and motivation to continue with their work whilst dealing with feelings of bereavement can be challenging. Providing people with the opportunity to grieve under such circumstances can be difficult. Offering support through talking, professional help, or rest days to reflect can help. Animal care professionals in diverse roles are often left wondering whether it was enough as we put our own feelings of grief to one side to ensure operational needs were not neglected. There is no fixed rule for navigating grief; there is no timetable for processing the death of a loved one. However, working in a professional capacity with animals can mean exposure to multiple losses. With that exposure comes an increased risk of empathic strain, depression, or symptoms of long-term grief. Striking a balance to cope and continue to function in a meaningful way is challenging, empathy, and are key components to working with animals, but allowing ourselves to become overwhelmed with grief, and enter dangerous territory and can cease to offer the effective care required by the role. Dealing with grief is personal, and different variables can affect the impact it has on us. Personal circumstances can influence our response to the loss of an animal, as can the circumstances surrounding the death and the relationship with the animal.

For animal professionals, there is also another aspect of grief that can often be uncomfortable to address, the sense of guilt. Guilt can be about the decision of euthanasia, or even the timing of euthanasia. Guilt can also be experienced when contemplating how the animal was cared for or treated. These feelings are unjustified but can be debilitating in the process of mourning an animal's death. The reality is that caring for animals is complex and nuanced. Rarely are the decisions made about their care the sole responsibility of one individual, so carrying the burden of guilt is—in theory—misplaced. However, it is something that is often tied to mourning the loss of animals.

Death and the sense of loss that accompanies it can, too, influence the relationships we have with others who loved and lost the same animal. Grief can be an incredibly unifying experience for a group of people, which can be wonderfully comforting. On the other hand, it can also be very isolating. The depth of loss felt can be different for everyone, and the degree of upset can be perceived as lesser or greater for different people. What is important is acceptance of the grief, recognition that how it is experienced, and how it is dealt with is different for all of us. The good news is that the mental and psychological wellbeing of animal care professionals is, as well as others in diverse roles in the animal care, education, conservation, and other related fields, increasingly, being recognised. Whether it is the death of a beloved animal, hearing the stories of grief and (species and ecological) losses, or the associated pressures that come from working in a profession that requires compassion and empathy, empathic strain can impact any one of us. As well as the rituals already described in remembering the passing of our animal friends, many animal facilities, be it zoos, wildlife shelters, or sanctuaries, are taking the time to recognise grief. In human health care, formal debriefing has been found to be beneficial for addressing and acknowledging loss and grief (e.g. Keene et al. 2010). This approach is now being adopted in many professional animal settings (Dow et al. 2019). It can provide those attending the opportunity to share their thoughts and memories of their dead animal friends, and more importantly, it can provide the opportunity to express personal feelings in a professional context.

Another way to approach mourning and remembering the loss of animals is to acknowledge that along with the grief felt, having the ability to relieve their suffering and make a difference to that life can be enormously satisfying. One of the ways of addressing our feelings of loss is to accept that their passing has taken a piece of us and acknowledge that every individual has enriched our life. It has truly been an honour to be involved with their lives. Taking the time to write and/or journal can be comforting (Lichtenthal et al. 2012). Photos, videos, and/or beautiful drawing of our favourite animals can serve as reminders of a life-changing experience and an incredible individual. Sometimes when animals pass away, we cannot be there in that moment, but contributing to an organisational and public obituary, and other acts can all contribute to being and feeling of the process of mourning and celebrating.

Finally, experiencing grief and sorrow might not be what we wish for when we wake up in the morning, but it is part and parcel of being human. Avoiding it is avoiding who we are. Accepting it as a part of the experience of being us is being who we are. It is learning what it is to be a being like us: to embrace the nature of us. Thus, it can be seen as a way of being fully human, of being authentic. And it is a way to develop compassion, empathy, and understanding in general in our lives. We shared "*As for man, his days are as grass: as a flower of the field, so he flourisheth. For the wind passeth over it, and it is gone; and the place thereof shall know it no more.*" We hope we have demonstrated that this is also true for animals. They are not *just* an animal, but that wonderful individual with a name, with her or his idiosyncratic ways. They will all die in the biological sense and in the existential sense. They, and we, will all die and that is sad and tragic but not something to shy away from or deny. Entering deep relationships with humans and animals is inevitably going to open us to grief and sorrow, but if we do not, we cannot care for each other—and we miss the joy of knowing each other as individual beings who each have this time alive on Earth to flourish.

References

Animals Asia. Funeral poem Spirit of Hope

Brando S, Buchanan-Smith HM (2018) The 24/7 approach to promoting optimal welfare for captive wild animals. Behav Process 156:83–95

Brando S, Rachinas-Lopes P, Goulart VDLR, Hart LA (2023) Understanding Job Satisfaction and Occupational Stressors of Distinctive Roles in Zoos and Aquariums. Animals 13(12):2018

Brown SE (2007) Companion animals as self-objects. Anthrozoös 20(4):329–343

Brown SE (2011) Theoretical concepts from self psychology applied to animal hoarding. Soc Anim 19(2):175–193

Davis H, Balfour D (1992) The inevitable bond: Examining scientist-animal interactions. Cambridge University Press, New York

Davis H, Gibson JA (2000) Can rabbits tell humans apart? Discrimination of individual humans and its implications for animal research. Comp Med 50(5):483–485

Davis H, Taylor AA, Norris C (1977) Preference for familiar humans by rats. Psychon Bull Rev 4:118–120

Davis H, Norris C, Taylor A (1998) Whether ewe know me or not: The discrimination of individual humans by sheep. Behav Process 43:27–32

Dow MQ, Chur-Hansen A, Hamood W, Edwards S (2019) Impact of dealing with bereaved clients on the psychological wellbeing of veterinarians. Aust Vet J 97(10):382–389

Hart LA, Mader B (1995) Pretense and hidden feelings in the humane society environment: a source of stress. Psychol Rep 77(2):554

Hart LA, Hart BL, Mader B (1990) Humane euthanasia and companion animal death: caring for the animal, the client, and the veterinarian. J Am Vet Med Assoc 197(10):1292–1299

Hart LA, Dorairaj K, Camacho S, Hart BL (2001) Nurturing older dogs: Attitudes and experiences of caregivers. J Am Anim Hosp Assoc 37(4):307–310

Heðinsdóttir K, Kondrup S, Röcklinsberg H, Gjerris M (2018) Can friends be copied? Ethical aspects of cloning dogs as companion animals. J Agric Environ Ethics 31(1):17–29

Hoage RJ, Deiss WA (eds) (1996) New worlds, new animals: from menagerie to zoological park in the nineteenth century. JHU Press, Baltimore

Hosey G, Melfi V (2012) Human–animal bonds between zoo professionals and the animals in their care. Zoo Biol 31(1):13–26

Iliff SA (2002) An additional "R": remembering the animals. ILAR J 43(1):38–47. https://doi.org/10.1093/ilar.43.1.38

Kalof L (2007) Looking at animals in human history. Reaktion Books, London

Kean H (2013) Human and animal space in historic 'pet'cemeteries in London. New York and Paris Animal death:21–42

Keene EA, Hutton N, Hall B, Rushton C (2010) Bereavement debriefing sessions: an intervention to support health care professionals in managing their grief after the death of a patient. Pediatr Nurs 36(4)

Kohut H (1971) The analysis of the self: A systematic approach to the psychoanalytic treatment of narcissistic personality disorders. International Universities Press, New York

Kohut H (1984) How does analysis cure? University of Chicago Press, Chicago

Lichtenthal WG, Neimeyer RA (2012) Directed journaling to facilitate meaning-making. Creative practices for counseling the bereaved, Techniques of grief therapy, pp 165–168

Liehrmann O, Crawley JA, Seltmann MW, Feillet S, Nyein UK, Aung HH et al (2021) Handler familiarity helps to improve working performance during novel situations in semi-captive Asian elephants. Sci Rep 11(1):15480

Lott DF (1988) Feeding wild animals: The urge, the interaction, and the consequences. Anthrozoös 1(4):255–257. https://doi.org/10.2752/089279388787058326

Martin RA, Melfi V (2016) A comparison of zoo animal behavior in the presence of familiar and unfamiliar people. J Appl Anim Welf Sci 19(3):234–244

Mitima-Verloop HB, Mooren TT, Boelen PA (2021) Facilitating grief: An exploration of the function of funerals and rituals in relation to grief reactions. Death Stud 45(9):735–745

Ogden LE (2012) Do animals have personality? The importance of individual differences. BioScience 62(6):533–537

Romanoff BD, Terenzio M (1998) Rituals and the grieving process. Death Stud 22(8):697–711

Salmi R, Jones CE, Carrigan J (2022) Who is there? Captive western gorillas distinguish human voices based on familiarity and nature of previous interactions. Anim Cogn 25(1):217–228

Serpell J (1996) In the company of animals: A study of human-animal relationships. Cambridge University Press

Taylor A, Davis H (1993) Acknowledging animals: A memorial service for teaching and research animals. Anthrozoös 6(4):221–225. https://doi.org/10.2752/089279393787002141

Taylor AA, Davis H (1996) The responses of llamas (*Lama glama*) to familiar and unfamiliar humans. Int J Comp Psychol 9(1):42–49

Taylor AA, Davis H (1998) Individual humans as discriminative stimuli for cattle (Bos taurus). Appl Anim Behav Sci 58:13–21

Taylor AA, Davis H, Boyle GJ (1998) Increased vigilance toward unfamiliar humans by harbor (*Phoca vitulina*) and gray (*Halichoerus grypus*) seals. Mar Mamm Sci 14(3):575–583

Veldkamp E (2009) The emergence of "pets as family" and the socio-historical development of pet funerals in Japan. Anthrozoös 22(4):333–346

Vigne JD (2011) The origins of animal domestication and husbandry: a major change in the history of humanity and the biosphere. C R Biol 334(3):171–181

Villalobos A, Kaplan L (2018) Canine and feline geriatric oncology: honoring the human-animal bond, 2nd edn. Wiley, Hoboken

Weller F (2015) The wild edge of sorrow. Rituals of renewal and the sacred work of grief. North Atlantic Books, Berkeley

The View from Beyond the Fence: Ageing Zoo Animals and Communicating with the Outside World

Philip Knowling

Mouss. Mouss is a 37-year-old male scarlet macaw (*Ara macao*) living at Zoo de Bordeaux Pessac in France. He was born in a small cage opposite the tigers and at the time, many years ago, all parrots would fly high in the sky above the zoo. Unfortunately, one of the parrots landed in the tiger enclosure and was not fast enough to take off in time. After this incident, the parrots lived in enclosed habitats and large trees. His personality is one of independence and exploration. He loves the large tree he lives in, ascending high into the canopy where he has a wide overview of the park. He likes to interact with the care staff and be carried around on a branch. His favourite food is apples, nuts, and treats such as coloured pellets. Although he is getting older, he is in very good

P. Knowling (✉)
Devon, England, UK

S. Brando, S. Chapman (eds.), *Optimal Wellbeing of Ageing Wild Animals in Human Care*, https://doi.org/10.1007/978-3-031-30659-4_15

health and is active. When Captian, (an *Ara ararauna*), was injured, he took him under his wing and cared for him as a son. For a while, he lived alone after the passing of two of his companions, but now he lives with his daughter Mocco with whom he has an interactive and friendly relationship.

Abstract

This chapter is written from the viewpoint of a zoo press officer and stresses the importance of keeping the public fully informed about the wellbeing of older animals. This approach brings three benefits: (1) keeping regular visitors and the general public up to date, (2) demonstrating care and commitment to the animals in human care, including the human–animal relationship of zoo staff with the animals, and (3) helping shape and maintain a facility's public image as a place of professionalism and progression. Older animals are often the best-known individuals in a facility and often have a special place in the hearts of visitors, which should be considered. Communicating about the welfare of ageing animals provides excellent educational opportunities, chances to inspire the public and shape opinions, and occasions to strengthen and maintain a facility's commitment to animal care and wellbeing. Working with journalists as well as the use of social media channels helps the zoo community reach a wider audience. Throughout, the words zoo or zoos will be used as shorthand for all places caring for wild animals, such as but not limited to zoos, aquariums, sanctuaries, safari parks, reserves, or facilities using a title of their own devising.

Keywords
Media · Information · Openness · Honesty · Transparency · Emotion · Reputation · Celebrate · Division · Controversy · Social media

1 Introduction: The Importance of Communicating

This chapter is about the people who are not directly caring for ageing wild animals, in other words, most people. Where exactly do they fit in, do they want to know what is going on, and do they need to know what is going on? It is important that zoos clearly and openly communicate, so that information is readily available to those searching for and interested in it. This includes proactively reaching out. More on this later in the chapter. As in most instances animal care is not carried out in private but in an institution that relies on visitors for both generating income and achieving their purpose, communication is key in all types of facilities. This chapter covers who needs to be kept informed and why a zoo's public reputation is both valuable and vulnerable. It will explore why people want to know so much about zoos and the animals in their care, and why in some cases they become so attached to them. It will look at ways of sharing stories and information, from the traditional news media perspective to how social media has increased in reach and scope, with all of its advantages and drawbacks. Perspectives on what you share with people and when will be discussed. There are ways of sharing that can help everyone, from, e.g., caregivers, veterinarians, and the organisation itself to the general public, including visitors, policymakers, and people who have (or do not yet have) opinions, including those not in favour of zoos. Why it is important to take charge of a story and what to do if it all goes horribly wrong will be covered in the last section of this chapter. Zoos exist because of people: because people are removed from nature, endangering species, threatening habitat, damaging ecosystems, and changing the climate. But zoos, including private institutions, exist because people want them to exist. Zoos give people things they need, things that are worth paying a fair bit to have, like comfort, wonder, and togetherness. There is a more detailed discussion of this later in this chapter. Zoos have and need a relationship with the outside world. It is an illusion to believe a facility does not have to think or be concerned about what people believe and say or consider the changing cultural climate. Education, research, engagement, and conservation, as well as animal wellbeing, are the remits of a contemporary zoo and communication plays a central role in all of these. Understanding public opinion, concerns, and desires is essential. Beyond the fence means just about everyone else: the visitors when they are not visiting; their friends and neighbours who never visit; and suppliers, sub-contractors, businesses, organisations, charities, and government departments who may or may not have a view on captive animals. Linking them are the news media who love stories, science, and drama, and social media platforms, where opinions rule.

2 Who Needs to Know

2.1 The External Audience

It matters what other people think or know (or think they know) about what goes on in a zoo. Zoo visitors support zoos through their interest and acceptance, and with their money, spent on tickets, zoo merchandise, memberships, and donations. Places that look after captive wild animals vary greatly according to geography, culture, and philosophy, but a typical zoo business model looks something like this. People come to the zoo and pay for a ticket which is not an insignificant amount of money. In return, the zoo provides them with a day out (the requirements for which can include car parking, refreshments, toilets, and play areas as well as animals and plants). A zoo generally makes enough money to care for animals, pay staff, invest in, and develop the animal and public facilities and carry out animal welfare, education, research, and conservation activities.

Zoos have a contract with the people who visit. People pay to come in and the zoo in return aims to provide a great day out. If the general public does not perceive zoos positively, zoos cannot build good reputations and ultimately the social licence to operate will be withdrawn. A zoo's reputation like any other organisation is a valuable asset. The zoo's good standing in the community and the sympathetic outlook of individuals and families are crucial both for achieving their remits and in challenging and sensitive situations, such as caring for, or the death of, an elderly animal. A good reputation based on integrity and trust supports a facility during a controversy, a story that appears to cast doubt on the zoo's ability to look after animals or do good science and education. It is important to note that the reputation of a facility is largely in the hands of people who never visit. These people can, to some degree, be positively influenced by the sharing of optimistic and engaging stories about work in progress or completed, including breeding successes, newly arrived animals, conservation projects, and how people can get involved. Good stories, honest stories, and stories that illustrate the values and the ethos of the organisation are key. Reputation may not be viewed as the number one concern of animal staff or people working in the restaurant or shop, but it is extremely important that a facility takes reputation seriously at all levels of the operation. Zoos work to care for individual animals and to save species; they are committed to the science and best practices of animal welfare, education, and conservation. It is important that a zoo, parent charity, or zoological society is seen as progressive, moral, open, and credible. People perceiving an organisation as untrustworthy is problematic. It does not take many disapproving people, who are or are not correct in what they think or say, to create a noise that social media will amplify, and the news media will pick up on. Dealing with people and animals honestly is essential.

Generally speaking, animals live longer and, some may say, better lives in zoos than in the wild (Tidière et al. 2016). Veterinary care in zoos has evolved and improved, with more preventative work and closer cooperation between animal carers and veterinary staff. When I started out in zoos, there was often an air of suspicion and even protectiveness amongst animal care staff when it came to communicating and working with veterinarians. In zoos today, animal care staff, which includes veterinarians, work closely together. Science has given us a much better understanding of species and their needs and preferences, from nutrition and behaviour to exercise and enclosure design, and considers a lifespan approach (see chapter "Holistic Approaches for Promoting Good wellbeing for Ageing Wild Animals").

Who in the public domain needs to know when an animal is ageing, is unwell, or dies? Potentially, everyone. The news media, social media, interested parties, uninterested parties, stakeholders, suppliers, partners, and the public should all have access to information if they want to find it. The news media will help you reach the public as well as social media, which is part news media and part public.

How much you say and how you say it depends on, e.g., the animal, facility type, and philosophy. Some zoo animals can be born, live out their lives, and die without ever being noticed, known to or singled out by the thousands or millions of people who visit, e.g., frogs, flamingos from a large flock, and fish. Any keeper can happily share all the names and relationships of the antelopes in their care, but they remain nameless and faceless to most of the guests or people beyond the fence.

Communicating how zoos care for all animals, from the ants to the alligators and from the zebra finches to the zebra, demonstrates that all animal care is equally important, and that welfare is taken seriously for all animals. Often, the techniques required to care for less known and obvious species are more interesting than for your general hoof stock. No one has to mimic rain in order to get a giraffe to mate, while they do with frogs. Most zoos today assign one or multiple roles to animals, as part of a population or a breeding programme, a public exhibit, or a conservation project. Tributes paid to an animal which was part of a studbook but who was generally unknown to visitors are as important as those for Nellie the elephant because she had been the star of the TV commercial for 20 years.

Zoo visitors and the wider public tend to know about the larger, charismatic animals, like giraffes and elephants, and a single iconic animal is more likely to be known to guests by name. This is also the case if a member of the public has adopted a specific animal. Ageing animals are often well known because they have been in the facility for a long time. Stories and videos and blogs about other animals are useful and interesting but stories, videos, and blogs about the animals recognised by guests are vital. Why? Because they so often become emblematic of a facility. In the UK, generations grew up on the marketing slogan "Visit the lions of Longleat", whilst Twycross Zoo is widely known for its primates. These animals represent the whole place in the minds of the general public; the health of Reggie the Rhino parallels and symbolises the health of the zoo. You cannot make these animals live forever, but you can make sure their stories are told with affection and respect.

2.2 The Internal Audience

Communicating with the audience inside the zoo is also important. A large institution has a spread of jobs, people, and workplaces. There are people with many qualifications and people with no qualifications at all. There are people for whom animal care is a vocation and people for whom working in a zoo is just a way to pay the bills. Reaching, addressing, and convincing them all is challenging but necessary. In fact, they need special consideration. Do not let your teams find out what is going on from your or their social media feed or a radio news bulletin. The zoo's philosophy, missions, and aspirations should be clearly communicated and give people a sense of purpose and direction. All news, ranging from good to bad, should be shared and known internally (at least a short time) before going public. The internal audience includes volunteers, who are part public and part zoo. Volunteers help for all sorts of reasons, including love for the zoo, and will rightly feel hurt if they are left out of communications. Senior managers must share plans to all involved, including those who pick up litter, or the volunteers on the fundraising stall. These people also want to know what is going on and why and they are the ones visitors are most likely to talk with.

3 Why People Want to Know

The general public is not of one mind. Some people love the idea of zoos, enjoy seeing animals in public facilities, and support wildlife institutions and their goals. There are also people who detest everything about animals in captivity. Then there are people who will change their minds easily and regularly, swayed by an anti-zoo report or a friend's favourable comment, but who will always be delighted by photos of a furry new arrival. It is complicated. This is because we hear from different parts of the public at different times. The anti-zoo lobby seems large because it is represented by various campaigning organisations, e.g. Born Free and Freedom for Animals, whilst zoo supporters are mainly individuals who like going to zoos. There are also people who change their views or have no views. This is why a zoo should always be striving to put the best and most honest version of itself out there for the public to experience.

There are a lot of things to consider in zoo communications. Zoo visitors, and people who never visit zoos, have opinions on everything from the size and state of animal enclosures to the species in those enclosures and their mental and physical wellbeing. The immediate response from a member of staff in a zoo might well be along the lines of "Who exactly is the expert here?", but this stance is not helpful. People have opinions, and they like to share them. Some people are genuinely worried (or delighted) by what they see and experience, others are cynically trying to get free tickets out of the customer care team, and some people want to challenge the very existence of zoos.

However, lots of people really like zoos. More than 700 million people visit zoos around the world annually. There are different reasons why people are pro zoo. People care about the concept of zoos, the animals they can see, or the wider concerns they represent, like the decline of species, the destruction of habitats, and wildlife conservation in general. We want and need people to engage with zoos because that is how they share information, build relationships, and shape hearts and minds. Engagement works in various ways. Zoos need to interest the public and cause them to visit. The public needs to feel free to ask any questions they might have and to receive full and honest answers.

Zoos can provide people with many things: a great day out, discovery, comfort, wonder, excitement, somewhere to keep the kids occupied for a few hours. Sometimes zoos give people a reason to complain; sometimes they give people a reason to live. There are individuals (your zoo may attract some) who rely on their local zoo as an absolute lifeline in times of poor mental health; although they can cause moments of antisocial behaviour, possibly towards staff or other visitors (personal experience of the chapter's author). Zoos can give you a good feeling about the world because they demonstrate what is being done to save the planet, or a bad one because they demonstrate how much there is to do which might make you feel hopeless.

Zoos should aim to inspire. The need to inspire the public should inform everything they do. This means telling good news stories, success stories, and positive stories and engaging people into action. Finding good stories or points of action can be easier said than done, but moments of innovation, occasions when staff and volunteers go that extra mile or when a scientific discovery illuminates a new way to do things, or new actions people can take to play their part. All these can inspire eager and open-minded visitors. The care given to ageing animals can most certainly provide moments of inspiration too. To some people zoos are not about the wild world, but simply another entertainment option, like the beach, the cinema, or shopping. Even if this is the case, zoos should take every opportunity to engage and inspire. Perhaps, as a result, the next time these people visit the beach instead of a zoo they will help with a clean-up or decide to volunteer with a local conservation project. Perhaps, they will pay attention to how they shop. Other people are interested in zoos for other reasons. The news media are looking for stories, both positive and negative; they are mostly not bothered, they just want content. A local journalist who loves the zoo might regularly call wanting to know what is going on but will happily tell a negative story if there is one to be told, because stories are their job (personal experience of the chapter's author). The anti-zoo movement is watching for any hint of neglect, scandal, or controversy. In summary, the outside world views zoos with delight, suspicion, devotion, indifference, rage, and ignorance. People are entitled to their opinions and to the wish to be better informed. They have a right to ask questions and they have a right to honest answers from their zoo. Good communication will please the positive, potentially sway some of the doubters, and might make little or no difference to zoo opponents. But how we communicate is important, and doing it right is vital.

4 How to Tell Them

Informing people about the wellbeing of an older animal is important. Channels for reaching the wider public include the traditional news media (newspapers, radio, and TV) and a steady to changing array of websites and social media platforms. Whilst the news media may be viewed as being the professionals and social media amateurs, there are people who make a living from social media and others who make no kind of living out of news. Websites straddle the divide, such as it is; these days the news media and social media overlap, mingle, and feed off and support each other.

Another way of reaching people is through face-to-face contact around the zoo. Informal public talks or timetabled educational presentations reach fewer people but are direct. Information is also shared in the guise of gossip; regular visitors, keen volunteers, and chatty staff will, deliberately or otherwise, share information, misinformation, rumour, and confusion. It may be unpredictable, but it can work for as well as against a facility. Word of mouth can be a respectable form of communication which should not be underestimated, so it is important that key people talk to staff and volunteers to provide the appropriate and correct information. This will also help all levels of staff, from the CEO to junior carers, to connect on at least an intermittent basis.

4.1 News Media

Whilst some zoo staff are of the opinion that all journalists are cold-hearted and calculating and will do anything to get their hands on a juicy story that would close the zoo for good, the majority of zoo professionals know that working with the news media can be engaging and mutually beneficial. Whilst one should be careful, there is really no "off the record"; one should not be terrified or suspicious without cause. Journalists are an important link between zoos and the public. For any facility still on the fence, it is time to drop being indifferent and uncommunicative and to become media savvy.

Some people may say journalists are biased, and they are, though not in the way they mean. People think journalists are biased in their politics, or in their view for or against zoos; it would be surprising if they were not biased, as journalists are human beings and human beings are biased. We are all biased: biased towards zoos or ice cream or a certain football team or type of music and towards people, ideas, places, nations. However, this is not the bias referred to here. Most journalists are biased in favour of a good story. Stories are their business, and good stories are good business. Some do not need the story to be true, though all should, and luckily most do. The news media deal in news by using stories. The news is not just stuff that is happening in the world; it is the stories of the people (or animals) to whom that stuff is happening. Most journalists (consciously or unconsciously) are biased towards drama. Like most other humans, they like a good plot, a twist, a surprise. In the modern world, news can be seen as entertainment, found at various points along the spectrum between the serious and the frivolous. News has to be worth reading, or watching, or hearing, and a bit of drama helps. Division also creates newsworthy content; opposing opinions make any news story work better and last longer.

It is worth sketching out how news happens. The news media and their methods vary between regions, countries, and cultures. Generally, local news outlets use journalists, employed or freelance, and other content creators who are looking for stories. They find them through contacts, tip-offs, the public, social media, word of mouth, and professional outlets such as press officers and public relations agencies. These last two are paid to supply stories on behalf of their clients or employers; others might have less obvious agendas. Local journalism is usually a mix of parochial news, local angles on national stories, and offbeat human interest. News agencies make a living selling stories, photos, and video clips to national print, broadcast, and online titles. Freelance press photographers and videographers take and sell photos and videos. Social media bypasses all of that, but also plugs back into it; social media can be a good source of stories for all kinds of journalists. You can think of journalists, social media, and the public pretty much as one. They all need and want to be informed and entertained. Journalists may ask different questions, more incisive questions, but they are asking them on behalf of their audience.

An institution should preferably have an established way of connecting to the outside world. This could be one office person who dabbles in social media, a dedicated press officer, an outside public relations agency, or a full communications team. There may be no plan and no inclination to have one, or there could be a jargon-ridden strategy with regular social media activity and news releases planned for months ahead. Some zoos even have their own YouTube channel or TV documentary show.

Wherever your institution sits on the communications spectrum, it is useful to have a formalised process, person, or team responsible for finding stories amongst the day-to-day goings-on of departments and sections that are worth sharing with the public.

Stories can be shared as photos, written news releases, blogs, vlogs, live spots on social media channels, static social media posts, or films made in-house. They will typically be about new animals arriving, successful births and hatchings, conservation projects, or education activities. They may feature care staff, researchers, volunteers, or students talking about an aspect of their role. Some stories are successful and attract attention; others are not and do not. It depends on timing, the other stories doing the rounds on the day, the choices and decisions of editors, and chance.

The aim is to put out interesting, honest, and trustworthy stories that promote the zoo and then respond to the resulting interest from outside, reinforcing the public image of the place and, most importantly, making people want to visit or support the zoo through other means. This is also the route through which approaches from external media, TV production company pitches, and other random queries should be channelled. The same system can be used to share bad news when necessary. Animal care staff need to know that they can trust the person who does this job to say the right thing in the right way and they in turn need to know they can trust animal care staff to provide the right information and the right context at the right time. The main currency of a press officer is the news release; how to write one of these will be covered later in the chapter. Journalists tend to look for clarity and simplicity for a black and white, goodies-and-baddies story that is easy to write and easy to understand. They work to deadlines and usually have too much to do. They are after a quick, clear, interesting story, or else an unfolding, conveniently episodic drama. Often, animal care is complicated, unclear, and long-drawn-out. Reconciling these is not always easy; journalists often up the drama and reduce the complexity, neither of which impresses scientists or care staff who know more details. These days, zoos are likely to face fewer professional journalists and more amateur observers. Social media and mobile phones have made everyone an instant commentator, whether we like it or not. Social media platforms share information, gossip, and falsehoods in seemingly equal measure, though it depends on what comes into your feed and what one chooses to read. Different platforms are used by different age groups and in different ways. News may not feature at all for some users. One of the great advantages, or disadvantages, of social media is the way information or rumour spreads, especially when it is something juicy and amazing and unlikely. Social media is about networks, making connections and spreading the word. The truth may not go viral because it seems dull and boring, but it is important to keep contributing to the conversation, and to be seen to be contributing, in an honest and transparent manner. Zoos and aquariums should work to reach as many people as possible, promoting a positive narrative reflecting the work they do and the goals they aim to achieve.

5 What to Share with the General Audience

What should one share with a general audience about the health of an older animal? As much information as possible, without getting bogged down in complicated veterinary or other jargon or potentially distressing descriptions. The illness, recovery, or death of animals should be communicated publicly, especially for those animals well known to the public. A zoo should provide updates, reassurance, information, and context. Celebrate the dedication and passion of all the staff. Highlight professional successes. If and when it comes to it, share the passing of a beloved, characterful animal.

An animal that is unwell is of course a cause for concern. The people caring for that animal will be focused on their work. But it is important that those people (veterinarians, animal caregivers, curators) find time to think of the wider community, and the people in their organisation who connect to the outside world. Animal staff involved and focused on the animal who is ill should also be aware of the wider audience and community outside the zoo. Lots of people beyond the fence will be keen for news and updates. Whilst this will not be the first concern of care staff, it is important that they make time to help colleagues for whom it is a concern. An animal who is sick and recovers is a cause for celebration or relief, so share this good news too. When veterinarians apply a new technique, or try novel treatments, that is a great story too. Plenty of things animal carers may take for granted can make good, engaging, offbeat stories for the news media, for example, using old, donated socks to keep the hands and feet of small monkeys warm during veterinary procedures, environmental enrichment, animal training, animal–visitor interaction programmes, and habitat modifications using recycled items; the ideas are endless. It is natural that the attention of the animal staff is focused on the animal. It is inevitable that the attention of senior managers is on the bigger picture, including the rest of the facility, visitors, and beyond. However, it is important to consider the wellbeing of the animal care staff who have worked with that animal every day for years or even, in the case of a long-lived animal, their whole career. Also important are the volunteers who helped prepare feeds; the educators who gave fond talks to delighted school groups; and anyone else who connected to the animal in one way or another. They need to be kept informed and can also contribute to the necessary public communications. Indeed, if they feel left out, they might even take to social media to say so.

This is the time to talk about the care and dedication shown by the team. Mention the people who have gone to great lengths to give the animal the finest care, the right environment, the very best quality of life. Show empathy with sad or disappointed guests, these people are your supporters, your constituency. Some animal care professionals can be shy when it comes to communicating about the care and welfare of animals. Their reluctance to talk about the subject outside their establishments may feed the false idea that zoos are secretive, unapproachable, and uncaring. Animal experts may know best, but they do not always know how to explain this. What are some of the experiences of people caring for older animals? Strained faces and tears in difficult meetings where tough quality of life and end-of-life decisions have to be made; staff up all night with their charges; special feeds and changes to nutrition; and alterations to dens and enclosures. This is what working as an animal carer means, and it communicates the very best qualities of the profession: the knowledge, the patience, the dedication, care, love, and the sheer hard work. How much emotion gets shared with the outside world is an individual choice. Whilst it is natural they are sad and it is healthy to acknowledge it, some may not want to be seen to be upset in public. Whilst the public can be deeply touched by their emotion, it needs to be a personal choice for a carer to go public with their feelings. Communications staff can only advise and guide. Care staff may well want to pay tribute to an animal they have poured their heart and soul into for 10 years. Handled sensitively, this can help build understanding and support in the wider community. Care staff will want to see the right things said publicly about the animal with which they, and possibly they alone, had such a close bond. They do not want to see press officers second-guessing their feelings. Therefore, it is paramount to create and maintain a culture of communication where staff can and do feel able to share, question, and trust. No one wants animal carers feeling reluctant to come forward and then feeling left out afterwards. Whilst it is important not to delay, as press officers must work quickly, animal staff need time to process their feelings. It is a delicate balance. At times of stress and emotion, it is the little things that can make or break a mood. When an animal is absent, someone will have the job of getting a sign made for a den window or paddock fence explaining why to visitors. This is no time for wordy, officious pronouncements. Write briefly and from the heart. Empathise. Keep it simple. Illness and old age can change the behaviour of individuals and the dynamics of a group. Some animals might be treated harshly by conspecifics as they change their place in the group. They might also look different, with changes in their coat or in their gait. Explaining these things to visitors, volunteers, and staff is an opportunity for moments of education and connection. This is particularly important when the care staff have found ways to keep the ageing animal living with his or her group members.

6 Showing the Way

6.1 Taking Charge

Be prepared to talk about sick and dying animals. News will get out, like it or not, whether you are ready or not. Some managers may not want to discuss these topics: I knew one curator who felt it was a sign that animal carers had failed. Whilst this is touchingly protective, it is unfortunately also myopic and unhelpful. The bigger and more well known the animal, the more likely it is that people will find out and the more important it is that the zoo makes the first move. It should be part of a communication plan, because if it is not, it is going to have to be in your emergency response plan or your business continuity plan. Talking to the outside world about the death of a key animal is as important as raising funds for a new exhibit or flagging up an event for schools. It is important not to try to suppress unofficial channels. It looks high-handed, and it would not work. Instead, be prepared to act quickly to take hold of the news agenda and make sure the zoo speaks first. This will help set the narrative going in the preferred direction and with the preferred tone.

It is impossible to control the flow of information in, around, and out of a zoo. Social media channels mean that any member of staff or regular visitor who is chatty and well-informed can share the truth (or what they know of it) or simply pure speculation and gossip. Senior people going walkabout and talking to staff and volunteers can help to set the record straight and will maintain a professional approach. It is key to have a clear and simple method for collecting and sharing official information, and an understanding amongst staff and volunteers that this is how it should be done. Whilst this may sound authoritarian, and could provoke people with contrary ideas into defiance, I think this really is the best approach. The news media and the public will pick up on any divisions. Diverging and dissenting claims will confuse and fuel a story, hence the importance of an organisation having a clear message. It is extremely unhelpful when someone inside the organisation feels so strongly about something that they break ranks and go off script. It is possible that experts from elsewhere (other institutions, animal welfare charities, and government agencies) might disagree with the story communicated, but this is something one cannot control. A way to counter it is to contact peers in other zoos and relevant organisations to find supportive interviewees.

A clear and simple method for collecting and sharing official information should ensure that the facts, the context, and the appropriate quotes are drawn together so they can be used by the right person. That person might need to work very quickly. Everyone in the zoo should help them to speak with a single voice that is fully informed, honest, and truthful. The focus of this effort will be a news release. A news release, i.e. press release, statement, and announcement, needs to gather a definitive version of the story, as understood at that moment, into one short, clear piece of writing that can then be used by any news media as the basis for their reporting. The details of the animal, his or her background, time spent at the zoo, and an explanation of what happened will all need to be included. Journalists love facts and figures, and you might be called upon to supply information that seems irrelevant, e.g. where and when the animal was born, when he or she came to the zoo, offspring, and so on. It may be beneficial to write a short summary for social media, although this can then link to the full story.

Be as honest as possible, keeping in mind that technical details of illness and decline can be too much for some people. They also need to be translated from veterinary jargon into plain language. If there is any doubt, simply explain that the full facts are currently not available. Reduce speculation as much as possible and make sure details are accurate. It can be useful to mention that there will be a post-mortem examination. Whilst a lot can be learned from an animal after death, the results can take a while to come back. This may lead to the opportunity for an update story at a later date, though quite often results are inconclusive.

Discussing the welfare of ageing wild animals in human care will sooner or later mean discussing the death of ageing wild animals in human care. There is no need to be afraid of communicating the death of an animal in a zoo. In fact, a facility may receive negative feedback from visitors, social media followers, and others when they do not. Animals in zoos die for a wide variety of reasons, including old age. It is paramount not to let fear of the anti-captive lobby denouncing the death as another reason to close all facilities and stop the celebration of the life of a much-loved animal.

It is important to take people on the care journey and not surprise them with a cold, clinical announcement stating that an animal of this species with that tricky Latin name died at 08:52 hours, followed by a list of long, difficult science-related words. Express sadness and empathy; a genuine quote or anecdote from an animal caregiver is always a wonderful way to share and connect. If there is time, signpost a death with news stories about the animal's illness, about treatments, and about how care staff are working hard to help. This might include installing ramps for when those awkward steps into the den get too much, for example. Stories and photos of health assessments highlight the caring and professional approach of zoos and show how animal experts also love the animals in their care. Whilst it may seem obvious and something to take for granted, it is crucial to show the wider public how much people working in zoos care.

I was the press officer at a zoo that lost one of its two elephants. Keeping two ageing female elephants, one Asian and one African, was regularly and widely criticised, and everyone knew it was far from ideal, but the elephants had been there for many years, every option had been considered, and changes were not on the cards. Not, that is, until one became seriously sick. The zoo assembled key people and planned. It was likely that the elephant had some months to live, depending on treatments and outcomes; the plan involved charting her progress through news stories and updates. At the time we did not report that she was going to die, as we really hoped she would not. We simply used stories about her care to show the way things were going. This approach was successful. The zoo received a lot of positive news coverage, including national exposure from a photo story about special padded boots made to measure to help cushion her sore feet. We set the tone for the coverage. When it came to breaking the news about the difficult decision to euthanise her, few people were surprised, and most were sympathetic.

Staff and volunteers will be sad when an animal dies. They will want to remember, each in their own way, alone and together, by talking, leaving tributes, or signing a book of condolence. Human beings deal with death in different ways, and this needs to be considered and respected. It is vital that management does not underestimate the impact that losing an animal has on carers. Some people might not talk about it openly, so managers may not realise that grief counselling may be needed. The potential for distraction and a loss of focus at these times should also be considered and a communication and support plan as indicated in chapter "Caring for Elderly Wild Animals: The Human Experience" will be fundamental. Euthanasia is a delicate subject. In some countries, euthanasia is illegal or ethically frowned upon, which can make things awkward for veterinarian and animal care staff. There can be dissent amongst staff as to when to euthanise. Care staff may want to be present at the time of euthanasia, or rather be somewhere else. Aspects concerning this are discussed in chapter "Euthanasia of Geriatric Zoo Animals: Decision-Making and Procedure". The idea of members of the public sending cards and flowers and messages of condolence is anathema to some zoo people and a sign of support and comfort to others. These actions can be seen as evidence of the strong link between people and animals, the very link zoos are trying to promote. Visitors see zoo animals in a way that is not the same as care staff, veterinarians, or zoo scientists. Respect the wishes of people, give them space, and let them, whether they are inside the zoo or outside, experience the death in their own way, with sensitivity and patience.

7 Points of Reference: Helping People to Understand

In the wild, animals may die, for example, of starvation, predation, or disease. Animals tend to live longer in modern zoos because experts consider quality of life across their lifespan. This presents a whole new challenge: how do we care for ageing animals under human management? Animal geriatrics is a fascinating subject, and one which can catch the imagination of the public as there is a clear parallel: people around the world are also enjoying longer lives.

The art and science of engagement, communication, and connection is to take people on a journey, learning (in this case) about a possibly complicated illness, a myriad of animal care issues and welfare considerations, group dynamics, and quality of life, all interwoven with the individual stories of the animals themselves. Having said that, the mantra should be Keep It Simple. Remain true and transparent and draw parallels where appropriate. Comparing the care of an animal to the care of a person, a companion animal, or horse can be a useful and recognisable way to connect with the audience. Caring for older animals can be likened to caring for older people, as many ageing animals face the same sort of problems that ageing humans face: mobility limitations, arthritis, dental disease, kidney disease, and neoplasia. Talk about what carers are doing for the animal: giving anti-inflammatories and supplements, providing physiotherapy through environmental enrichment, and increasing warmth and comfort. Some animals will get what amounts to end-of-life care, as people might receive in a hospice. All these are things many people can relate to.

Share stories about old age health assessments. These might need to be edited for clarity and understanding for non-veterinary professionals, but details of urine checks, blood tests, blood pressure, radiographs, and heart monitoring, as well as psychological and affective state monitoring, all demonstrate the remarkable level of care in zoos. Action taken as a result of health checks can also be used in communication and educational messages. Nutritional supplements to help with stiff joints, pain killers, or anti-inflammatories to improve mobility, lowering high steps and providing gentle gradients for older limbs, making bedding deeper and softer, and providing extra warmth in winter are all things people will recognise as good and considerate care. Help the public to understand; look for ways in which people can connect readily and empathise easily. When an animal dies, empathise with people who were fond of it. Some zoo visitors get emotionally invested and will need particular consideration and respect. Professional detachment translated into newspeak becomes cold and uncaring and fails to reflect the caring and loving attitudes of so many care professionals.

Whilst anthropomorphism, the idea of attributing human qualities and behaviours to animals, can be challenging for people both inside and outside zoos, used wisely it can help non-zoo people connect with amazing animals. Look at how the news media love videos and photos of animals with pumpkins at Halloween or presents at Christmas. Many animals often do behave in ways that parallel human actions; for example, animals play, squabble, mourn, and socialise. It is a fascinating and useful angle, and if approached with caution it can really help communicate conservation and education ideas. After an animal dies, social media may be full of emotional rhetoric and anthropomorphic hyperbole: well-informed messages and tributes from animal carers can both reset the balance and show how much they cared. The names, personalities, and stories of individual animals can engage people and bring attention onto species and habitats. Animals housed in zoos genuinely are ambassadors for their species in the wild, inasmuch as an ambassador is a diplomatic resident representative whose role is to facilitate connections between different groups. Another way to engage people is through their experiences of domestic animals. Many will have kept animals on a farm or had companion animals. Anyone who grew up on a farm, who had a pony and a pet chicken, or who raised a lamb knows that death is a part of life. They know what it is like to see a beloved animal grow old, get ill, and die. They will have experienced that moment when it was up to them to do the right thing and recognise that age and illness had taken away quality of life. Care staff can help draw parallels for people by openly and honestly sharing the personality traits and characteristics of an animal. Occasional crankiness and moodiness, along with curiosity and sociability, shows character. When people recognise themselves in the personalities of animals, they can feel more connected and more willing to invest themselves. What are perceived as "negative" traits can make for good connections and recognition. If an animal was grumpy or greedy at times, well, so are people. Nothing connects the public with a zoo animal like the fond comments of a dedicated keeper who misses that animal like an old friend/enemy/rival/colleague.

8 What Do You Do If Things Do Not Go According to Plan?

What if the zoo has made a mistake? What if husbandry has been less than perfect, if there has been an accident that is identifiably someone's fault, or if errors have contributed to the injury or death of an animal? What if a visitor filmed it all on their phone?

First and foremost, do not panic. If there is bad news, never attempt to hide it. A cover-up uncovered later is ten times as bad. Speak up, be honest, be open. Take ownership as an organisation and/or as a senior animal expert. This is the time to rely on the reputation for integrity and transparency you have built with the public. An honest approach now will make a positive difference and may make the whole crisis quicker and easier to solve. Above all, honesty is the professional response.

Journalists and other people will ask difficult questions, such as "How did this happen?", "Who is to blame?", and "Is this zoo safe?" Be prepared to answer a wide variety of questions. One good idea is to gather key people and brainstorm for various tough situations before they happen; another is to do a table-top exercise, a sort of role-play game where staff respond to problems set by colleagues and revealed in real time. This sort of thing should be part of your emergency response planning. For example, there may be questions about whether this illness or that issue occurs in wild populations, or whether it is a problem only for captive animals. Mostly, because of a lack of data, the answer must be "We do not know". At other times, people might ask why the problem was not spotted earlier, not knowing just how well wild animals can conceal the outward signs of disease. Answer questions to the best of your ability in an open and honest manner, do not build barricades, even if you are feeling defensive. Humility is essential. Find out what went wrong and set about fixing it. The reputation of the zoo might be dented in the short term, but dealing with problems genuinely will help reduce the damage and make repair easier. A sensible organisation will keep the public fully informed about the health of older animals. Communication brings understanding and support; it reminds people of all the great work zoos do and what important objectives they have. It helps keep the public image positive and progressive. Talking about the welfare of ageing animals, and their death, offers opportunities to educate and inspire both zoo supporters and the wider public. Used wisely, the news media and social media channels can help you get honest and caring messages to a wide and receptive audience.

Reference

Tidière M, Gaillard JM, Berger V, Müller DW, Bingaman Lackey L, Gimenez O et al (2016) Comparative analyses of longevity and senescence reveal variable survival benefits of living in zoos across mammals. Sci Rep 6:36361

Index

S. Brando, S. Chapman (eds.), *Optimal Wellbeing of Ageing Wild Animals in Human Care*, https://doi.org/10.1007/978-3-031-30659-4